Advances in Neurobiology

Volume 25

Series Editor
Arne Schousboe, Department of Drug Design & Pharmacology
University of Copenhagen, Copenhagen, Denmark

More information about this series at http://www.springer.com/series/8787

Emanuel DiCicco-Bloom • James H. Millonig
Editors

Neurodevelopmental Disorders

Employing iPSC Technologies to Define and Treat Childhood Brain Diseases

Springer

Editors
Emanuel DiCicco-Bloom
Department of Neuroscience and Cell Biology/Pediatrics, Rutgers Robert Wood Johnson Medical School
Rutgers University
Piscataway, NJ, USA

James H. Millonig
Department of Neuroscience and Cell Biology, Center for Advanced Biotechnology and Medicine, Rutgers Robert Wood Johnson Medical School, Rutgers University
Piscataway, NJ, USA

ISSN 2190-5215 ISSN 2190-5223 (electronic)
Advances in Neurobiology
ISBN 978-3-030-45492-0 ISBN 978-3-030-45493-7 (eBook)
https://doi.org/10.1007/978-3-030-45493-7

This Springer imprint is published by the registered company Springer Nature Switzerland AG
The registered company address is: Gewerbestrasse 11, 6330 Cham, Switzerland

Contents

About the Editors

Emanuel DiCicco-Bloom, MD, is a Professor of Neuroscience and Cell Biology and Pediatrics (Child Neurology and Neurodevelopmental Disabilities) at Rutgers Robert Wood Johnson Medical School (RWJMS) in New Jersey and a member of graduate programs in Cell and Developmental Biology, Neuroscience, and Toxicology at Rutgers University. He graduated *summa cum laude* from Princeton University, received his M.D. from then Cornell University Medical College, and trained in Pediatrics and Neurology at New York Hospital-Cornell Medical Center, joining RWJMS in 1990.

Dr. DiCicco-Bloom has broad experience performing basic and translational research on neurodevelopmental disorders including animal models and human induced pluripotent stem cells (iPSCs). As an active child neurologist, his research focuses on defining molecular and cellular pathways that regulate the production of neuronal cells (neurogenesis) during brain development, and how related abnormalities contribute to developmental disorders including autism. He investigates how growth signals, genetic factors, and environmental toxins impact cell proliferation, survival, and fate determination during brain development using rat and mouse neural stem cells in culture and in vivo, with a focus on the cerebral cortex, cerebellum, and hippocampus. To begin defining mechanisms that are more directly relevant to human disorders, recent collaborative studies with Co-Editor James H. Millonig, Ph.D. have focused on creating human induced neural stem cells (NPCs) from people with autism, to determine their neurobiological signatures. Significantly, by comparing NPCs from two forms of autism, including idiopathic and genetically defined (CNV 16p11.2 Deletion Syndrome), these studies reveal a common neurobiological phenotype consisting of reduced neural process outgrowth and cell migration in autism compared to control NPCs. Moreover, dysregulation of the mTOR signaling pathway appears to be central to these phenotypes, as pathway manipulation both corrects autism deficits and recreates abnormalities in relevant controls. This new era of exploring neuropsychiatric conditions in human neurons may provide more relevant cellular and molecular pathways on which to target therapeutic interventions that may be "personalized" to the needs of the specific individual.

Dr. DiCicco-Bloom has long served the autism scientific and advocacy communities, providing scientific expertise to federal agencies including NIH, DOD, NSF, and IACC, and disease advocacy organizations including National Alliance for Autism Research, Autism Science Foundation, Autism Speaks, Autism Tissue Program, Brain Canada, Rett Syndrome Foundation, and Simons Foundation. He Chaired the Scientific Program Committee of the 2008 International Meeting for Autism Research (IMFAR) meeting (London) and Co-Chaired 2010 (Philadelphia), and has long served the IMFAR Program Committee. For the Society for Neuroscience, he has served on many committees (PECC, GPA, Audit, Rigor and Reproducibility Working Group) and as a Councilor. Currently, he is the Chair of the NIH Developmental Brain Disorders Study Section, is a member of the DOD Autism Research Program and the American Brain Coalition (ABC) Board of Directors, and serves as Scientific Advisor to the Eagles Autism Foundation. He serves on the editorial boards of Autism Research, Molecular Autism, and other developmental neuroscience journals, and has authored numerous research articles and book chapters in neuroscience, child neurology, and psychiatry.

James H. Millonig, PhD, is an associate professor in the Department of Neuroscience and Cell Biology and resident member in the Center for Advanced Biotechnology and Medicine at Robert Wood Johnson Medical School—Rutgers University.

Dr. Millonig graduated from The University of Rochester *magna cum laude* with a B.S. in Biochemistry. After college, he received an M.Sc. degree from Oxford University, University College studying *B. subtilis* sporulation. Dr. Millonig matriculated in the Princeton University's Molecular Biology PhD program and performed his thesis research in the lab of Shirley Tilghman Ph.D., *President Emerita* of Princeton University. During his Ph.D., he was trained in genetics, mouse development, and molecular biology. Subsequent post-doctoral research, using mouse genetic approaches to study cerebellar development, was performed in Mary E. Hatten's laboratory at The Rockefeller University. He joined the faculty at Rutgers in 1999.

Dr. Millonig has studied CNS development throughout his career applying molecular genetic approaches to understand basic mechanisms and disease. His work in collaboration with Emanuel DiCicco-Bloom M.D. and Linda Brzustowicz M.D. demonstrated the genetic association of *ENGRAILED2* (*EN2*) with autism spectrum disorder and defined autism-like behaviors in mouse models that could be reversed by pharmacological treatment. His lab also identified the orphan GPCR, Gpr161, as an unknown, important regulator of neural tube closure and lens development. Forward mouse genetic approaches demonstrated that Gpr161 functions through the retinoic acid (RA) pathway. His lab is currently studying rare human mutations in the RA metabolic pathway and Gpr161 and their impact on development. More recently, Dr. Millonig and Dr. DiCicco-Bloom have applied iPSC-based approaches to study idiopathic and 16p11.2 Deletion Syndrome. They have defined neurodevelopmental phenotypes in culture and are applying numerous omic strategies to define the downstream signaling and cell biological pathways affected in autism. He is the recipient of many research awards including Basil O'Connor Starter Research Award, NARSAD Young Investigator Award, and the Thomas A Edison Patent Award for medical diagnostics. He has authored over 100 abstracts, papers, and grants related to autism and neurodevelopmental disorders. He has been an active reviewer for many scientific journals including Molecular Psychiatry, Translational Psychiatry, and eLife. For the NIH, he has been an *ad hoc* member of Clinical Neuroplasticity and Neurotransmitters and Genetics of Health and Disease, was a standing member of the NIH Developmental Brain Disorders Study Section (2010–16), and has also contributed to NIH Loan Repayment Program, BRAIN, and NIH Director's Early Independence Awards initiatives.

He is currently the Director of the Rutgers—RWJMS—Princeton University MD/PhD program and is the senior associate dean in the Rutgers School of Graduate Studies. He was co-PI on an NIH Director's Fund Broadening Experiences Scientific Training (BEST) training grant. In these roles, Dr. Millonig has developed numerous educational initiatives including the BEST program called iJOBS, which includes all Rutgers campuses across NJ that have reached 836 trainees or 16,258 person-hours in 4 years.

Modeling Neurodevelopmental Deficits in Tuberous Sclerosis Complex with Stem Cell Derived Neural Precursors and Neurons

Maria Sundberg and Mustafa Sahin

1 Background of Tuberous Sclerosis Complex

1.1 Neurology of TSC

TSC is an autosomal dominant, genetic disorder that profoundly affects development of different organs, including the central nervous system (CNS). TSC is classified as a rare disorder with a prevalence of 1:6000 live births worldwide [1–3]. The genes affecting the occurrence of TSC are *TSC1* and *TSC2*, encoding for proteins that form a complex regulating the mechanistic target of the rapamycin (mTOR)-pathway [4, 5].

During the development of the CNS, TSC is characterized by formation of cortical and cerebellar tubers, low grade astrocytic tumors known as subependymal giant cell astrocytomas (SEGAs) and small non-cancerous lesions known as subependymal nodules (SENs). Epilepsy is very common within TSC patients affecting >80–90% of the population [6]. Thus, TSC is considered to be one of the most common genetic causes of epilepsy. Clinically, epilepsy affects the lives of the vast majority of TSC patients, and without effective treatment it can negatively affect the patient's brain development. In general, the majority of treatments used for epilepsy suppress seizures in the short-term but do not provide long lasting cure for the deficits within the neural circuitry, and the mechanisms that cause epileptogenesis in TSC patients are still under investigation. Autism spectrum disorder (ASD) is diagnosed in approximately 40–60% of patients with TSC [7, 8]. Previous studies have indicated that children with idiopathic ASD and children with TSC and ASD have similar behavioral and cognitive deficits, indicating common circuit-level dysfunction [9, 10]. In addition, although several cognitive- and behavioral-therapies exist

M. Sundberg · M. Sahin (✉)
Department of Neurology, F.M. Kirby Center for Neurobiology, Boston Children's Hospital, Harvard Medical School, Boston, MA, USA
e-mail: Maria.sundberg@childrens.harvard.edu; Mustafa.sahin@childrens.harvard.edu

E. DiCicco-Bloom, J. H. Millonig (eds.), *Neurodevelopmental Disorders*, Advances in Neurobiology 25, https://doi.org/10.1007/978-3-030-45493-7_1

for TSC patients with cognitive impairment and autism, there are currently no treatments that permanently reverse these deficits. Thus, novel approaches are crucially needed to provide more effective and safe treatments for TSC patients.

In recent years several research studies have shown that human pluripotent stem cells, such as human embryonic stem cells (hESCs) and human induced pluripotent stem cells (hiPSCs), are an ideal tool for characterization of human neuronal development in vitro in neurodevelopmental and neurodegenerative disorders [11–19]. In addition, TSC-patient derived hiPSC-derived neurons can be used for disease phenotyping at the cellular and genetic level and for screening of compounds for the development of new drug interventions. Although TSC is a multi-organ disorder that affects development of the eye, heart, lung, kidneys, skin, and brain, here we are going to concentrate on the neurological manifestations of TSC and describe recent advances in the human stem cell derived neuronal models of TSC. We are also going to introduce Tsc1/2-deficient animal models whose characterization has supported the establishment and validation of the disease phenotyping assays for human stem cell derived neuronal cells in vitro.

1.2 mTORC1 and mTORC2

mTOR is a conserved serine/threonine kinase that is formed by two functionally distinct complexes: mTORC1 and mTORC2 [20]. Hyperactivation of mTORC1 greatly affects the development of neuronal cells and is a key factor associated with neurodevelopmental deficits in TSC. mTORC1 complex consists of several subunits, including nonobligate protein proline-rich AKT1 substrate 40 kDa (PRAS40), DEP domain containing mTOR-interacting protein (DEPTOR), regulatory associated protein of mTOR complex 1 (RAPTOR), mammalian lethal with sec13 protein 8 (mLST8), telomere maintenance 2 (TEL2), and TELO interacting protein 1 (TTI1). mTORC1 phosphorylates eukaryotic translation initiation factor 4E binding protein (EIF4EBP1) and protein kinase S6-kinase1 (S6K) which affects mRNA translation, cell growth, cell proliferation, and differentiation [20]. In the regulation of mTORC1, TSC1 and TSC2 are accompanied by TBC1D7 [21], which is a member of the TBC-domain containing proteins and an important subunit of the tuberous sclerosis TSC1-TSC2 complex. TSC2 functions as a GTPase activating protein (GAP), which regulates hydrolysis of GTP into GDP on the GTPase Ras homolog enriched in the brain (RHEB). RHEB-GTP activates mTORC1 and stimulates phosphorylation of 4EBP1 and S6K [22]. Mutations in either *TSC1* or *TSC2* can cause loss of function of the TSC1-TSC2-protein complex causing increased RHEB-GTP activation leading to mTORC1 overactivation in cells [22].

mTORC2 complex also consists of several components, including telomere maintenance 2 (TEL2), TELO interacting protein 1 (TTI1), protein observed with rictor (PROTOR), DEPTOR, mammalian stress-activated protein kinase 1 (mSin1), mLST8, and rapamycin-insensitive companion to mTOR (RICTOR) [20]. The mTORC2 regulates cellular cytoskeleton organization and affects phosphorylation of AKT1 by phosphorylation of Ser-473. mTORC2 is insensitive for short-term rapamycin treatment, but it is affected by long-term exposure to high doses of rapamycin. mTORC2 is activated by the TSC1-TSC2 complex along with several growth factors and nutrients such as insulin. mTORC2 is involved in cell growth, cytoskeleton organization, and activation of protein kinases. mTORC2 has also been involved in morphological and functional development of neurons [23, 24]. A summary of the mTORC1 and mTORC2 pathways and interacting molecules are presented in Fig. 1.

To develop effective treatment options for TSC, much research has been done with human pluripotent stem cell derived neuronal cells in vitro. In the next section, we will discuss the current research advances that have been made with disease phenotyping of TSC-deficient human pluripotent stem cell derived neuronal cells and how the mTOR-pathway inhibitors affect development of human neurons in vitro.

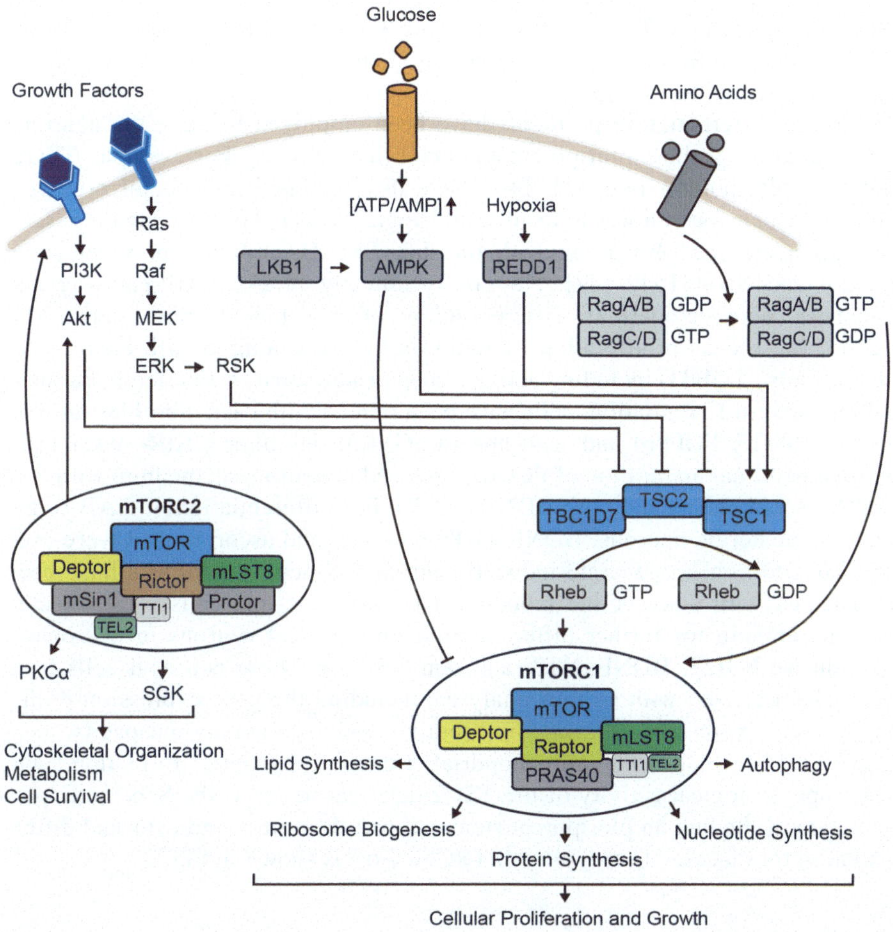

Fig. 1 Schematic presentation of the regulation of the mTORC1 and mTORC2 activation in the cell

2 Neural Stem Cell Development in TSC

2.1 *TSC2 Deficiency in hESCs and hiPSCs*

To study the neural stem cell development in humans, hESCs and hiPSCs are an ideal source of pluripotent stem cells that can be differentiated in vitro into all three germ layer cell types of the human body [25, 26]. Several neuronal differentiation protocols have previously been developed to drive these cells towards neuronal fate [27, 28]. To study specifically the neurodevelopmental deficits of TSC in vitro, a genetically engineered hESC-derived neuronal cell model has been created with zinc-finger nucleases to induce mutations in one *TSC2* allele ($TSC2^{+/-}$) or both *TSC2* alleles ($TSC2^{-/-}$) [29]. In addition, our group has created an allelic series of *TSC2* mutation in isogenic hiPSCs lines [17] by correction of the $TSC2^{+/-}$ mutation in TSC-patient derived hiPSCs, using the CRISPR-Cas9 gene editing method [30–32], or by inducing a mutation into second allele of the TSC2-patient hiPSCs ($TSC2^{+/-}$) to generate a biallelic deletion in *TSC2* ($TSC2^{-/-}$) with the TALEN method [33].

Several differentiation protocols have been employed to drive the neuronal differentiation of these pluripotent stem cell lines in vitro. For example, TSC2-deficient pluripotent stem cells have been differentiated into neural precursor cells in suspension cultures to form neural aggregates [17] or by using the rosette formation method [29] along with the dual-SMAD-inhibition protocol using noggin and SB-431542 [27]. Noggin specifically blocks BMP-pathway and enhances neuroectodermal differentiation of pluripotent stem cells, and SB-431542 blocks proliferation by inhibiting transforming growth factor beta-superfamily (TGFβ3) including activin receptor like kinases. For cerebellar neural precursor differentiation, cells have been treated with basic fibroblast growth factors (bFGF, FGF8b) and with specific GSK3β inhibitor CHIR-99021 [17]. Further neuronal maturation of PCs was induced in neurobasal medium with B27 and N2 supplements and with BDNF and T3. For differentiation of GABAergic and glutamatergic neurons, BDNF, GDNF, cAMP, and ascorbic acid were used [29]. In addition, to generate induced neurons (iN-neurons) hiPSCs have been transfected with doxycycline inducible NGN2-vector, and cells were selected with puromycin for further differentiation into cortical neurons in neurobasal medium with B27, BDNF, NT3, and laminin [34]. These neuronal cells have been characterized with various analyses, including the gene expression profiling, protein expression profiling, proliferation and differentiation capacity, morphological development, mitochondrial function, synaptic formation, and electrophysiological activity of the TSC2-deficient neural cells. Schematic presentation of the human pluripotent stem cell line derivation and neuronal differentiation for disease phenotyping of TSC in vitro is shown in Fig. 2.

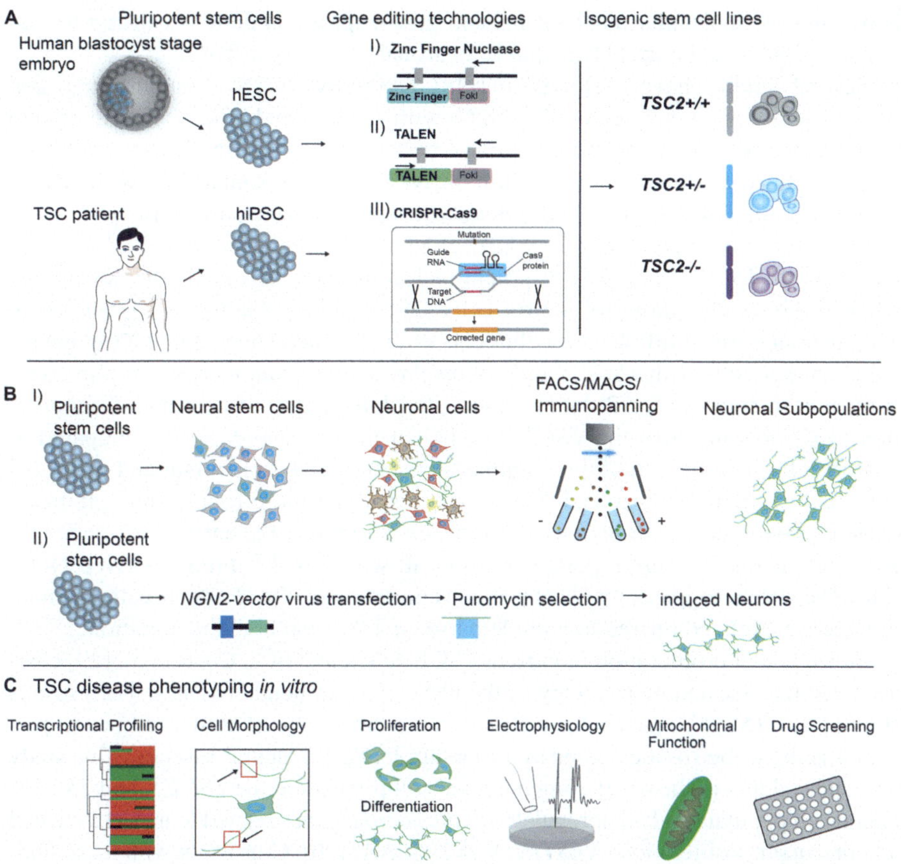

Fig. 2 (**a**) Schematic presentation of human pluripotent stem cell line derivation and gene editing technologies. (**b**) Neuronal differentiation of human pluripotent stem cell lines. (**c**) TSC disease phenotyping assays in vitro

2.2 *Effects of mTORC1 Pathway Activation in Human Neural Stem Cell Proliferation*

During CNS development SENs and SEGAs are typically formed in the TSC-patient brain. These non-malignant tumors consist of giant cells and dysplastic cells that originate from abnormally differentiated neural stem cells (NSCs) during development. Several mouse models have been previously created with *Tsc1* or *Tsc2* mutations in Nestin-Cre-positive NSCs or Emx1-Cre-positive NSCs [35–37]. Although these mice have abnormal cortical lamination and cellular abnormalities that resembled SENs in the cortex, these mouse models have not reliably recapitulated formation of SEGAs [35–37]. Thus, human stem cell models are useful for answering questions surrounding abnormal differentiation of TSC-patient derived

NSCs and the molecular mechanisms and gene-expression deficits causing formation of SEGAs or abnormal function of neuronal networks in TSC.

Several studies have evaluated mTORC1 activity in TSC2-mutant stem cell derived neural precursor cells (NSC/NPCs with *TSC2*$^{+/-}$ and *TSC2*$^{-/-}$) using protein expression analyses of phosphorylation of ribosomal protein S6, 4EBP1, and AKT [17, 29, 38]. According to these studies, the NPCs derived from hESCs with heterozygous loss of TSC2 (*TSC2*$^{+/-}$) did not show significant increase in the pS6 levels compared to control cells (*TSC2*$^{+/+}$) [29]. However during further neuronal differentiation both *TSC2*$^{+/-}$ and *TSC2*$^{-/-}$ neuronal cells had increased pS6 levels and p4EBP1 levels compared to control cells [29]. Similar results were observed in TSC-patient derived hiPSC-cerebellar cell studies, which showed that early neuronal precursor cells with *TSC2*$^{+/-}$ did not display a significant increase in pS6 levels compared to control (*TSC2*$^{+/+}$). However, after later stages of neuronal differentiation (>32 days in vitro) the *TSC2*$^{+/-}$ cells had increased pS6 levels compared to control cells [17]. In both of these human stem cell derived-precursor and -neuronal cell studies, the homozygous *TSC2*$^{-/-}$ cells displayed more robust and significant disease phenotypes, with increased pS6 levels already at the early NSC stage [17, 29]. Similar results were reported by Li et al. with *TSC2*$^{+/-}$ hiPSC-derived NPCs [38]. These data suggests that heterozygous mutation of TSC2 affects cellular function less severely compared to complete loss of TSC2 via biallelic mutation.

To study cell proliferation in vitro in TSC2-deficient NSCs, Costa et al. [29] characterized the size and morphology of the hESC-derived neural rosettes. Their results showed that *TSC2*$^{-/-}$ rosettes were larger due to the increased number of neural precursor cells in the rosettes compared to control *TSC2*$^{+/+}$ neural rosettes. This study also showed that in the developing neuronal cell population the *TSC2*$^{+/-}$ and *TSC2*$^{-/-}$ neural cell populations had upregulated expression of genes related to neural stem cell and precursor proliferation (*NESTIN, SOX2, PCNA*) [29]. Consistent with these findings, increased proliferation of TSC-patient hiPSC-derived NPCs was also described with increased BrdU-incorporation in TSC2-deficient cells or with increased expression of proliferation marker Ki67 in TSC2-deficient cerebellar precursor cell populations [17, 38]. These proliferation deficits were further characterized by detection of increased expression of CYCLIN D1 [17], which is a key regulator of cellular proliferation in NSCs [39, 40]. Abnormal regulation of CYCLIND1/CYCLIND2 has been described as an important factor in the development of megalencephaly–polymicrogyria–polydactyly–hydrocephalus syndrome [41, 42]. Similarly to the pathophysiology of TSC, megalencephaly–polymicrogyria–polydactyly–hydrocephalus syndrome results in increased growth of neuronal tissue and formation of enlarged balloon cells in the patient brain [41, 42]. In addition, abnormally increased neuronal proliferation and brain size have been described in children with ASD [43, 44]. Thus, the abnormal neural precursor cell proliferation via CYCLIND1/CYCLIND2 and mTORC1 pathway dysregulation creates an environment which predisposes the brain towards tuber formation and abnormal brain growth. The human stem cell derived neural precursor models described here, effectively recapitulated these cellular abnormalities and molecular deficits in vitro. Interestingly, we and other researchers have shown that mTORC1-pathway inhibition with rapamycin treatment rescued both the hESC-derived NSCs and the hiPSC-derived NPC proliferation deficits in vitro [17, 29, 38].

These results suggest that the early developmental abnormalities seen in TSC-patient neural cells may be reversible with correctly timed inhibition of mTORC1.

2.3 Differentiation Capacity and Morphological Development of TSC-Deficient Stem Cell Derived NPCs

Abnormal cellular proliferation of TSC-deficient neural precursor cell populations also suggests imbalance in neuronal cell differentiation capacity of the cells. Previously created *Tsc1 or Tsc2* deficient rodent models have displayed enhanced development of neural progenitors and premature differentiation and degeneration of neurons, as well as increased astroglial differentiation of the neural progenitors [36, 45, 46]. We and others have compared these findings to human pluripotent stem cell derived neuronal data [17, 29, 38]. Decreased expression of post-mitotic neurons that express neuronal proteins HuC/D were discovered in *TSC2*$^{-/-}$ and *TSC2*$^{+/-}$ hESC-derived neural cell populations at day 14 of differentiation compared to control neural cell populations [29]. Also, decreased expression of neural cell adhesion molecule (NCAM) was detected in *TSC2*$^{-/-}$ hiPSC-derived neural precursor cell populations at day 16 and day 30 of differentiation [17]. In line with this finding, increased astroglial differentiation was detected in these cellular populations in both hESC-derived TSC2-deficient (*TSC2*$^{+/-}$, *TSC2*$^{-/-}$) neural cell populations [29] and TSC-patient derived hiPSC-neural cell populations (*TSC2*$^{+/-}$, *TSC2*$^{-/-}$) [17, 38] during the differentiation process shown with increased GFAP expression.

In addition to an altered neural cell fate, the mTORC1 pathway overactivation in TSC leads to increased protein synthesis and cellular growth, which dysregulates normal morphological development of the neurons, as shown in Tsc1/2 deficient rodent models with increased branching and number of axons [36, 45, 47]. Consistent with these studies, the human neuronal cells derived from TSC patients or TSC2-mutant hESCs displayed enlarged soma size and abnormal dendritic branching with an increased number of neurites compared to control neurons [17, 29, 38]. These in vitro findings with human stem cell derived neural cells are also consistent with in vivo findings from TSC-patient brains where abnormal mTORC1-pathway activation causes neural cell overgrowth, abnormal neuronal lamination, tuber formation, and increased glial cell differentiation and SEGA formation [48–50].

2.4 Abnormal Purkinje Cell Differentiation of TSC2-Deficient hiPSCs

Cerebellar damage and deficits, including loss of Purkinje cells (PCs), have been associated with occurrence of autism [51–54]. Also, mouse models with loss of *Tsc1* or *Tsc2* specifically in the PCs display autistic-like behavioral deficits, including abnormal social behavior, ultrasonic vocalizations, cellular spine development, and PC functionality [46, 55]. Thus, to study the molecular deficits predisposing autism, it is important to take a closer look at how stem cells give rise to the differentiation of PCs and which pathways are important for their normal development and function in the cerebellum.

To create PCs in vitro, mouse embryonic stem cells have been differentiated towards cerebellar lineage with a high concentration of bFGF and insulin [56]. This study described a method of differentiation of Neph3-positive (KIRREL2) precursor cells that were sorted with FACS for enrichment of L7 (PCP2) positive PCs, and these cells were co-cultured with mouse embryonic granule cell progenitors for final PC maturation in vitro [56]. Transcription factors that were upregulated in the cerebellar progenitors were *Wnt1*, *Fgf8*, *En2*, *Lhx5*, *Corl2* (*SKOR2*), which are important for PC precursor development [56]. Similarly, hESCs have been differentiated into PCs using high bFGF and insulin in vitro, and co-culturing the cells with mouse cerebellar cells or cerebellar tissue sections [57, 58]. This protocol also included a cell sorting step that facilitated enrichment of KIRREL2+ PC precursors in vitro [57]. These methods have been used to differentiate spinocerebellar ataxia patient hiPSCs into PCs in vitro [59]. However, reliable repetition of these methods with different hESCs or hiPSC-lines has been challenging. Thus, our group has recently developed an improved differentiation protocol for derivation of human PCs from hiPSCs in vitro [17]. In this new protocol, hiPSCs are first cultured on suspension spheres to form neural precursor aggregates, in the presence of noggin and SB-43752 that induces dual SMAD-inhibition and drives the cells towards neural lineage by inhibiting the BMP4 and TGFβ3-pathways [27]. To further induce the midbrain/hindbrain boundary development, the cells were differentiated with a specific GSK3β-pathway inhibitor, CHIR-99021 [17]. In addition, transcription factors FGF8b and bFGF were added to the cells to guide them to express cerebellar lineage markers [17]. With this method, the hiPSCs were differentiated efficiently towards cerebellar lineage, and they expressed the transcription factors *EN1*, *EN2*, *HOX1*, *GBX2* and proteins such as KIRREL2, SKOR2, LHX1A [17]. At day 30 of differentiation > 20% of the cells expressed PCP2. The developing PC populations were sorted with THY1+ selection, and co-cultured in the presence of mouse cerebellar granule neurons towards PCs. After 120 days of differentiation 60–90% of cells were PCP2 positive, which indicated efficient human stem cell derived PC production in vitro [17].

To further analyze the differentiation capacity of TSC2-mutant patient derived hiPSCs into PCs, the cerebellar lineage marker expression profiles were characterized during the differentiation process. The TSC2-deficient hiPSCs displayed decreased *GBX2* expression at early stages of cerebellar differentiation, and the expression levels of proteins SKOR2 and LHX1A were also downregulated. Increased astroglial derivation of the precursor cells was also detected, as described above [17, 29, 60]. This study also showed that post-mitotic neuronal markers THY1 and PCP2 were reduced during *TSC2*$^{-/-}$ hiPSC-derived cerebellar precursor cell differentiation into PCs in vitro [17]. Morphological abnormalities of THY1+ TSC2-deficient PCs were detected with increased soma size and increased number of neurites compared to control PCs. These deficits were mTORC1-dependent, and they were rescued with long-term rapamycin treatment in vitro. These results suggest that developmental deficits of TSC2-mutant PCs could contribute to the development of ASD in patients with TSC.

3 Gliosis and Myelination in TSC

3.1 Astrocyte Differentiation from TSC-Deficient Neural Precursor Cells

During spontaneous differentiation of TSC2-deficient pluripotent stem cell derived NPCs towards neuronal lineage, the TSC2-mutant cells have significantly lower capacity to differentiate into neurons and increased capacity to differentiate towards astroglial cells [60]. Gene expression analyses of TSC2-deficient hESC-derived NPCs revealed that TSC2-mutation increased activation of inflammatory pathways and increased cellular metabolism in these cells. At the gene- and protein expression level the TSC2-deficient neural cells had increased synthesis of angiogenic factors that are important inducers of vasculogenesis in the cells [60]. Interestingly, this study also described that inhibition of mTOR-pathway hyperactivation with rapamycin rescued the translational deficits, but did not affect the mRNA expression levels in the TSC2-mutant cells [60]. Grabrole et al. detected upregulation of genes regulating glial fate, *NF1B* and *NF1X,* and downregulation of the pro-neuronal marker *MASH1* in TSC2-mutant NPCs compared to control cells. During further differentiation of the TSC2-deficient NPCs, the expression of astrocyte markers *GFAP and* phospho-STAT3, and glial precursor markers CD44 and CD184, were significantly increased in TSC2-mutant neural cell populations compared to control populations [60]. Similarly, a previous study of TSC-patient hiPSC-derived neural cell populations has discovered that heterozygous mutation of TSC2 was sufficient to alter the neural differentiation capacity of the NPCs towards an astroglial cell fate [38]. This study also detected increased expression of proliferative BrdU+ astroglial cells and increased mTORC1-pathway activation in astrocytes in the TSC-patient derived cell populations compared to control cell populations [38]. Increased astrocyte differentiation capacity has also been described in TSC-patient hiPSC-derived cerebellar precursor cell populations with increased CD44 and GFAP expression during differentiation [17]. Taken together, these studies indicate that loss of function of TSC2 during differentiation of the human NPCs leads to increased astroglial precursor cell differentiation and decreased neuronal differentiation and maturation in vitro [17, 29, 38, 60]. Increased astroglial differentiation of the TSC-deficient NPC populations is also consistent with clinical data showing increased formation of astrocytomas and tubers with increased number of astroglial cells in the TSC-patient brain [61].

To gain a greater understanding of the gene expression changes of the human TSC2-deficient astrocyte enriched cell populations, researchers compared their dataset with two microarray datasets from previous studies of mouse models of reactive astrogliosis; ischemic-stroke-model and LPS-mediated neuroinflammation model [60, 62]. This comparison revealed significant overlap of the differently expressed genes among *TSC2*$^{-/-}$ human neural cells versus the mouse reactive astrocyte-models. Further analyses of the enriched gene sets revealed that pathways related to metabolic and inflammatory responses were upregulated in the TSC2-deficient hESC-derived astrocyte enriched cell cultures [60].

Strong correlation has also been detected between gene ontology analyses of the hESC-derived TSC2-deficient neural cell populations compared to microarray gene expression analyses of SEGAs and cortical tubers of TSC patients [60, 63]. Pathways related to neo-angiogenesis and inflammation (interferon-induced guanylate binding proteins *GBP1* and *GBP2*) are significantly enriched in the TSC2-deficient neural cell populations and in cortical tubers and SEGAs. Interestingly, increased expression of angiogenic markers, vascular endothelial growth factor (VEGF) and platelet derived growth factor (PDGF), at the gene- and protein-level in TSC2-deficient neural cultures suggests that TSC2-deficient tuber forming cells have capacity to induce vascularization during development of the SEGAs [60]. These data suggest that development of anti-angiogenic treatment options for TSC with VEGF-receptor blocking drugs could be used to prevent vascularization of tumors and to inhibit growth of astrocytomas during differentiation of the TSC-deficient patient cells. Moreover, increased inflammatory signaling in astrocytes has previously been linked to mTOR-pathway hyperactivation and occurrence of epilepsy in a *Tsc1-GFAP*-conditional knock-out mouse [64]. Treatment of these mice with anti-inflammatory agent epicatechin-3-gallate partially improved their viability and reduced the number of seizures [64]. It is possible that combining existing mTORC1-pathway inhibitors together with anti-inflammatory drugs and anti-angiogenic drugs could provide more efficacious treatments for TSC patients. Such combined treatments could work simultaneously to inhibit reactive astrocyte activation, hyperexcitability, and vascularization of the SEGAs in the TSC-patient brain.

3.2 Development of Oligodendrocytes and Myelination Deficits in TSC

Oligodendrocytes are myelin forming cells of the central nervous system. Previous mouse studies have shown that the mTOR-pathway has an important role in oligodendrocyte development [65], and mTOR affects myelination through regulation of myelination production related protein synthesis, lipid metabolism, and cholesterol consumption in oligodendrocytes [66, 67]. In TSC patients, hyperactivated mTOR signaling dysregulates myelination of the axons in the cortex and large white matter tracts, which has been detected in neuroimaging studies [68–70]. In line with these clinical findings, mTORC1-pathway hyperactivation in Tsc1/2-deficient mice affects differentiation of glial precursor cells leading to increased astroglial differentiation and altered differentiation of oligodendrocytes, causing myelination deficits in the

CNS [71, 72]. A mouse model of specific loss of *Tsc2* in oligodendrocyte precursor cells displayed reduced numbers of Olig2+ oligodendrocytes in the cortex and corpus callosum compared to a control mouse brain. In addition, the Tsc2-deficient mouse oligodendrocyte precursors are drawn towards an astrocyte cell fate instead of maturation into oligodendrocytes [71]. This finding is in line with previous mouse studies, where hyperactivation of that mTORC1-pathway in Tsc1-deficient oligodendrocytes leads to hypomyelination of axons [72]. At the molecular level, the deficient mTORC1-pathway hyperactivation and loss of Tsc1/2 in mouse oligodendrocytes have been shown to reduce the expression of transcription factors that regulate cholesterol consumption and lipid production, which leads to deficient myelin production [71, 72]. Interestingly, mTORC1 plays a more critical role in the myelination regulation than mTORC2 alone [72, 73]. This has been shown in raptor (mTORC1) mutant mouse oligodendrocytes where myelin protein expression and lipogenesis were downregulated significantly compared to control cells, whereas in rictor (mTORC2) mutant oligodendrocytes myelination or lipogenesis were not altered as significantly compared to control cells [72].

As described above, altered myelination in TSC appears to be a multifactorial process. To summarize, firstly Tsc1 or Tsc2 deficiency regulates glial precursor fate decisions leading to decreased number of oligodendrocytes, and by negatively affecting expression of myelin producing transcription factors in the oligodendrocytes [71, 72]. The second cause of myelination deficits in TSC is due to non-cell autonomous effects, where abnormalities in the TSC-deficient neuronal cells lead to abnormal development of oligodendrocytes and deficits in myelin production [74, 75]. The molecular mechanism behind this effect has previously been described in a mouse model of specific loss of Tsc1 in cortical neurons [74, 75] and with TSC-patient hiPSC-derived TSC2-deficient neurons [74]. Mouse model showed decreased myelination of the CNS due to upregulation of an inhibitory signaling mediator known as connective tissue growth factor (CTGF). Increased expression of CTGF was also detected in TSC2-deficient patient hiPSC-derived neurons [74]. These myelination deficits were partially rescued by deletion of CTGF in the neurons [74].

To study oligodendrocyte development in human stem cell models, several differentiation protocols have been developed to derive oligodendrocytes from pluripotent stem cells [76–80]. These methods include several different growth factors that have been detected originally in mouse embryonic oligodendrocyte development and specification, such as SHH, bFGF, PDGF-AA, IGF-1, RA, T3, and NT3 [81]. The differentiation of human pluripotent stem cells to oligodendrocytes usually takes around 75–120 days in vitro, and includes both suspension culturing of glial-spheres and adherent culturing of the maturing oligodendrocyte precursor cells [76–80]. TSC-patient hiPSC derived oligodendrocyte precursors cells have shown increased proliferation capacity and decreased maturation in co-cultures with patient neurons, and rapamycin treatment rescued these deficits. These TSC-patient hiPSC derived oligodendrocyte models are valuable for characterization of the molecular pathways driving the myelination deficits in the TSC-patient brain [82].

4 Cellular Homeostasis and Viability in TSC

4.1 *Mitochondrial Dysfunction in TSC*

Mitochondria are cell organelles that regulate cell respiration and energy consumption via production of ATP. Normal mitochondrial function is important for cells and especially for differentiating neuronal cells that require a lot of energy for synapse formation and for development of electrophysiologically active neuronal networks [83]. In TSC, the hyperactivated mTOR-pathway dysregulates the normal neuronal development and function, and it has been detected that mitochondrial function is impaired in Tsc1 and Tsc2-deficient hippocampal and cortical neurons [83, 84]. Specifically, oxidative stress was increased in the Tsc-deficient neurons [83, 84]. In addition, in the Tsc2-deficient neuronal axons the number of mitochondria was decreased compared to control axons. Mitochondrial respiration was also altered in the Tsc2-deficient rodent neurons [83]. These disease phenotypes were also studied in TSC-patient derived hiPSCs that were differentiated into cortical neurons with the NGN2-overexpression iN-protocol [34], both from TSC-patient derived hiPSCs ($TSC2^{+/-}$) and from the TALEN induced biallelic-mutant hiPSCS ($TSC2^{-/-}$), and compared to parental control cells ($TSC2^{+/+}$) [83]. In contrast to rodent data, no differences in mitochondrial respiration were detected in TSC2-deficient patient derived cortical neurons ($TSC2^{+/-}$) compared to control neurons ($TSC2^{+/+}$). However, the $TSC2^{-/-}$ hiPSC-derived cortical neurons had significant reduction in the mitochondrial function compared to control neurons ($TSC2^{+/+}$) [83]. Further staining and intensity analyses of the mitochondrial membrane potential with tetramethyl-rhodamine-ethyl-ester dye showed significant reduction in the membrane potential of the $TSC2^{-/-}$ hiPSC-neurons compared to control neurons ($TSC2^{+/+}$), although significant changes were not detected in the patient derived cortical neurons ($TSC2^{+/-}$) compared to control neurons. In a more detailed analysis, this study also discovered accumulation of mitochondria in the soma of the TSC2-deficient human neurons ($TSC2^{-/-}$). In addition, there was abnormal axonal transport of the mitochondria back to the soma in the Tsc2-deficient neurons, leading to decreased presynaptic mitochondria in these neurons [83]. These results suggest that dysfunctional mitochondria and decreased energy production at synapses may contribute to the impaired synaptic development in TSC-deficient neurons.

4.2 *Autophagy in TSC*

Autophagy is a cellular process that maintains catabolic processes in the body, by breaking down dysfunctional and damaged cells and their organelles and by releasing degraded molecular components in order for new cells to reform. Growth factor depletion, nutrient depletion, oxidative stress of mitochondria, endoplasmic reticulum-stress, toxic accumulation, and infection can induce autophagy in cells

[85]. Autophagy regulates cell death via interaction with necrotic and apoptotic pathways. Autophagy also functions as a main regulator of recycling damaged or aged mitochondria in the cells, in a process called mitophagy [85]. Autophagy activation may lead to apoptosis and cell death in neurodegenerative [86, 87] and neurodevelopmental diseases [88], where abnormal protein accumulation and organelle dysfunction are present. Typically, mTOR-pathway activation suppresses autophagy in the cells [89]. In brains of ASD patients, it was reported that a hyperactivated mTOR-pathway reduced autophagy activation[90]. This study also showed that deficient autophagy activation affected spine pruning and caused increased spine density in ASD patient neurons compared to control neurons [90]. Related to these findings, impaired autophagy activation has been detected in both TSC-patient brain tissues and in mouse models with Tsc1 mutations, where dysfunctional autophagocytosis is linked to increased epileptogenesis in TSC [91]. In Tsc2-deficient mouse embryonic fibroblasts, autophagy activation was decreased compared to control cells, leading to autophagy substrate p62 accumulation. In addition, combinatorial treatment of $TSC2^{-/-}$ tumors with mTORC1-inhibitor and autophagy inhibitor blocked tumorigenesis more effectively than treatment with either one alone in vivo [92]. Our group has shown that autophagy activation was regulated in opposite directions in the Tsc2 deficient fibroblasts and in the Tsc2-deficient neurons [93]. Tsc2-deficient rat cortical neurons displayed accumulation of the autophagic marker LC3-II and of the autophagy substrate p62 [93]. There was also impaired recruitment of autophagosomes to damaged axonal mitochondria in *Tsc2*-deficient neurons [83]. Similar findings were found in the TSC-patient hiPSC-derived cerebellar cultures, where mitochondrial reactive oxygen species (mROS) and LC3II expression were significantly increased in the $TSC2^{-/-}$ hiPSC-derived PC precursors compared to control ($TSC2^{+/+}$) PC precursors [17]. This may be due to accumulation of damaged mitochondria and increased oxidative stress in the cells, which may lead to increased apoptosis of the TSC2-deficient developing human cerebellar neurons. This molecular deficit is in line with previous data showing reduced volume of the cerebellum of TSC patients [94] and reduced PC number in ASD affected brains [53, 54].

5 Autism and Epilepsy in TSC

5.1 MTOR-Pathway in Autism and Epilepsy

Common features of autism include repetitive behaviors, restricted interests, and a prominent deficit presenting as abnormal and impaired social interactions [95–97]. Previous studies have shown that there exists a prevalent correlation between epilepsy and ASD [98–100]. In general, over 30% of autistic patients also have epilepsy [98–100]. Strikingly, from the TSC patients with deficient mTOR-pathway regulation 80–90% have epilepsy [6], and autism has been described to be present in approximately 40–60% of TSC patients [7, 8]. Additionally, deficient expression of another mTOR-pathway suppressor called phosphatase and tensin homolog (PTEN) has been associated with development of epilepsy and autism in both patient and mice models in vivo [101–104]. Also, TSC2 deficient hiPSC-derived NGN2-induced cortical neurons with biallelic loss of TSC2 have shown hyperactive network formation, synchronization, and increased mTOR-pathway activation, which were rescued with rapamycin treatment [105]. Taken together, this data suggests that there might exist shared pathological mechanisms that lead to development of both epilepsy and autism, and this mechanism may be linked to hyperactivation of the mTOR-pathway.

5.2 *Tsc-Deficient Rodent Models of Epilepsy and Autism*

Previously, researchers have created several rodent models to study the molecular link between epilepsy and autism. Here we are briefly introducing a few of them as an example of in vivo models. One of these models is a *Tsc2*$^{+/-}$ haploinsufficient rat model that displayed a reduced tendency for novel objects and social exploration in open field studies [106]. When both naïve and *Tsc2*$^{+/-}$ rats were exposed to kainic acid to induce epilepsy, both groups showed reduced social interaction due to increased anxiety [106]. These results suggest that epilepsy in both *Tsc2*$^{+/-}$ mutation and healthy background caused increased tendency for development of autistic-like social deficits in behavior [106]. These deficits were rescued with treatment of the animals with mTORC1-inhibitor rapamycin [107]. Moreover, our research group has previously generated a mouse model of Tsc2 with conditional expression of Tsc2 hypomorphic and *Tsc2*-null alleles together with synapsin-I-cre (SynIcre), which resulted in restricted expression of Tsc2 mutation only in the neurons of the mouse brain [108]. In the Tsc2 mutant mouse brain, the cortical neurons were larger than control neurons, and this difference was related to increased mTORC1-pathway activation and pS6 expression with increased protein synthesis [108, 109]. Importantly, these mice demonstrated behavioral deficits including altered social interaction and ASD-like behavior, anxiety, and epilepsy [108, 109].

Similar to Tsc2-deficient mouse models, researchers have also created a mouse model with specific loss of Tsc1 in neurons and showed that these mice displayed seizures, deficits in neuronal myelination, and enlarged cortical and hippocampal soma size in the brain [75]. Specific mTOR-inhibitors rapamycin and everolimus were sufficient to rescue these disease phenotypes in the mouse model of neuronal loss of Tsc1 [110]. In addition, another mouse model of *Tsc1* gene mutation in serotonergic neurons has displayed autistic-like behaviors and epilepsy related to insufficient neurotransmission of 5-HT. In this study, seizures increased hyperactivation of the mTOR-pathway by increasing pS6 levels in the cells [111]. Hyperactivation of the mTOR-pathway alone without epilepsy also caused autistic-like behaviors in the mouse model of *Tsc1*$^{flox/flox}$*;Slc6a4-cre*, where Tsc-deficiency is specific to 5-HT neurons [111]. Previously, our lab created a mouse model with specific loss of *Tsc1* in Purkinje cells (PCs), which displayed autistic-like behavioral deficits, including decreased interests in social novelty and increased self-grooming habits, abnormal vocalizations, and deficits in motor learning tasks [46]. At the cellular level the Tsc1-mutant PCs had enlarged soma size and increased pS6 expression, increased spine density, and increased cellular stress compared to control PCs [46]. These mice did not have seizures, but this study showed that cerebellar deficits caused by loss of Tsc1 and mTORC1-pathway activation in PCs are connected to the development of ASD, and mTORC1-pathway inhibition with rapamycin could partially rescue these deficits [46]. Similar results were obtained using Tsc2-knockout specifically in PCs [55]. In summary, these Tsc1/Tsc2 mice models provide evidence that neuron specific loss of Tsc1 or Tsc2 causes mTORC1-hyperactivation leading to both epilepsy and ASD-like phenotypes suggesting a shared pathological origin of these neurological deficits in TSC.

Although these rodent models have been valuable in studying the link between epilepsy and autistic-like behaviors in vivo, these models do not provide insights into the molecular and genetic deficits that drive disease development and manifestation in TSC-patient derived neurons. Thus, in the following sections, we will review the molecular signaling studies in TSC related epilepsy and autism in diagnostically and therapeutically relevant human stem cell derived neuronal cell models.

5.3 Genes Associated with Neural Differentiation and Synaptic Development in TSC2-Deficient Human Neurons

To characterize the effects of TSC2 mutations on synaptic development of human neurons, researchers have previously established an allelic series of TSC2-mutations in isogenic hESC-lines with control *TSC2*$^{+/+}$ and zinc-finger-nuclease-induced *TSC2*$^{+/-}$ and *TSC2*$^{-/-}$ [29]. This cellular model described deficient neuronal differentiation and altered synaptogenesis in human stem cells with decreased expression of a group of genes related to synaptic transmission and autism [29], including: *CNTNAP2*, *NLGN3*, *KCC2*, and *RBFOX1*. More specifically, the *CNTNAP2* gene codes contactin-associated protein-like 2, which belongs to the neurexin family of proteins, and functions as a cell adhesion molecule and receptor in the nervous system. The neurexin family of proteins are expressed in myelinated axons, and they regulate neuron and glia interactions during development of the nervous system. Deficits in the expression of this gene have been associated with several neurodevelopmental disorders, which are highly associated with autism and intellectual disability, epilepsy, ADHD, schizophrenia, and Tourette syndrome [112]. NLGN3 is a neuroligin belonging to a family of post-synaptic adhesion molecules that regulate formation, maturation, and function of synapses. Previous studies have indicated that mutations in *NLGN3* and *NLGN4* are associated with autism [113] and induced point-mutations in *NLGN3* altered synaptic functionality in mice hippocampal and cortical neurons [114]. KCC2 is a potassium chloride cotransporter 2 that is coded by the SLC12A5 gene and operates as a transporter of chloride in mature neurons, in addition to regulating neuronal inhibition [115, 116]. Decreased KCC2 expression reduces activation of GABA receptors by increasing intracellular Cl^- levels and depolarization of the cells, which leads to decreased inhibitory signaling of the GABAergic cells. In the studies of Rett-syndrome, researchers have shown that stem cell derived neurons from Rett-patients had deficient MeCP2 expression which caused loss of KCC2 expression and delayed developmental switch of GABAergic cells from excitatory to inhibitory [117]. Interestingly, previous studies have also described that loss of function of KCC2 causes infantile epilepsy. Downregulation of KCC2 expression is also associated with idiopathic and acquired epilepsy and neurodevelopmental disorders with seizures [115]. Thus, the downregulation of *KCC2* in TSC2-mutant human stem cell derived neurons [29] could cause reduced inhibitory signaling of the GABAergic neurons and lead to hyperactivation of excitatory neurons resulting in seizures in TSC patients. *RBFOX1* gene expression has also been shown to be downregulated in TSC2 deficient human neurons compared to control neurons [29]. RBFOX1 also known as Ataxin-2-binding protein 1 (A2BP1) or FOX1 is coded by the *RBFOX1* gene which is located in chromosome region 16p13 in humans [118]. Several studies have shown that deficient RBFOX1 expression is highly associated with autism, and other neurodevelopmental disorders, including epilepsy [119–123].

In addition, detailed gene expression analyses of TSC2 deficient hESC-derived neural precursor cells have discovered downregulation of genes related to neuronal maturation and synapse formation, with increased expression of genes related to astroglial differentiation, and increased expression of genes related to inflammation and metabolism [60]. This is a relevant finding, since previous pathological studies have shown that in the cortical tubers of TSC patients, inflammatory genes were also enriched, whereas synaptic transmission related genes were decreased [63], and both of these datasets contained interferon-inducible genes *GBP1* and *GBP2* [60, 63]. When these in vitro data of TSC2-mutant hESC-derived neural cells was compared to microarray data of SEGAs from TSC patients, correlation in the gene expression profiles was even more significant. Angiogenic factors and inflammation were increased in both groups, whereas the neuronal expression was decreased significantly within both of these groups of genes. In general, the TSC2-deficiency in hESC-derived neural populations drives the cells towards an astrocyte cell fate during the differentiation process, and thus the gene expression profiles were also more strongly correlated with the astrocytomas than with the datasets from cortical tuber biopsies [60]. When one compares this hESC-derived neuronal cell population data to transcriptional gene expression profiling of hiPSC-derived TSC2-deficient PCs with TSC2-gene dosage dependent weighted co-expression analyses [17], there is significant overlap within the gene expression profiles [17, 60].

Gene ontology analyses of the TSC-patient hiPSC-derived PC precursors revealed that genes whose expression were increased in $TSC2^{-/-}$ cells compared to control cells were related to mitochondria function, protein transport, and autophagy. In contrast, the group of downregulated genes in the $TSC2^{-/-}$ cells were genes related to mRNA transportation and production [17]. This led to further comparison of genes expressed in TSC-patient hiPSC-derived PCs [17] to the gene expression data of FMRP targets in two previous transcriptional gene expression studies [124, 125]. FMRP is coded by *Fragile-X-mental-retardation 1* gene *FMR1*. FMRP functions as an mRNA binding protein and regulator of mRNA trafficking within axons together with motor proteins, and it also mediates protein translation, synaptic protein expression, and synapse formation in axons [124–131]. In addition, several genes related to the development of ASD are enriched in the FMRP target gene group. When the TSC2-deficient hiPSC-derived PC data was compared to these previous gene expression analyses [124, 125], it was discovered that FMRP target genes and target genes of FXR1 were significantly enriched in the co-expression group of downregulated genes in the $TSC2^{-/-}$ hiPSC-derived PCs [17]. These results suggest that both the mTOR-pathway and FMRP regulates the same genetic pathways that are involved in the neuronal dysfunction predisposing to ASD and FXS.

5.4 Synaptic Function and Excitability of TSC2-Deficient Human Neurons

Whole cell patch-clamping is a conventional method that has been utilized for several decades for characterization of functional properties of individual neurons in vitro. The first study to describe the effects of loss of function of TSC2 in the functionality of hESC-derived neurons, discovered that TSC2-deficient neurons ($TSC2^{+/-}$ and $TSC2^{-/-}$) displayed decreased excitability compared to control neurons ($TSC2^{+/-}$) [29]. This study also described reduced glutamatergic synapse formation and function in $TSC2^{-/-}$ neurons compared to control $TSC2^{+/+}$ cells, with decreased frequency of spontaneous excitatory post-synaptic currents (sEPSCs) and miniature excitatory post-synaptic currents (mEPSCs) [29]. These functional deficits were rescued with long-term rapamycin treatment in vitro, which confirmed involvement of mTOR-pathway regulation [29]. This functional data is in line with previously described altered functional and behavioral deficits in the SynCre-Tsc2-mouse model [108, 109]. Importantly, the $TSC2^{-/-}$ hESC-derived neuronal cells displayed similar functional deficits to dysplastic and enlarged neurons in cortical tubers, by being significantly hypoexcitable compared to healthy control neuronal cells [29]. These in vitro findings support previous clinical studies, which have described that formation of cortical tubers predispose the brain towards development of epilepsy in TSC patients by disturbing the formation of neuronal network connectivity and functionality during brain development [132, 133].

In line with these studies, our group has developed a TSC-patient hiPSC-derived PC model, which displayed significant synaptic deficits and electrophysiological abnormalities in the TSC2-deficient PCs compared to control PCs in vitro [17]. According to this study, the TSC-patient hiPSC-derived PCs with specific TSC2-mutations showed significantly increased capacitance and input resistance compared to control PCs [17]. This finding was consistent with the increased cellular size of TSC2-deficient cells compared to control cells. Interestingly, the TSC2-deficient patient PCs and $TSC2^{-/-}$ PCs had lower firing rates compared to control cells. In addition, TSC2-deficient patient PCs and $TSC2^{-/-}$ PCs were significantly hypoexcitable compared to control PCs. Although it is important to note that these cells are only modeling human cerebellum development in vitro, the detected disease phenotypes in patient neurons with specific loss of TSC2 are consistent with the functional phenotypes detected previously in Tsc1/2-deficient mice cerebellar PCs [46, 55]. Functional analyses of Tsc1-mutant PCs in the mouse cerebellum slices have shown reduced spontaneous firing rates in the $Tsc1^{+/-}$ and $Tsc1^{-/-}$ PCs compared to control PCs $Tsc1^{+/+}$ [46].

At the molecular level, TSC2-deficient hiPSC-derived PCs have reduced expression of pre- and post-synaptic proteins: synaptophysin and PSD95, and PC specific glutamate receptor δ2 [17]. Importantly, we also detected decreased expression of FMRP protein, in TSC2-deficient hiPSC-derived PCs, which affects synaptic development and functionality of the cells [17]. TSC2-deficient hiPSC-derived GABAergic PCs also had decreased synaptic transmission, which was detected by decreased frequency of mEPSCs in $TSC2^{-/-}$ PCs compared to control $TSC2^{+/+}$ hiPSC-derived PCs [17]. These functional phenotypes of human PCs are regulated by mTORC1 pathway hyperactivation, since long-term rapamycin treatment rescued these deficits in TSC-patient hiPSC-derived PCs [17]. TSC and FXS are two similar neurodevelopmental disorders that lead to neuronal deficits and ASD. Thus, these studies suggest that there might exist a synergy in mTORC1 pathway regulation and FMRP expression in TSC and FXS. In general, decreased inhibitory neuron activity in the development of the neural network can cause imbalance between inhibitory and excitatory signaling and predispose to dysfunctional neural circuitry formation in TSC and ASD. Although more research needs to be done to characterize the detailed molecular mechanisms affecting the neuronal development and network functionality during inhibitory and excitatory neuron differentiation in TSC-patients brain, here we have summarized the disease phenotypes detected in the TSC2-deficient pluripotent stem cell derived neuronal cells in vitro: Fig. 3, Table 1.

6 Future Prospects

6.1 *Pharmacological Treatment of Neurological Deficits in TSC*

To date, several preclinical animal studies have been conducted with mTOR-inhibitors rapamycin and everolimus to rescue the behavioral, functional, and cellular deficits in TSC [110, 134]. These studies show that rapamycin treatment increased the lifespan of the animals and decreased occurrence of seizures in the mice [110, 134]. In addition, in clinical studies with TSC patients, everolimus reduced the frequency of the seizures and improved the quality of life of the patients, and the treatment was well tolerated [135]. Other clinical studies have shown that

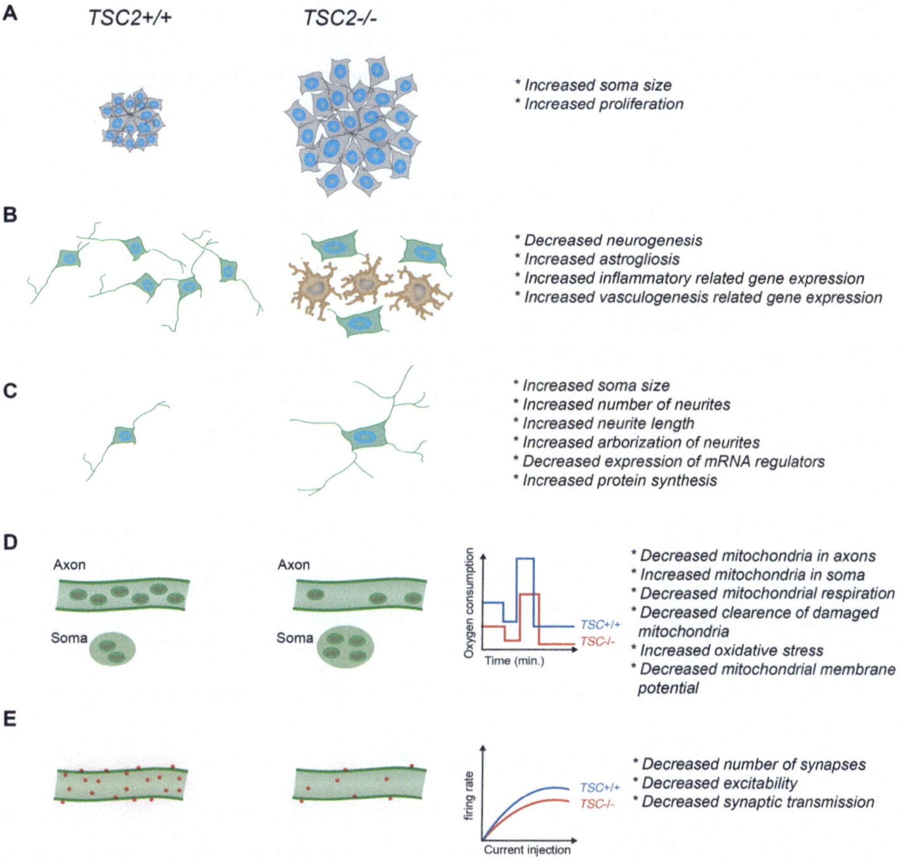

Fig. 3 Schematic presentation of TSC2-deficient cellular phenotypes in vitro. (**a**) Control and TSC2-deficient neural precursor cells. (**b**) Control and TSC2-deficient developing neuronal cultures. (**c**) Control and TSC2-deficient maturing neurons, (**d**) Mitochondrial development and function in control and TSC2-deficient neurons, (**e**) Synapse development and electrophysiological activity of the control and TSC2-deficient neurons

long-term everolimus treatment reduced seizure frequency in ≥50% TSC patient populations [136, 137] and > 50% of the TSC patients became seizure free [137]. Related to this, it has been reported that treatment of TSC patients with sirolimus (rapamycin) decreased frequency of seizures by 41% [138], although significant improvements were not discovered in this study, and cognitive development of the patients did not improve [138]. Taken together, these studies indicate that mTOR-inhibitors may have beneficial effects for the treatment of epilepsy in TSC patients, although more controlled, long-term clinical studies are needed to confirm these findings.

Clinical studies have also shown that mTOR-inhibitors are effective in slowing the growth of SEGAs and renal angiomyolipoma in TSC patients [139, 140]. Although mTOR-inhibitors possess clear advantageous effects for TSC patients,

Table 1 Human pluripotent stem cell derived neural cell cultures as a platform to study TSC

TSC-mutant human stem cells	Stem cell derived neural cell types	Gene deficits in TSC2-mutant human neural cells	Cellular deficits in the TSC2-mutant human neural cells	References
– Isogenic hESCs (*TSC2+/+*) with zinc-finger nuclease-induced allelic series of: *TSC2+/–* *TSC2–/–*	– NPCs and neural rosettes – Mixed population of GABAergic neurons and glutamatergic neurons	– Increased expression of genes regulating NSC proliferation – Downregulation of genes regulating neuronal maturation and synapse formation – Downregulation of genes related to synaptic transmission – Dysregulation of ASD-related genes	– Hyperactivation of mTORC1 pathway in NPCs and neurons – Increased rosette area and abnormal cellular structure – Increased soma size – Increased dendritic arborization – Reduced firing rate – Reduced frequency of sEPSCs and mEPSCs – Rapamycin rescued the early proliferation deficits and soma size in NPCs – Rapamycin rescued the synaptic deficits and functionality of neurons	Costa et al. [29], *Cell Rep.*
– Isogenic hESCs (*TSC2+/+*) with zinc-finger nuclease-induced allelic series of: *TSC2+/–* *TSC2–/–*	– NPCs – Mixed population of neurons and astrocytes	– Increased astroglial genes – Decreased glutamatergic and GABAergic neuron markers – Increased inflammatory induced genes – Increased expression of angiogenic genes	– Reduced neuronal maturation – Increased astrogliosis – Increased active inflammatory response – Increased metabolic activity – Increased ribosome occupancy and protein synthesis – Increased angiogenic protein expression – Rapamycin rescued the translational deficits	Grabole et al. [60], *Genome Med.*

(continued)

Table 1 (continued)

TSC-mutant human stem cells	Stem cell derived neural cell types	Gene deficits in TSC2-mutant human neural cells	Cellular deficits in the TSC2-mutant human neural cells	References
– TSC/ASD-patient derived hiPSCs (*TSC2+/–*) vs. parental control or sex-matched control hiPSCs (*TSC2+/+*) – Isogenic TSC-patient derived hiPSCs (*TSC2+/–*), with allelic series of: *TSC2+/+* (CRISPR-cas9-corrected patient line),*TSC2–/–* (TALEN induced mutation in the second TSC2 allele of patient-derived cell line)	– Cerebellar precursor cells – THY1+ Purkinje cells	– Increased mitochondrial gene expression – Increased autophagy gene expression – Increased expression of genes related to protein transport and translation – Decreased nuclear mRNA expression and mRNA regulator genes – Decreased post-transcriptional regulator genes – Decreased expression of FMRP target genes	– Hyperactivation of mTORC1- pathway in NPCs and PCs – Increased NPC proliferation – Reduced neuronal differentiation – Increased astrocyte differentiation – Increased mROS – Increased soma size of PCs – Increased number of neurites per cell – Decreased synaptic marker expression – Reduced firing rate – Reduced excitability – Reduced frequency of mEPSCs – Rapamycin rescued the early proliferation and differentiation deficits of the NPCs – Rapamycin rescued the synaptic deficits and functionality of the neurons	Sundberg et al. [17], *Mol Psych.*
– TSC/ASD-patient derived hiPSCs (*TSC2+/–*) vs. parental control hiPSCs (*TSC2+/+*) – TSC-patient *TSC2+/–*, parental control *TSC2+/+* and *TSC2–/–* (TALEN-induced mutation into second TSC2 allele of patient-derived cell line)	– NGN2-induced cortical neurons	– No gene expression data from human neurons	– Decreased number of mitochondria in axons – Increased accumulation of mitochondria in the soma – Decreased mitochondrial membrane potential – Decreased mitochondrial respiration	Ebrahimi-Fakhari et al. [83], *Cell Rep.*

(continued)

Table 1 (continued)

TSC-mutant human stem cells	Stem cell derived neural cell types	Gene deficits in TSC2-mutant human neural cells	Cellular deficits in the TSC2-mutant human neural cells	References
– TSC/ASD-patient derived hiPSCs (*TSC2+/–*) vs. parental control hiPSCs (*TSC2+/+*)	– Neuronal cells differentiated with dual-SMAD-protocol	– No gene expression data from human neurons	– Increased expression of negative regulator of neuronal myelination CTGF	Ercan et al. [74], *J Exp Med.*
– TSC-patient derived hiPSC (*TSC2+/–)* vs. unaffected non-related sex-matched control hiPSCs (*TSC2+/+*)	– NPCs – Mixed population of neurons and astrocytes	– Sequencing data from hiPSC-NPCs confirmed the presence of heterozygous TSC2 mutation in patient derived cells	– Hyperactivation of mTORC1 pathway in NPCs, neurons and astrocytes – Increased NPC proliferation – Increased astrocyte precursor cell proliferation – abnormal neurite development and increased branching – Increased astrogliosis – Rapamycin rescued the proliferation deficits	Li et al. [38], *Stem Cell Reports*

safety issues must be assessed in more detail, to ensure reliable long-term usage of these drugs in future clinical studies. Rapamycin is an immunosuppressant drug and long-term use of immunosuppressants may alter patients' risk of infections [139]. This is important to consider, since previous clinical and rodent studies have shown that neurological deficits reoccur if the drug treatment is terminated [134, 139, 141]. Since mTOR-pathway activation affects several cellular pathways during development, including cell proliferation and neuronal differentiation, further studies are needed to optimize the administered dosage of mTOR-inhibitors in children with TSC to avoid adverse effects with regard to brain development, motor skills, and learning capacity.

In addition to these ongoing clinical trials of mTOR-inhibitors, the neurological symptoms of TSC are treated with commonly used antiepileptic drugs. Antiepileptic drugs are utilized to reduce the overactivation of the neurons in the brain by modulating transmembrane ion channels. The problem with these common antiepileptic drugs is that they do not target specifically the mTOR-pathway, which is aberrant in the TSC-patient derived neurons. Thus, a key aspect in seeking new treatments for TSC is to develop effective, safe, and stable inhibitors of mTOR-pathway overactivation that can rescue the neuronal dysfunction in the TSC patient brain.

6.2 Drug Screening Assays in TSC2-Deficient Human Neurons

As described in this chapter, the possibility to rescue the differentiation and functional deficits of TSC-patient derived neural cells and embryonic stem cell derived neurons in vitro [17, 29] opens up new opportunities for the development of novel drug screening assays for human neurons. TSC-patient derived neurons will facilitate screening of small molecule libraries and FDA-approved drug-libraries in a high throughput manner and will facilitate selection of new compounds that have high target specificity and efficacy. TSC-patient derived neurons are also useful for testing the safety, and optimal concentration of the novel compounds in vitro, prior to entering into preclinical animal studies or clinical trials with TSC patients. These compounds can be tested in vitro for cellular toxicity, and for detection of the minimum effective concentration that could improve the function of the patient derived neuronal networks. In recent years, several new technologies have been established for these phenotypic compound-library-screens, such as high throughput imaging platforms, cytokine release assays, cellular metabolism assays, and high-resolution multi-electrode-array-platforms for assessment of electrophysiological activity. We anticipate that these drug screening assays will eventually facilitate development of new treatment options for TSC related neurological dysfunctions.

Acknowledgements Owing to limited space, we have not quoted all literature in the field, and we apologize to those whose articles are not referenced. We would like to thank Denise McGinnis for critical reading of the manuscript. We would like to thank Ville Kujala for design and preparation of the images. The Sahin lab has received grant funding from the US National Institutes of Health (NIH) (U01-NS082320, U01-NS092595, and U54-HD090255), US Department of Defense W81XWH-15-1-0189, Nancy Lurie Marks Family Foundation, Autism Speaks, TS Alliance, National Ataxia Foundation, Harvard Stem Cell Institute, Tommy Fuss Center, Roche, Novartis, Pfizer and LAM Therapeutics.

References

1. Feliciano, D. M., Lin, T. V., Hartman, N. W., Bartley, C. M., Kubera, C., Hsieh, L., et al. (2013). A circuitry and biochemical basis for tuberous sclerosis symptoms: From epilepsy to neurocognitive deficits. *International Journal of Developmental Neuroscience, 31*, 667–678.
2. Orlova, K. A., & Crino, P. B. (2010). The tuberous sclerosis complex. *Annals of the New York Academy of Sciences, 1184*, 87–105.

3. Sahin, M., Henske, E. P., Manning, B. D., Ess, K. C., Bissler, J. J., Klann, E., et al. (2016). Advances and future directions for tuberous sclerosis complex research: recommendations from the 2015 strategic planning conference. *Pediatric Neurology, 60*, 1–12.
4. Crino, P. B., Nathanson, K. L., & Henske, E. P. (2006). The tuberous sclerosis complex. *The New England Journal of Medicine, 355*, 1345–1356.
5. Lipton, J. O., & Sahin, M. (2014). The neurology of mTOR. *Neuron, 84*, 275–291.
6. Chu-Shore, C. J., Major, P., Camposano, S., Muzykewicz, D., & Thiele, E. A. (2010). The natural history of epilepsy in tuberous sclerosis complex. *Epilepsia, 51*, 1236–1241.
7. Jeste, S. S., Sahin, M., Bolton, P., Ploubidis, G. B., & Humphrey, A. (2008). Characterization of autism in young children with tuberous sclerosis complex. *Journal of Child Neurology, 23*, 520–525.
8. Richards, C., Jones, C., Groves, L., Moss, J., & Oliver, C. (2015). Prevalence of autism spectrum disorder phenomenology in genetic disorders: A systematic review and meta-analysis. *Lancet Psychiatry, 2*, 909–916.
9. Bruining, H., Eijkemans, M. J., Kas, M. J., Curran, S. R., Vorstman, J. A., & Bolton, P. F. (2014). Behavioral signatures related to genetic disorders in autism. *Molecular Autism, 5*, 11.
10. Jeste, S. S., Varcin, K. J., Hellemann, G. S., Gulsrud, A. C., Bhatt, R., Kasari, C., et al. (2016). Symptom profiles of autism spectrum disorder in tuberous sclerosis complex. *Neurology, 87*, 766–772.
11. Chamberlain, S. J., Chen, P. F., Ng, K. Y., Bourgois-Rocha, F., Lemtiri-Chlieh, F., Levine, E. S., et al. (2010). Induced pluripotent stem cell models of the genomic imprinting disorders Angelman and Prader-Willi syndromes. *Proceedings of the National Academy of Sciences of the United States of America, 107*, 17668–17673.
12. Chiu, F. L., Lin, J. T., Chuang, C. Y., Chien, T., Chen, C. M., Chen, K. H., et al. (2015). Elucidating the role of the A2A adenosine receptor in neurodegeneration using neurons derived from Huntington's disease iPSCs. *Human Molecular Genetics, 24*, 6066–6079.
13. Cooper, O., Seo, H., Andrabi, S., Guardia-Laguarta, C., Graziotto, J., Sundberg, M., et al. (2012). Pharmacological rescue of mitochondrial deficits in iPSC-derived neural cells from patients with familial Parkinson's disease. *Science Translational Medicine, 4*, 141ra190.
14. Marchetto, M. C., Carromeu, C., Acab, A., Yu, D., Yeo, G. W., Mu, Y., et al. (2010). A model for neural development and treatment of Rett syndrome using human induced pluripotent stem cells. *Cell, 143*, 527–539.
15. Pasca, S. P., Portmann, T., Voineagu, I., Yazawa, M., Shcheglovitov, A., Pasca, A. M., et al. (2011). Using iPSC-derived neurons to uncover cellular phenotypes associated with Timothy syndrome. *Nature Medicine, 17*, 1657–1662.
16. Sundberg, M., Bogetofte, H., Lawson, T., Jansson, J., Smith, G., Astradsson, A., et al. (2013). Improved cell therapy protocols for Parkinson's disease based on differentiation efficiency and safety of hESC-, hiPSC-, and non-human primate iPSC-derived dopaminergic neurons. *Stem Cells, 31*, 1548–1562.
17. Sundberg, M., Tochitsky, I., Buchholz, D. E., Winden, K., Kujala, V., Kapur, K., et al. (2018). Purkinje cells derived from TSC patients display hypoexcitability and synaptic deficits associated with reduced FMRP levels and reversed by rapamycin. *Molecular Psychiatry, 23*(11), 2167–2183.
18. Wainger, B. J., Kiskinis, E., Mellin, C., Wiskow, O., Han, S. S., Sandoe, J., et al. (2014). Intrinsic membrane hyperexcitability of amyotrophic lateral sclerosis patient-derived motor neurons. *Cell Reports, 7*, 1–11.
19. Woodard, C. M., Campos, B. A., Kuo, S. H., Nirenberg, M. J., Nestor, M. W., Zimmer, M., et al. (2014). iPSC-derived dopamine neurons reveal differences between monozygotic twins discordant for Parkinson's disease. *Cell Reports, 9*, 1173–1182.
20. Saxton, R. A., & Sabatini, D. M. (2017). mTOR signaling in growth, metabolism, and disease. *Cell, 169*, 361–371.

21. Dibble, C. C., Elis, W., Menon, S., Qin, W., Klekota, J., Asara, J. M., et al. (2012). TBC1D7 is a third subunit of the TSC1-TSC2 complex upstream of mTORC1. *Molecular Cell, 47*, 535–546.
22. Inoki, K., Li, Y., Xu, T., & Guan, K. L. (2003). Rheb GTPase is a direct target of TSC2 GAP activity and regulates mTOR signaling. *Genes & Development, 17*, 1829–1834.
23. Huang, W., Zhu, P. J., Zhang, S., Zhou, H., Stoica, L., Galiano, M., et al. (2013). mTORC2 controls actin polymerization required for consolidation of long-term memory. *Nature Neuroscience, 16*, 441–448.
24. Thomanetz, V., Angliker, N., Cloetta, D., Lustenberger, R. M., Schweighauser, M., Oliveri, F., et al. (2013). Ablation of the mTORC2 component rictor in brain or Purkinje cells affects size and neuron morphology. *The Journal of Cell Biology, 201*, 293–308.
25. Takahashi, K., Tanabe, K., Ohnuki, M., Narita, M., Ichisaka, T., Tomoda, K., et al. (2007). Induction of pluripotent stem cells from adult human fibroblasts by defined factors. *Cell, 131*, 861–872.
26. Thomson, J. A., Itskovitz-Eldor, J., Shapiro, S. S., Waknitz, M. A., Swiergiel, J. J., Marshall, V. S., et al. (1998). Embryonic stem cell lines derived from human blastocysts. *Science, 282*, 1145–1147.
27. Chambers, S. M., Fasano, C. A., Papapetrou, E. P., Tomishima, M., Sadelain, M., & Studer, L. (2009). Highly efficient neural conversion of human ES and iPS cells by dual inhibition of SMAD signaling. *Nature Biotechnology, 27*, 275–280.
28. Gerrard, L., Rodgers, L., & Cui, W. (2005). Differentiation of human embryonic stem cells to neural lineages in adherent culture by blocking bone morphogenetic protein signaling. *Stem Cells, 23*, 1234–1241.
29. Costa, V., Aigner, S., Vukcevic, M., Sauter, E., Behr, K., Ebeling, M., et al. (2016). mTORC1 inhibition corrects neurodevelopmental and synaptic alterations in a human stem cell model of tuberous sclerosis. *Cell Reports, 15*, 86–95.
30. Cho, S. W., Kim, S., Kim, J. M., & Kim, J. S. (2013). Targeted genome engineering in human cells with the Cas9 RNA-guided endonuclease. *Nature Biotechnology, 31*, 230–232.
31. Horii, T., Tamura, D., Morita, S., Kimura, M., & Hatada, I. (2013). Generation of an ICF syndrome model by efficient genome editing of human induced pluripotent stem cells using the CRISPR system. *International Journal of Molecular Sciences, 14*, 19774–19781.
32. Hou, Z., Zhang, Y., Propson, N. E., Howden, S. E., Chu, L. F., Sontheimer, E. J., et al. (2013). Efficient genome engineering in human pluripotent stem cells using Cas9 from Neisseria meningitidis. *Proceedings of the National Academy of Sciences of the United States of America, 110*, 15644–15649.
33. Mussolino, C., Morbitzer, R., Lutge, F., Dannemann, N., Lahaye, T., & Cathomen, T. (2011). A novel TALE nuclease scaffold enables high genome editing activity in combination with low toxicity. *Nucleic Acids Research, 39*, 9283–9293.
34. Zhang, Y., Pak, C., Han, Y., Ahlenius, H., Zhang, Z., Chanda, S., et al. (2013). Rapid single-step induction of functional neurons from human pluripotent stem cells. *Neuron, 78*, 785–798.
35. Feliciano, D. M., Quon, J. L., Su, T., Taylor, M. M., & Bordey, A. (2012). Postnatal neurogenesis generates heterotopias, olfactory micronodules and cortical infiltration following single-cell Tsc1 deletion. *Human Molecular Genetics, 21*, 799–810.
36. Magri, L., Cambiaghi, M., Cominelli, M., Alfaro-Cervello, C., Cursi, M., Pala, M., et al. (2011). Sustained activation of mTOR pathway in embryonic neural stem cells leads to development of tuberous sclerosis complex-associated lesions. *Cell Stem Cell, 9*, 447–462.
37. Zhou, J., Shrikhande, G., Xu, J., McKay, R. M., Burns, D. K., Johnson, J. E., et al. (2011). Tsc1 mutant neural stem/progenitor cells exhibit migration deficits and give rise to subependymal lesions in the lateral ventricle. *Genes & Development, 25*, 1595–1600.
38. Li, Y., Cao, J., Chen, M., Li, J., Sun, Y., Zhang, Y., et al. (2017). Abnormal neural progenitor cells differentiated from induced pluripotent stem cells partially mimicked development of TSC2 neurological abnormalities. *Stem Cell Reports, 8*, 883–893.

39. Ma, J., Yu, Z., Qu, W., Tang, Y., Zhan, Y., Ding, C., et al. (2010). Proliferation and differentiation of neural stem cells are selectively regulated by knockout of cyclin D1. *Journal of Molecular Neuroscience, 42*, 35–43.
40. Sundberg, M., Savola, S., Hienola, A., Korhonen, L., & Lindholm, D. (2006). Glucocorticoid hormones decrease proliferation of embryonic neural stem cells through ubiquitin-mediated degradation of cyclin D1. *The Journal of Neuroscience, 26*, 5402–5410.
41. Mirzaa, G., Dodge, N. N., Glass, I., Day, C., Gripp, K., Nicholson, L., et al. (2004). Megalencephaly and perisylvian polymicrogyria with postaxial polydactyly and hydrocephalus: A rare brain malformation syndrome associated with mental retardation and seizures. *Neuropediatrics, 35*, 353–359.
42. Mirzaa, G. M., Parry, D. A., Fry, A. E., Giamanco, K. A., Schwartzentruber, J., Vanstone, M., et al. (2014). De novo CCND2 mutations leading to stabilization of cyclin D2 cause megalencephaly-polymicrogyria-polydactyly-hydrocephalus syndrome. *Nature Genetics, 46*, 510–515.
43. Courchesne, E., Mouton, P. R., Calhoun, M. E., Semendeferi, K., Ahrens-Barbeau, C., Hallet, M. J., et al. (2011). Neuron number and size in prefrontal cortex of children with autism. *JAMA, 306*, 2001–2010.
44. Redcay, E., & Courchesne, E. (2005). When is the brain enlarged in autism? A meta-analysis of all brain size reports. *Biological Psychiatry, 58*, 1–9.
45. Crowell, B., Lee, G. H., Nikolaeva, I., Dal Pozzo, V., & D'Arcangelo, G. (2015). Complex neurological phenotype in mutant mice lacking Tsc2 in excitatory neurons of the developing forebrain(123). *Eneuro, 2*. https://doi.org/10.1523/ENEURO.0046-15.2015
46. Tsai, P. T., Hull, C., Chu, Y., Greene-Colozzi, E., Sadowski, A. R., Leech, J. M., et al. (2012). Autistic-like behaviour and cerebellar dysfunction in Purkinje cell Tsc1 mutant mice. *Nature, 488*, 647–651.
47. Choi, Y. J., Di Nardo, A., Kramvis, I., Meikle, L., Kwiatkowski, D. J., Sahin, M., et al. (2008). Tuberous sclerosis complex proteins control axon formation. *Genes & Development, 22*, 2485–2495.
48. Buccoliero, A. M., Franchi, A., Castiglione, F., Gheri, C. F., Mussa, F., Giordano, F., et al. (2009). Subependymal giant cell astrocytoma (SEGA): Is it an astrocytoma? Morphological, immunohistochemical and ultrastructural study. *Neuropathology, 29*, 25–30.
49. Grajkowska, W., Kotulska, K., Jurkiewicz, E., & Matyja, E. (2010). Brain lesions in tuberous sclerosis complex. Review. *Folia Neuropathologica, 48*, 139–149.
50. Grajkowska, W., Kotulska, K., Jurkiewicz, E., Roszkowski, M., Daszkiewicz, P., Jozwiak, S., et al. (2011). Subependymal giant cell astrocytomas with atypical histological features mimicking malignant gliomas. *Folia Neuropathologica, 49*, 39–46.
51. Bailey, A., Luthert, P., Dean, A., Harding, B., Janota, I., Montgomery, M., et al. (1998). A clinicopathological study of autism. *Brain, 121*(Pt 5), 889–905.
52. Limperopoulos, C., Bassan, H., Gauvreau, K., Robertson Jr., R. L., Sullivan, N. R., Benson, C. B., et al. (2007). Does cerebellar injury in premature infants contribute to the high prevalence of long-term cognitive, learning, and behavioral disability in survivors? *Pediatrics, 120*, 584–593.
53. Skefos, J., Cummings, C., Enzer, K., Holiday, J., Weed, K., Levy, E., et al. (2014). Regional alterations in purkinje cell density in patients with autism. *PLoS One, 9*, e81255.
54. Whitney, E. R., Kemper, T. L., Bauman, M. L., Rosene, D. L., & Blatt, G. J. (2008). Cerebellar Purkinje cells are reduced in a subpopulation of autistic brains: A stereological experiment using calbindin-D28k. *Cerebellum, 7*, 406–416.
55. Reith, R. M., McKenna, J., Wu, H., Hashmi, S. S., Cho, S. H., Dash, P. K., et al. (2013). Loss of Tsc2 in Purkinje cells is associated with autistic-like behavior in a mouse model of tuberous sclerosis complex. *Neurobiology of Disease, 51*, 93–103.
56. Muguruma, K., Nishiyama, A., Ono, Y., Miyawaki, H., Mizuhara, E., Hori, S., et al. (2010). Ontogeny-recapitulating generation and tissue integration of ES cell-derived Purkinje cells. *Nature Neuroscience, 13*, 1171–1180.

57. Muguruma, K., Nishiyama, A., Kawakami, H., Hashimoto, K., & Sasai, Y. (2015). Self-organization of polarized cerebellar tissue in 3D culture of human pluripotent stem cells. *Cell Reports, 10*, 537–550.
58. Wang, S., Wang, B., Pan, N., Fu, L., Wang, C., Song, G., et al. (2015). Differentiation of human induced pluripotent stem cells to mature functional Purkinje neurons. *Scientific Reports, 5*, 9232.
59. Morino, H., Matsuda, Y., Muguruma, K., Miyamoto, R., Ohsawa, R., Ohtake, T., et al. (2015). A mutation in the low voltage-gated calcium channel CACNA1G alters the physiological properties of the channel, causing spinocerebellar ataxia. *Molecular Brain, 8*, 89.
60. Grabole, N., Zhang, J. D., Aigner, S., Ruderisch, N., Costa, V., Weber, F. C., et al. (2016). Genomic analysis of the molecular neuropathology of tuberous sclerosis using a human stem cell model. *Genome Medicine, 8*, 94.
61. Wong, M., & Crino, P. B. (2012). Tuberous sclerosis and epilepsy: Role of astrocytes. *Glia, 60*, 1244–1250.
62. Zamanian, J. L., Xu, L., Foo, L. C., Nouri, N., Zhou, L., Giffard, R. G., et al. (2012). Genomic analysis of reactive astrogliosis. *The Journal of Neuroscience, 32*, 6391–6410.
63. Boer, K., Crino, P. B., Gorter, J. A., Nellist, M., Jansen, F. E., Spliet, W. G., et al. (2010). Gene expression analysis of tuberous sclerosis complex cortical tubers reveals increased expression of adhesion and inflammatory factors. *Brain Pathology, 20*, 704–719.
64. Zhang, B., Zou, J., Rensing, N. R., Yang, M., & Wong, M. (2015). Inflammatory mechanisms contribute to the neurological manifestations of tuberous sclerosis complex. *Neurobiology of Disease, 80*, 70–79.
65. Tyler, W. A., Gangoli, N., Gokina, P., Kim, H. A., Covey, M., Levison, S. W., et al. (2009). Activation of the mammalian target of rapamycin (mTOR) is essential for oligodendrocyte differentiation. *The Journal of Neuroscience, 29*, 6367–6378.
66. Guardiola-Diaz, H. M., Ishii, A., & Bansal, R. (2012). Erk1/2 MAPK and mTOR signaling sequentially regulates progression through distinct stages of oligodendrocyte differentiation. *Glia, 60*, 476–486.
67. Tyler, W. A., Jain, M. R., Cifelli, S. E., Li, Q., Ku, L., Feng, Y., et al. (2011). Proteomic identification of novel targets regulated by the mammalian target of rapamycin pathway during oligodendrocyte differentiation. *Glia, 59*, 1754–1769.
68. Arulrajah, S., Ertan, G., Jordan, L., Tekes, A., Khaykin, E., Izbudak, I., et al. (2009). Magnetic resonance imaging and diffusion-weighted imaging of normal-appearing white matter in children and young adults with tuberous sclerosis complex. *Neuroradiology, 51*, 781–786.
69. Makki, M. I., Chugani, D. C., Janisse, J., & Chugani, H. T. (2007). Characteristics of abnormal diffusivity in normal-appearing white matter investigated with diffusion tensor MR imaging in tuberous sclerosis complex. *AJNR. American Journal of Neuroradiology, 28*, 1662–1667.
70. Peters, J. M., Sahin, M., Vogel-Farley, V. K., Jeste, S. S., Nelson 3rd, C. A., Gregas, M. C., et al. (2012). Loss of white matter microstructural integrity is associated with adverse neurological outcome in tuberous sclerosis complex. *Academic Radiology, 19*, 17–25.
71. Carson, R. P., Kelm, N. D., West, K. L., Does, M. D., Fu, C., Weaver, G., et al. (2015). Hypomyelination following deletion of Tsc2 in oligodendrocyte precursors. *Annals of Clinical Translational Neurology, 2*, 1041–1054.
72. Lebrun-Julien, F., Bachmann, L., Norrmen, C., Trotzmuller, M., Kofeler, H., Ruegg, M. A., et al. (2014). Balanced mTORC1 activity in oligodendrocytes is required for accurate CNS myelination. *The Journal of Neuroscience, 34*, 8432–8448.
73. Bercury, K. K., Dai, J., Sachs, H. H., Ahrendsen, J. T., Wood, T. L., & Macklin, W. B. (2014). Conditional ablation of raptor or rictor has differential impact on oligodendrocyte differentiation and CNS myelination. *The Journal of Neuroscience, 34*, 4466–4480.
74. Ercan, E., Han, J. M., Di Nardo, A., Winden, K., Han, M. J., Hoyo, L., et al. (2017). Neuronal CTGF/CCN2 negatively regulates myelination in a mouse model of tuberous sclerosis complex. *The Journal of Experimental Medicine, 214*, 681–697.

75. Meikle, L., Talos, D. M., Onda, H., Pollizzi, K., Rotenberg, A., Sahin, M., et al. (2007). A mouse model of tuberous sclerosis: Neuronal loss of Tsc1 causes dysplastic and ectopic neurons, reduced myelination, seizure activity, and limited survival. *The Journal of Neuroscience, 27*, 5546–5558.
76. Douvaras, P., Wang, J., Zimmer, M., Hanchuk, S., O'Bara, M. A., Sadiq, S., et al. (2014). Efficient generation of myelinating oligodendrocytes from primary progressive multiple sclerosis patients by induced pluripotent stem cells. *Stem Cell Reports, 3*, 250–259.
77. Nistor, G. I., Totoiu, M. O., Haque, N., Carpenter, M. K., & Keirstead, H. S. (2005). Human embryonic stem cells differentiate into oligodendrocytes in high purity and myelinate after spinal cord transplantation. *Glia, 49*, 385–396.
78. Stacpoole, S. R., Spitzer, S., Bilican, B., Compston, A., Karadottir, R., Chandran, S., et al. (2013). High yields of oligodendrocyte lineage cells from human embryonic stem cells at physiological oxygen tensions for evaluation of translational biology. *Stem Cell Reports, 1*, 437–450.
79. Sundberg, M., Hyysalo, A., Skottman, H., Shin, S., Vemuri, M., Suuronen, R., et al. (2011). A xeno-free culturing protocol for pluripotent stem oligodendrocyte precursor cell production. *Regenerative Medicine, 6*, 449–460.
80. Wang, S., Bates, J., Li, X., Schanz, S., Chandler-Militello, D., Levine, C., et al. (2013). Human iPSC-derived oligodendrocyte progenitor cells can myelinate and rescue a mouse model of congenital hypomyelination. *Cell Stem Cell, 12*, 252–264.
81. Goldman, S. A., & Kuypers, N. J. (2015). How to make an oligodendrocyte. *Development, 142*, 3983–3995.
82. Nadadhur, A. G., Alsaqati, M., Gasparotto, L., Cornelissen-Steijger, P., van Hugte, E., Dooves, S., et al. (2019). Neuron-Glia interactions increase neuronal phenotypes in tuberous sclerosis complex patient iPSC-Derived models. *Stem Cell Reports, 12*(1), 42–56.
83. Ebrahimi-Fakhari, D., Saffari, A., Wahlster, L., DiNardo, A., Turner, D., Lewis Jr., T. L., et al. (2016). Impaired mitochondrial dynamics and mitophagy in neuronal models of tuberous sclerosis complex. *Cell Reports, 17*, 2162.
84. Di Nardo, A., Kramvis, I., Cho, N., Sadowski, A., Meikle, L., Kwiatkowski, D. J., et al. (2009). Tuberous sclerosis complex activity is required to control neuronal stress responses in an mTOR-dependent manner. *The Journal of Neuroscience, 29*, 5926–5937.
85. Sheng, R., & Qin, Z. H. (2015). The divergent roles of autophagy in ischemia and preconditioning. *Acta Pharmacologica Sinica, 36*, 411–420.
86. Pickrell, A. M., & Youle, R. J. (2015). The roles of PINK1, parkin, and mitochondrial fidelity in Parkinson's disease. *Neuron, 85*, 257–273.
87. Tan, C. C., Yu, J. T., Tan, M. S., Jiang, T., Zhu, X. C., & Tan, L. (2014). Autophagy in aging and neurodegenerative diseases: Implications for pathogenesis and therapy. *Neurobiology of Aging, 35*, 941–957.
88. Lee, K. M., Hwang, S. K., & Lee, J. A. (2013). Neuronal autophagy and neurodevelopmental disorders. *Experimental Neurobiology, 22*, 133–142.
89. Kim, J., Kundu, M., Viollet, B., & Guan, K. L. (2011). AMPK and mTOR regulate autophagy through direct phosphorylation of Ulk1. *Nature Cell Biology, 13*, 132–141.
90. Tang, G., Gudsnuk, K., Kuo, S. H., Cotrina, M. L., Rosoklija, G., Sosunov, A., et al. (2014). Loss of mTOR-dependent macroautophagy causes autistic-like synaptic pruning deficits. *Neuron, 83*, 1131–1143.
91. McMahon, J., Huang, X., Yang, J., Komatsu, M., Yue, Z., Qian, J., et al. (2012). Impaired autophagy in neurons after disinhibition of mammalian target of rapamycin and its contribution to epileptogenesis. *The Journal of Neuroscience, 32*, 15704–15714.
92. Parkhitko, A., Myachina, F., Morrison, T. A., Hindi, K. M., Auricchio, N., Karbowniczek, M., et al. (2011). Tumorigenesis in tuberous sclerosis complex is autophagy and p62/sequestosome 1 (SQSTM1)-dependent. *Proceedings of the National Academy of Sciences of the United States of America, 108*, 12455–12460.

93. Di Nardo, A., Wertz, M. H., Kwiatkowski, E., Tsai, P. T., Leech, J. D., Greene-Colozzi, E., et al. (2014). Neuronal Tsc1/2 complex controls autophagy through AMPK-dependent regulation of ULK1. *Human Molecular Genetics, 23*, 3865–3874.
94. Weisenfeld, N. I., Peters, J. M., Tsai, P. T., Prabhu, S. P., Dies, K. A., Sahin, M., et al. (2013). A magnetic resonance imaging study of cerebellar volume in tuberous sclerosis complex. *Pediatric Neurology, 48*, 105–110.
95. Constantino, J. N. (2011). The quantitative nature of autistic social impairment. *Pediatric Research, 69*, 55R–62R.
96. Happe, F., & Ronald, A. (2008). The 'fractionable autism triad': A review of evidence from behavioural, genetic, cognitive and neural research. *Neuropsychology Review, 18*, 287–304.
97. Sundberg, M., & Sahin, M. (2015). Cerebellar development and autism Spectrum disorder in tuberous sclerosis complex. *Journal of Child Neurology, 30*, 1954–1962.
98. Ekinci, O., Arman, A. R., Isik, U., Bez, Y., & Berkem, M. (2010). EEG abnormalities and epilepsy in autistic spectrum disorders: Clinical and familial correlates. *Epilepsy & Behavior, 17*, 178–182.
99. Parmeggiani, A., Barcia, G., Posar, A., Raimondi, E., Santucci, M., & Scaduto, M. C. (2010). Epilepsy and EEG paroxysmal abnormalities in autism spectrum disorders. *Brain Dev, 32*, 783–789.
100. Viscidi, E. W., Triche, E. W., Pescosolido, M. F., McLean, R. L., Joseph, R. M., Spence, S. J., et al. (2013). Clinical characteristics of children with autism spectrum disorder and co-occurring epilepsy. *PLoS One, 8*, e67797.
101. Backman, S. A., Stambolic, V., Suzuki, A., Haight, J., Elia, A., Pretorius, J., et al. (2001). Deletion of Pten in mouse brain causes seizures, ataxia and defects in soma size resembling Lhermitte-Duclos disease. *Nature Genetics, 29*, 396–403.
102. Goffin, A., Hoefsloot, L. H., Bosgoed, E., Swillen, A., & Fryns, J. P. (2001). PTEN mutation in a family with Cowden syndrome and autism. *American Journal of Medical Genetics, 105*, 521–524.
103. Kwon, C. H., Luikart, B. W., Powell, C. M., Zhou, J., Matheny, S. A., Zhang, W., et al. (2006). Pten regulates neuronal arborization and social interaction in mice. *Neuron, 50*, 377–388.
104. Takeuchi, K., Gertner, M. J., Zhou, J., Parada, L. F., Bennett, M. V., & Zukin, R. S. (2013). Dysregulation of synaptic plasticity precedes appearance of morphological defects in a Pten conditional knockout mouse model of autism. *Proceedings of the National Academy of Sciences of the United States of America, 110*, 4738–4743.
105. Winden, K. D., Sundberg, M., Yang, C., Wafa, S. M. A., Dwyer, S., Chen, P. F., et al. (2019). Biallelic mutations in TSC2 lead to abnormalities associated with cortical tubers in human iPSC-Derived neurons. *J Neurosci, 39*(47), 9294–9305.
106. Waltereit, R., Japs, B., Schneider, M., de Vries, P. J., & Bartsch, D. (2011). Epilepsy and Tsc2 haploinsufficiency lead to autistic-like social deficit behaviors in rats. *Behavior Genetics, 41*, 364–372.
107. Schneider, M., de Vries, P. J., Schonig, K., Rossner, V., & Waltereit, R. (2017). mTOR inhibitor reverses autistic-like social deficit behaviours in adult rats with both Tsc2 haploinsufficiency and developmental status epilepticus. *European Archives of Psychiatry and Clinical Neuroscience, 267*, 455–463.
108. Yuan, E., Tsai, P. T., Greene-Colozzi, E., Sahin, M., Kwiatkowski, D. J., & Malinowska, I. A. (2012). Graded loss of tuberin in an allelic series of brain models of TSC correlates with survival, and biochemical, histological and behavioral features. *Human Molecular Genetics, 21*, 4286–4300.
109. Kelly, E., Schaeffer, S. M., Dhamne, S. C., Lipton, J. O., Lindemann, L., Honer, M., et al. (2018). mGluR5 modulation of behavioral and epileptic phenotypes in a mouse model of tuberous sclerosis complex. *Neuropsychopharmacology, 43*, 1457–1465.
110. Meikle, L., Pollizzi, K., Egnor, A., Kramvis, I., Lane, H., Sahin, M., et al. (2008). Response of a neuronal model of tuberous sclerosis to mammalian target of rapamycin (mTOR) inhibi-

tors: Effects on mTORC1 and Akt signaling lead to improved survival and function. *The Journal of Neuroscience, 28*, 5422–5432.
111. McMahon, J. J., Yu, W., Yang, J., Feng, H., Helm, M., McMahon, E., et al. (2015). Seizure-dependent mTOR activation in 5-HT neurons promotes autism-like behaviors in mice. *Neurobiology of Disease, 73*, 296–306.
112. Singh, S. K., & Eroglu, C. (2013). Neuroligins provide molecular links between syndromic and nonsyndromic autism. *Science Signaling, 6*, re4.
113. Yan, J., Oliveira, G., Coutinho, A., Yang, C., Feng, J., Katz, C., et al. (2005). Analysis of the neuroligin 3 and 4 genes in autism and other neuropsychiatric patients. *Molecular Psychiatry, 10*, 329–332.
114. Etherton, M., Foldy, C., Sharma, M., Tabuchi, K., Liu, X., Shamloo, M., et al. (2011). Autism-linked neuroligin-3 R451C mutation differentially alters hippocampal and cortical synaptic function. *Proceedings of the National Academy of Sciences of the United States of America, 108*, 13764–13769.
115. Kahle, K. T., Khanna, A. R., Duan, J., Staley, K. J., Delpire, E., & Poduri, A. (2016). The KCC2 Cotransporter and human epilepsy: Getting excited about inhibition. *The Neuroscientist, 22*, 555–562.
116. Moore, Y. E., Kelley, M. R., Brandon, N. J., Deeb, T. Z., & Moss, S. J. (2017). Seizing control of KCC2: A new therapeutic target for epilepsy. *Trends in Neurosciences, 40*, 555–571.
117. Tang, X., Kim, J., Zhou, L., Wengert, E., Zhang, L., Wu, Z., et al. (2016). KCC2 rescues functional deficits in human neurons derived from patients with Rett syndrome. *Proceedings of the National Academy of Sciences of the United States of America, 113*, 751–756.
118. Shibata, H., Huynh, D. P., & Pulst, S. M. (2000). A novel protein with RNA-binding motifs interacts with ataxin-2. *Human Molecular Genetics, 9*, 1303–1313.
119. Bhalla, K., Phillips, H. A., Crawford, J., McKenzie, O. L., Mulley, J. C., Eyre, H., et al. (2004). The de novo chromosome 16 translocations of two patients with abnormal phenotypes (mental retardation and epilepsy) disrupt the A2BP1 gene. *Journal of Human Genetics, 49*, 308–311.
120. Bucan, M., Abrahams, B. S., Wang, K., Glessner, J. T., Herman, E. I., Sonnenblick, L. I., et al. (2009). Genome-wide analyses of exonic copy number variants in a family-based study point to novel autism susceptibility genes. *PLoS Genetics, 5*, e1000536.
121. Gai, X., Xie, H. M., Perin, J. C., Takahashi, N., Murphy, K., Wenocur, A. S., et al. (2012). Rare structural variation of synapse and neurotransmission genes in autism. *Molecular Psychiatry, 17*, 402–411.
122. Philippe, A., Martinez, M., Guilloud-Bataille, M., Gillberg, C., Rastam, M., Sponheim, E., et al. (1999). Genome-wide scan for autism susceptibility genes. Paris Autism Research International Sibpair Study. *Human Molecular Genetics, 8*, 805–812.
123. Pinto, D., Pagnamenta, A. T., Klei, L., Anney, R., Merico, D., Regan, R., et al. (2010). Functional impact of global rare copy number variation in autism spectrum disorders. *Nature, 466*, 368–372.
124. Ascano Jr., M., Mukherjee, N., Bandaru, P., Miller, J. B., Nusbaum, J. D., Corcoran, D. L., et al. (2012). FMRP targets distinct mRNA sequence elements to regulate protein expression. *Nature, 492*, 382–386.
125. Darnell, J. C., Van Driesche, S. J., Zhang, C., Hung, K. Y., Mele, A., Fraser, C. E., et al. (2011). FMRP stalls ribosomal translocation on mRNAs linked to synaptic function and autism. *Cell, 146*, 247–261.
126. Corbin, F., Bouillon, M., Fortin, A., Morin, S., Rousseau, F., & Khandjian, E. W. (1997). The fragile X mental retardation protein is associated with poly(A)+ mRNA in actively translating polyribosomes. *Human Molecular Genetics, 6*, 1465–1472.
127. Davidovic, L., Jaglin, X. H., Lepagnol-Bestel, A. M., Tremblay, S., Simonneau, M., Bardoni, B., et al. (2007). The fragile X mental retardation protein is a molecular adaptor between the neurospecific KIF3C kinesin and dendritic RNA granules. *Human Molecular Genetics, 16*, 3047–3058.

128. Dictenberg, J. B., Swanger, S. A., Antar, L. N., Singer, R. H., & Bassell, G. J. (2008). A direct role for FMRP in activity-dependent dendritic mRNA transport links filopodial-spine morphogenesis to fragile X syndrome. *Developmental Cell, 14*, 926–939.
129. Feng, Y., Absher, D., Eberhart, D. E., Brown, V., Malter, H. E., & Warren, S. T. (1997). FMRP associates with polyribosomes as an mRNP, and the I304N mutation of severe fragile X syndrome abolishes this association. *Molecular Cell, 1*, 109–118.
130. Fridell, R. A., Benson, R. E., Hua, J., Bogerd, H. P., & Cullen, B. R. (1996). A nuclear role for the fragile X mental retardation protein. *The EMBO Journal, 15*, 5408–5414.
131. Ling, S. C., Fahrner, P. S., Greenough, W. T., & Gelfand, V. I. (2004). Transport of Drosophila fragile X mental retardation protein-containing ribonucleoprotein granules by kinesin-1 and cytoplasmic dynein. *Proceedings of the National Academy of Sciences of the United States of America, 101*, 17428–17433.
132. Doherty, C., Goh, S., Young Poussaint, T., Erdag, N., & Thiele, E. A. (2005). Prognostic significance of tuber count and location in tuberous sclerosis complex. *Journal of Child Neurology, 20*, 837–841.
133. Jansen, F. E., Vincken, K. L., Algra, A., Anbeek, P., Braams, O., Nellist, M., et al. (2008). Cognitive impairment in tuberous sclerosis complex is a multifactorial condition. *Neurology, 70*, 916–923.
134. Zeng, L. H., Xu, L., Gutmann, D. H., & Wong, M. (2008). Rapamycin prevents epilepsy in a mouse model of tuberous sclerosis complex. *Annals of Neurology, 63*, 444–453.
135. Krueger, D. A., Wilfong, A. A., Holland-Bouley, K., Anderson, A. E., Agricola, K., Tudor, C., et al. (2013). Everolimus treatment of refractory epilepsy in tuberous sclerosis complex. *Annals of Neurology, 74*, 679–687.
136. Krueger, D. A., Wilfong, A. A., Mays, M., Talley, C. M., Agricola, K., Tudor, C., et al. (2016). Long-term treatment of epilepsy with everolimus in tuberous sclerosis. *Neurology, 87*, 2408–2415.
137. Samueli, S., Abraham, K., Dressler, A., Groppel, G., Muhlebner-Fahrngruber, A., Scholl, T., et al. (2016). Efficacy and safety of Everolimus in children with TSC - associated epilepsy - pilot data from an open single-center prospective study. *Orphanet Journal of Rare Diseases, 11*, 145.
138. Overwater, I. E., Rietman, A. B., Bindels-de Heus, K., Looman, C. W., Rizopoulos, D., Sibindi, T. M., et al. (2016). Sirolimus for epilepsy in children with tuberous sclerosis complex: A randomized controlled trial. *Neurology, 87*, 1011–1018.
139. Bissler, J. J., McCormack, F. X., Young, L. R., Elwing, J. M., Chuck, G., Leonard, J. M., et al. (2008). Sirolimus for angiomyolipoma in tuberous sclerosis complex or lymphangioleiomyomatosis. *The New England Journal of Medicine, 358*, 140–151.
140. Franz, D. N., Leonard, J., Tudor, C., Chuck, G., Care, M., Sethuraman, G., et al. (2006). Rapamycin causes regression of astrocytomas in tuberous sclerosis complex. *Annals of Neurology, 59*, 490–498.
141. Buckmaster, P. S., Ingram, E. A., & Wen, X. (2009). Inhibition of the mammalian target of rapamycin signaling pathway suppresses dentate granule cell axon sprouting in a rodent model of temporal lobe epilepsy. *The Journal of Neuroscience, 29*, 8259–8269.

Advances in Human Stem Cells and Genome Editing to Understand and Develop Treatment for Fragile X Syndrome

Xinyu Zhao and Anita Bhattacharyya

1 Introduction

Fragile X syndrome (FXS) is the most common inherited cause of intellectual disability affecting about 1 in 4000 boys and 1 in 7000 girls [1]. FXS patients exhibit learning deficits and IQs between 20 and 60, increased incidence of seizures and reduced motor coordination as well as hyperactivity, attention deficit disorder, and autistic-like behavior [2]. Because more than one-third of FXS individuals also meet the diagnostic criteria for autism [3–9], FXS is the largest single genetic contributor to autism. In addition, FXS features are not limited to the nervous system and include connective tissue dysplasia, facial dysmorphia, hyper-extensible joints, mitral valve prolapse, and macro-orchidism.

FXS is caused by a mutation in a single gene, the Fragile X Mental Retardation Gene 1 (*FMR1*), resulting in a lack of the encoded Fragile X Mental Retardation Protein (FMRP) [10, 11]. Animal models have been useful in studying characteristics and potential mechanisms underlying characteristics associated with FXS. FXS models have shown that FMRP is an RNA binding protein that binds specific mRNAs to control their location and protein translation [12]. This function implies that FMRP plays a crucial role in neuronal development, function, and synaptic

X. Zhao (✉)
Waisman Center, University of Wisconsin-Madison, Madison, WI, USA

Department of Neuroscience, University of Wisconsin-Madison, School of Medicine and Public Health, Madison, WI, USA
e-mail: Xinyu.zhao@wisc.edu

A. Bhattacharyya (✉)
Waisman Center, University of Wisconsin-Madison, Madison, WI, USA

Department of Cell and Regenerative Biology, University of Wisconsin-Madison, School of Medicine and Public Health, Madison, WI, USA
e-mail: bhattacharyy@waisman.wisc.edu

E. DiCicco-Bloom, J. H. Millonig (eds.), *Neurodevelopmental Disorders*, Advances in Neurobiology 25, https://doi.org/10.1007/978-3-030-45493-7_2

plasticity [13, 14]. The absence of FMRP results in increased protein synthesis, leading to enhanced signaling in a number of intracellular pathways, including the mTOR, mGLuR5, ERK, Gsk3β, PI3K, and insulin pathways. Mouse models have revealed that neural cells that lack FMRP exhibit neurogenesis and neuronal maturation deficits [15]. Further, altered synaptic plasticity has been established in the FXS mouse model and is thought to be due to the lack of FMRP's role as a negative regulator of translation [13]. Data from mouse models has informed our understanding of FXS and several drug trials have been instituted in FXS patients as a direct result of targets identified in animal studies.

However, how the loss of FMRP manifests in the human nervous system remains unclear.

Although higher density but immature long and thin neuronal dendritic spines are consistently found in FXS patients' brains [16–18], the underlying mechanisms of this phenotype are not well defined. The failure of FXS clinical trials based on animal models demands better understanding of pathogenesis of FXS in human models. Hence, it is important to define the cellular and molecular deficits in human FXS neural cells so that therapeutics to affect neural development and function in FXS can be designed more intelligently.

2 Unique and Complex Genetics of FXS

2.1 CGG Expansion of the Human FMR1 *Gene*

In humans, the causal mutation in FXS is a trinucleotide CGG repeat expansion in the 5′ untranslated region of the *FMR1* gene leading to epigenetic silencing. CGG trinucleotide repeats are normally present in the *FMR1* gene of all humans. The number of CGG repeats in the human *FMR1* gene is polymorphic with 30–31 as the mode and more than 90% of the human population has fewer than 40 CGGs [19, 20]. In people with the near modal CGG repeat numbers, the CGG repeat usually remains stable through generations [21]. However, due to reasons that are not understood, the CGG repeats sometimes expand through the germ line leading to CGG repeat lengths between 55 and 200, termed premutation. In premutation cells, *FMR1* gene transcription is enhanced with protein levels unchanged or reduced and leads to pathological conditions called Fragile X premutation with an increased risk of developing fragile X-related primary ovarian insufficiency (FXPOI) and fragile X-associated tremor/ataxia syndrome (FXTAS) in older age [5]. The premutation CGG repeats are unstable and may expand during meiosis in germ cells. When the mutational expansion of the CGG repeats exceeds 200 in humans, it leads to methylation of the repeats and the *FMR1* promoter, chromatin condensation, and a loss of FMRP protein expression. It remains unclear how the variable CGG repeat length in the *FMR1* gene affects *FMR1* gene transcription and why having more than 200 repeats leads to DNA methylation and gene silencing.

2.2 *Epigenetic Silencing of the Human* FMR1 *Gene*

Epigenetic mechanisms, mediated by DNA methylation, histone modification, and non-coding RNAs, are known to play significant roles in regulating stem cells and development as well as adult neuroplasticity [22]. DNA methylation is catalyzed by methyl transferases, including de novo Dnmt3a and Dnmt3b that add methyl groups onto unmethylated DNA, and Dnmt1 that recognizes hemi-methylated DNA and maintains DNA methylation. A majority of genomic DNA methylation, particularly in the brain, is at cytosine residues in the context of CpG dinucleotide (mCG), with additional methylation at non-CpG sites. Active DNA demethylation involves multi-step chemical reactions by several groups of proteins, including TET proteins and the production of 5-hydroxymethylation of cytosine (5hmC) [23]. The amino (N)-terminal tails of core histones are subject to a variety of covalent modifications including acetylation, methylation, ubiquitination, phosphorylation, ribosylation, SUMOylation, etc. The combination of these histone modifications is called "the histone code" and binding of modified histones to specific genomic regions controls the activation or repression of the associated genes [24]. These histone modifications are catalyzed by enzymes such as histone acetyltransferases (HATs), histone deacetylases (HDACs), histone methyltransferases (HMTs), and histone demethylases (HdMTs). A large portion of the genome is transcribed into non-protein-coding RNA called non-coding RNA (ncRNA). Recently, the involvement of long ncRNAs (lncRNAs) is increasingly recognized as an important aspect of regulation [22].

All three epigenetic mechanisms are involved in *FMR1* gene expression. DNA methylation of expanded CGG in the *FMR1* gene is the major, if not the only reason leading to *FMR1* gene inactivation in FXS. Associated with striking DNA methylation changes, the histone marks in the *FMR1* gene locus also shift from the active to the repressive state. The human *FMR1* gene locus also encodes several lncRNAs [25, 26]. Further, other genes such as *FAM 11A* [27] and several long non-coding RNAs such as antisense *FMR1* (AS*FMR1*) [28] and FMR6 [25] in the locus are also methylated and silenced in addition to *FMR1*. However, it is unclear how much these elements contribute to FXS and whether they are involved in the gene inactivation process.

The critical importance of *FMR1* methylation in gene silencing is illustrated by the existence of individuals who are mosaic for *FMR1* gene methylation and have partial FMRP expression [29]. Rare males with *FMR1* full-length CGG expansion mutations show no or only mild symptoms because their CGG and *FMR1* gene are unmethylated [30–33].

It is important to emphasize that none of the animal models, including all mouse models engineered to mimic the human CGG expansion mutation in the *FMR1* gene, show methylation and silencing [34]. Therefore, epigenetic mechanisms in human and animal models are different and preclude the ability to study epigenetic mechanisms of *FMR1* silencing in animal models of FXS.

3 Human Pluripotent Stem Cells Models for FXS

3.1 Unique Features of Human Genetics and Neural Development Require Human FXS Models

Animal models are limited in revealing some of the most fundamental aspects of development, genetics, pathology, and disease mechanisms that are unique to humans. Therefore, it is necessary to use human cells to define underlying mechanisms that lead to FXS characteristics that affect the nervous system.

Differences between the formation and structure of the brain in mice and humans present challenges to understanding the mechanisms of abnormal brain development and function in FXS. The neocortex in mammals is involved in higher cognitive functions that distinguish humans from rodents and the anatomical basis for these differences is accomplished through complex and evolutionarily emergent differences in cortical expansion, arealization, and connectivity. The evolutionary expansion of the cerebral cortex is reliant on increased cell numbers that results in an increase in the numbers of neural progenitors [35–40]. The presence of an additional outer subventricular zone (oSVZ) enables more proliferative neuronal progenitors that, in turn, differentiate into more neurons [41–45].

Global transcriptomic analyses of developing primate and human brain have identified unique molecular mechanisms that regulate the expansion of the human cortex [46–48]. One example of a human specific gene is ARHGAP11B, encoding Rho GTPase activating protein 11B, that promotes basal progenitor amplification and neocortex expansion and may have contributed to the evolutionary expansion of the human neocortex [49, 50]. Recently, the introduction of this gene into an animal model enabled induced hallmarks of primate neocortical expansion [51], emphasizing the importance of this human specific gene in human cortical development. Thus, there are additional unique genetic mechanisms in human in addition to the regulation of *FMR1*.

The human brain is more reliant on the role of interneurons and astrocytes and so FXS mouse models may not adequately reveal differences in these particular systems. For example, human interneuron development occurs over a protracted period of time and integrates unique mechanisms to generate more numerous and more elaborate interneurons [43, 44, 52–55]. *FMR1* is expressed in all neurons and in astrocytes and so human specific differences in interneurons and astrocytes may be affected by FMRP.

These data and others highlight significant cellular and molecular differences in human and mouse cortical development that may hinder understanding of disorders that affect cortical development [56]. Thus, it is important to study the cause and consequences of *FMR1* silencing in neural development and function in the human context.

3.2 *Human Pluripotent Stem Cells as a Platform to Understand the Pathogenesis of FXS*

Human pluripotent stem cells (hPSCs), both human embryonic stem cells (hESCs) and induced PSCs (hiPSCs), offer a model system to reveal cellular and molecular events underlying normal and abnormal human development. hESCs are isolated from preimplantation embryos [57]. Human iPSCs are reprogrammed from somatic cells by forced expression of stem cell genes and have the characteristics of hESCs [58, 59]. Patient-derived iPSCs provide a paradigm to understand FXS in a human context.

FXS hESCs were first isolated from preimplantation embryos carrying the FXS mutation by Verlinsky in 2005 [60]. The heritability of FXS enables the identification of affected embryos through preimplantation genetic diagnosis during the in vitro fertilization process. hESCs can be isolated from the inner cell mass (ICM) of these embryos for research purposes [61–64].

Multiple hESC lines have been derived from embryos diagnosed with FXS [65–68]. Importantly, FXS hESCs retain the full-length mutation of the *FMR1* gene and, although the gene was initially reported to be unmethylated and expressed in these cells [66], more recent evidence indicates that the epigenetic silencing in FXS hESCs occurs spontaneously [65]. These results also suggest that the initial embryonic ICM cells may have different methylation of the *FMR1* gene at the time of stem cell derivation. Therefore, epigenetic silencing can occur in the undifferentiated state.

The ability to generate hiPSCs from somatic cells of FXS individuals has enabled the generation of FXS hiPSC lines from patient fibroblasts [68–75]. Without exception, the methylated, silenced *FMR1* mutation in the patient fibroblasts is retained through the reprogramming process [76]. Reprogramming also causes a rare unmethylated full mutation in patient somatic cells to be silenced [72, 77]. Reactivation of the FMR1 gene has been shown to occur when FXS patient-derived iPSCs are treated to recapitulate ground state naïve pluripotency in which DNA methylation is reduced [78]. These naïve FXS pluripotent stem cells are useful to analyze the mechanisms of gene silencing while FXS iPSCs are useful to test mechanisms of epigenetic reactivation and for studying the effect of *FMR1* loss as cells differentiate from the undifferentiated state.

FMRP is expressed in all neural cells [79–81], so the characterization of FMRP loss in different cell types is needed to better understand FXS. While the focus of FXS neurobiology has been on excitatory neurons, inhibitory neurons are also dysfunctional in FXS [82–87]. Astrocytes express FMRP and mounting evidence from mouse suggests that the lack of FMRP in astrocytes may be detrimental to neuronal function [88–96]. In addition, FMRP expression in oligodendrocytes has been shown to correlate with myelin defects in mouse models [97, 98]. Human neuroimaging studies suggest that individuals with FXS have white matter defects that may link development and function of oligodendrocytes to FXS neuropathology [99, 100]. Thus, the loss of FMRP in all neural cell types likely contributes to neuropathology in FXS.

To take advantage of the power of hPSCs to model human brain development and to define the steps that go awry in FXS, it is critical to differentiate hPSCs into the specific neural subtypes that are affected in FXS [101]. Disease-relevant cell types differentiated from human FXS PSCs can be used to unveil cellular and molecular events underlying consequences of FMRP loss on neural development in a human genetic background. Additionally, pure populations of specific neural cell types derived from patient hiPSCs are useful for the rapid screening of pharmaceutical compounds to accelerate drug discovery and advance treatments. The power of human PSCs to model human brain development lies in the ability to generate specific neural cell types in vitro over a long period of time that corresponds to in vivo development, thus recapitulating many of the developmental steps. Methods to generate specific neuronal and glial subtypes from human PSCs that follow in vivo developmental principles and timing have been developed [102–116]. In addition, some neuronal subtypes can be directly differentiated from somatic cells or hPSCs [117–121]. These neurons, often termed induced neurons (iNs), are functional neurons but their generation bypasses development, where FMRP function is likely important.

Analyses of neurons derived from FXS hPSCs indicate that the epigenetic mutation is preserved, although the expression of FMRP decreases through development from PSC to neurons. The phenotypes of neural progenitor cells differentiated from these cells vary among reports, with some providing evidence that neurogenesis is aberrant [122] and others reporting that neurogenesis is unaffected [73, 123]. The gene expression patterns of hPSC-derived FXS neurons suggest defects in neuronal differentiation [124] and maturation [74] similar to what has been shown in mouse models ([125–127]. FXS patients have grossly normal brain formation, suggesting that neurogenesis is not obviously affected. The prevailing hypothesis is that FXS patients have synaptic and plasticity defects and some have been demonstrated in FXS mouse models. Neurons from FXS PSCs do exhibit functional deficits [73, 128–131], and reduced pre-synaptic vesicle recycling [132]. Recent assessment of synaptic function and plasticity in FXS human neurons shows loss of homeostatic plasticity in FXS neurons [131]. Thus, neurons generated from FXS hPSCs are beginning to provide insight into early events in neuronal development that are affected by loss of FMRP. It will be critical for future studies to define the underlying mechanisms and consequences of deficits at the cellular level.

3.3 Human PSCs for Understanding FMR1 *Silencing in FXS*

Several potential mechanisms underlying *FMR1* gene inactivation have been investigated and proposed. Since no animal model enables the study of epigenetic changes of human FXS, early studies used human non-neural cells, including fibroblasts and lymphocytes, to assess epigenetic signatures of active and repressed *FMR1* genes in human FXS. DNA footprinting studies have shown that *FMR1* gene repression is correlated with the absence of transcription factor binding in cells derived from FXS individuals [133–135]. Using "3C" chromosome conformation analysis of a 170 kb locus encompassing the human *FMR1* gene, Gheldof et al. discovered a significant difference in chromosome conformation between *FMR1*-expressing versus non-expressing cells [136], suggesting that silencing of the *FMR1* gene is associated with broader changes at the chromosome level than previously anticipated. Chromatin immunoprecipitation studies have shown that levels of acetylated H3 and H4 that are associated with actively transcribed genes reduced in FXS cells [137] and the levels of acetylated H3 on the *FMR1* promoter are inversely correlated with repeat sizes [138]. In addition, H3K4 methylation is decreased in FXS cells, whereas H3K9 methylation is increased, consistent with the inactive status of *FMR1* gene in FXS cells. These changes in histone modifications are restricted to *FMR1* promoter regions [136]. However, studies of full mutation males with an unmethylated *FMR1* promoter show an increase in deacetylated H3 and H4 and methylated H3K9 [139, 140], suggesting that histone modification might be independent of DNA methylation in *FMR1* gene silencing. The seemingly contradictory messages from these studies using various non-neural cell types demonstrate complex mechanisms regulate *FMR1* gene expression and demand the analysis of human brain-relevant cell types.

The successful development of FXS hPSCs and efficient differentiation of these cells into neural progenitors and neurons has allowed in-depth investigation of *FMR1* gene silencing in human neural cells. The silencing of *FMR1* gene in hESCs is associated with loss of active chromatin markers including H3K4me2 and gain of H3K9me3 [65]. Another study using hESCs suggests that changes in histone markers precede DNA methylation changes [66]. Further, CGG containing *FMR1* mRNA was found to inhibit its own expression. Demethylation of the *FMR1* promoter leads to increased repressive chromatin marker H3K27 methylation binding to the promoter that is dependent on the presence of mutant mRNA [141]. Recently, *FMR1* silencing was shown to be mediated by *FMR1* mRNA containing long CGG repeats [142]. Therefore, several mechanisms may be at play in silencing *FMR1* expression during human neural development.

4 Human PSCS for Targeting *FMR1* Gene Reactivation for Treating FXS

4.1 Rationale for FMR1 *Gene Restoration as a Potential Therapy*

Since the coding sequence of the silenced *FMR1* gene is normal in FXS individuals, a possible therapeutic strategy is to restore the transcription of *FMR1* in FXS. Support for this strategy comes from reports of males with *FMR1* full-length CGG expansion mutations who show no or mild symptoms because their *FMR1* gene is unmethylated [30–33, 139, 140, 143]. These studies show that an unmethylated *FMR1* gene carries out normal functions resulting in near normal intelligence instead of intellectual disability associated with FXS and provides justification for *FMR1* gene reactivation. In addition, adeno-associated virus (AAV) has been used to deliver exogenous FMRP into adult or neonatal *Fmr1* KO mouse brains. AAV-FMRP injected into the hippocampus of 5-week old mice restores hippocampal synaptic plasticity and reduces behavioral symptoms of *Fmr1* KO mice [144, 145]. Further, mouse genetic studies provide more confidence in the notion of therapeutic gene reactivation. Restoration of *FMR1* in adult-born new neurons using inducible genetics restores several adult neurogenesis-dependent learning and memory in mice [125]. These data suggest that the neuronal developmental deficits seen in FMRP-deficient neurons may be reversible.

4.2 *FMR1* Gene Restoration Strategies by Targeting Known Pathways

The identification of effective methods to reactivate *FMR1* gene and restore FMRP expression has been extremely challenging. A number of studies have shown that treatment of human FXS lymphoblastoid cell lines with a DNA methyltransferase (DNMT) inhibitor 5-azacytidine (5azaC) or 5-azadeoxycytidine (5azadC) results in partial reactivation of the *FMR1* gene and FMRP expression. Upon treatment, the *FMR1* promoter becomes passively unmethylated through cell division [146, 147]. Importantly, the increase in *FMR1* mRNA production is associated with increased active chromatin marker binding and decreased repressive chromatin marker binding to the *FMR1* promoter [141, 147–149]. In contrast, methotrexate, a folate antagonist that acts by inhibiting dihydrofolate reductase (DHFR) and has some DNA methylation inhibition activity does not reduce DNA methylation in the *FMR1* promoter but leads to some mRNA expression but not protein expression [148]. These results suggest that reversing DNA methylation of CGG repeats might be a promising method for gene restoration therapy.

Chemicals affecting histone modification have also been explored for *FMR1* reactivation. Most studies have so far focused on Class I, II, and IV HDAC inhibitors include butyrate and trichostatin A (TSA). Using human FXS lymphoblastoid cell lines, Chiurazzi et al. have shown that treatment with HDAC inhibitors, phenylbutyrate, sodium butyrate, and TSA, leads to moderately increased *FMR1* gene transcription, yet less than compared to the effect of 5azaC or 5-azadC [150]. However other studies show no *FMR1* transcription after TSA treatment [137, 138] or VPA treatment [149]. Interestingly, combined treatment with 5dazaC and HDAC inhibitor leads to two- fivefold higher reactivation compared to 5azaC treatment alone [150]. Recently, more effective reactivation, comparable to that by 5azaC, has been achieved by using splitomicin (SPT), an inhibitor of SIRT1 a class III HDAC [141, 151]. Knockdown of SIRT1 in either lymphoblastoid cell lines or FXS patient-derived fibroblasts leads to increased deacetylation of H4K16 and increased *FMR1* gene transcription without significantly affecting DNA methylation. This study suggests that inhibition of certain key HDACs may be able to reactivate the *FMR1* gene without altering DNA methylation.

4.3 *Search for Novel* FMR1 *Gene Restoration Strategies Using High-Throughput Screening of Chemicals*

The limited success of using known epigenetic reagents to reactivate *FMR1* genes has prompted studies to explore novel chemical reagents and molecules as well as genetic strategies. A major challenge is to establish a screening method that can effectively and efficiently report *FMR1* gene expression. Several reports of screening technologies have been published that rely on antibodies to detect FMRP expression. In one study, human FXS-derived iPSCs were differentiated into neural progenitor cells (NPCs), immunostained with an FMRP antibody, and analyzed by high content imaging for FMRP levels [75]. Using this system, 50,000 compounds covering epigenetic targets and known FMRP regulated pathways were screened and several compounds were identified (identity not revealed) that induced weak reactivation [75]. In another screen, FMRP antibodies were used to establish a time-resolved fluorescence resonance energy transfer (TR-FRET) dual antibody assay to increase specificity. They also used human iPSC-derived NPCs and screened ~5000 compounds including a FDA-approved drug library. Six hits were identified to enhance *FMR1* gene transcription modestly, although no significant FMRP was detected [152]. Interestingly, one of the identified compounds is SB216763 which we have previously found to rescue learning deficits in *FMR1*-null mice through enhancing Wnt signaling [126]. Yet, none of the compounds identified so far can reactivate *FMR1* expression to near normal levels, necessitating new and better strategies.

With advances in gene-editing technology combined with hPSCs, the ability to use in vitro "disease in a dish" models to conduct large-scale, high-throughput drug screens to find new treatments for neurological and psychiatric diseases has greatly expanded. Human PSCs can be derived from human patients and modified using CRISPR/Cas9 gene editing to insert reporter genes for screening purposes. Antibody-based assays for high-throughput screening (HTS) of compounds for reactivation of *FMR1* show poor sensitivity and are expensive and/or difficult to scale up. Luciferase-based HTS has been shown to be highly sensitive and can be scaled up [153]. Therefore, we created luciferase reporter lines of FX-iPSCs that have great potential for large-scale HTS of compounds to detect *FMR1* activation [154–156].

4.4 FMR1 *Gene Reactivation by Genetic Targeting of CGG Repeats*

The discovery and rapid development of various gene-editing technologies has opened new avenues for gene correction-based therapies. The zinc finger protease (ZFN) and transcription activator-like effector nucleases (TALENs) methods are promising but are difficult to use and have relatively low specificity [157].

Nevertheless, proof of concept experiments show that TALEN can be used to correct AT-rich repeats in FATS a common fragile site in mice [158]. The newer CRISPR/Cas9-based gene editing method is significantly easier to use and exhibits much higher specificity [159]. CRISPR/Cas9-based gene editing has enabled the dissection of molecular aspects of *FMR1* silencing and reactivation. Park et al. used CRISPR/Cas9 to create a single double strand break (DSB) near the CGG repeat followed by non-homologous end joining (NHEJ) to delete the CGG repeat in the silenced *FMR1* gene in FXS-derived iPSCs and demonstrated activation of *FMR1* gene [160]. Although the efficiency of this deletion is extremely low, the results of this work suggest that deletion of the silenced CGG repeat is a promising gene reactivation strategy for FXS. One major issue of this work is that NHEJ generates random deletion and so the deletion is not restricted to the CGG repeat. In an alternative approach, Xie et al. used CRISPR/Cas9-mediated double strand break to cut out only the CGG repeats from the mutant *FMR1* gene in FXS patient iPSCs [161]. This more specific deletion of the CGG repeats resulted in FMRP expression in only 20% of FXS iPSC clones. Interestingly, the clones with FMRP expression also had DNA demethylation of CpG sites upstream of the deleted CGG repeats, whereas the clones without FMRP expression did not, suggesting that deletion of CGG repeats alone is not sufficient to reactivate the gene. The factor(s) contributes the different *FMR1* gene reactivation and DNA methylation in these CGG repeat-deleted clones remains unknown.

Targeting epigenetic modifiers or transcriptional activators to the CGG repeats is also a potential method for reactivating the *FMR1* gene [70]. This idea became possible with the development of mutant Cas9 (or dead Cas9, dCas9) that can bind DNA in a guide-RNA dependent manner without cutting the DNA sequence [162]. dCas9 has been used to create fusion proteins and molecules to deliver transcriptional activators, transcriptional repressors, epigenome modifiers, fluorescent proteins, RNA tags, dimerization domains, RNA editing machinery, etc. to specific DNA sequences in the genome [163]. However, sequence-specific activation of a fully silenced FMR1 gene remains challenging. Haenfler et al. [164] targeted dCas9 fused with different versions of the VP16 transcriptional activation domains to the promoter or the CGG repeats of the human *FMR1* gene. They found that dCas9-VP192 (16 tandem repeats of VP16 transactivator [165]) and targeting the CGG repeat, instead of other parts of the promoter, leads to the highest *FMR1* gene transcription in FXS ESCs and NPCs. However, the reactivation of *FMR1* transcription is moderate, especially in FXS hESCs with full silencing, and no protein FMRP expression was detected in any cells. Liu et al. used another approach by using dCas9-Tet1 fusion protein to target the catalytic domain of Tet1, an enzyme in DNA demethylation pathway, to the CGG repeats of *FMR1* gene in FXS iPSCs. Demethylation and reactivation of the *FMR1* gene occurred to ~90% of normal levels, demonstrating substantial and impressive reactivation [166]. Even more exciting, the method worked on post-mitotic human neurons in culture, providing hope for future reactivation strategies in the brain [166]. A major concern for this study, as well as the Haenfler et al., is the targeting to the CGG repeat, rather than sequences specific to the *FMR1* gene, leads to gene activation. CGG repeats are present in other genes and thus may lead to aberrant transcription of other genes. To address the off target effect, the authors performed Cas9 ChIP-seq to identify dCas9-tet1 binding and Cas9 ChIP-BS-seq to identify DNA methylation changes of dCas9-Tet1 targets in reactivated cells and concluded that the off target effects are minimal and can be further reduced by lower the levels of dCas9-Tet1. Further analyses and careful monitoring of other genes are necessary in future.

In addition to the issues with efficacy and specificity, CRISPR /Cas9- based methods also face other challenges. Cas9 protein is very large and difficult to deliver and distribute through solid tissues including the brain. Nanoparticle-based delivery of CRISPR/Cas9 has been used to reduce the levels of FMRP targets in mouse brains, but the efficiency of targeting remains low [167]. In addition, a substantial number of humans have pre-existing T cells towards the most widely used Cas9 homolog derived from *S. pyogenes* (SpCas9), conferring immunity to Cas9 based delivery. In fact about 12% children have an asymptomatic colonization of the facial mucosa with *S. pyogenes* [168]. Despite these limitations, the fast advancement in gene editing and delivery methods promises that reactivation of the *FMR1* gene using genetic methods may become feasible in the near future.

5 Summary: Challenges and Perspectives

While hPSCs provide an unparalleled tool to investigate the pathogenesis of FXS and develop novel human-relevant therapies, major challenges remain. For example, a number of factors introduce variability that affect the ability to compare data from multiple studies. Variability can be introduced through patient differences, iPSC reprogramming methods, and neuronal differentiation paradigms. Some of these problems can be overcome by using cells from enough different individuals to enable statistically meaningful results. Importantly, engineered isogenic cell lines with either *FMR1* gene deletion in control cells or *FMR1* gene restoration in FXS cells can limit the impact of genetic background. Although hPSCs and gene editing methods have greatly advanced our ability to study human diseases, whether these cell culture-based models can sufficiently mimic human brain development is unclear. Thus, how to best model FXS remains controversial. For a single-gene disease, FXS patients exhibit surprising variability in phenotypes, which likely results from a combination of genetic, biological, psychosocial, and environmental risk factors. How do we evaluate and model the impact of these factors?

FMR1 gene reactivation remains an attractive option for treating FXS at its root cause. However, small molecules that can specifically and effectively reactivate the silenced *FMR1* genes have yet to be discovered. Current genome editing methods have a number of issues that hinders the ability to be effective treatments for human FXS. Although mouse studies have shown that restoration of FMRP in juvenile and young adult mice restores neuronal and behavioral functions, whether such functional restoration will work for humans past the developmental period remains unknown. This question demands further studies using both animal models, especially studies using non-human primate with more complex brains. These challenges are daunting. However, we are hopeful that the rapid scientific development in neuroscience, genetics, stem cell biology, and bioengineering will solve these problems in the coming years and provide hope for individuals and families living with FXS.

References

1. Coffee, B., Keith, K., Albizua, I., Malone, T., Mowrey, J., Sherman, S. L., et al. (2009). Incidence of fragile X syndrome by newborn screening for methylated FMR1 DNA. *American Journal of Human Genetics, 85*, 503–514.
2. Hagerman, R. J., & Hagerman, P. J. (2002). *Fragile X syndrome* (Vol. 2). Baltimore, MD: Johns Hopkins University Press.
3. Cohen, I. L., Sudhalter, V., Pfadt, A., Jenkins, E. C., Brown, W. T., & Vietze, P. M. (1991). Why are autism and the fragile-X syndrome associated? Conceptual and methodological issues. *American Journal of Human Genetics, 48*, 195–202.
4. Fisch, G. S., Cohen, I. L., Wolf, E. G., Brown, W. T., Jenkins, E. C., & Gross, A. (1986). Autism and the fragile X syndrome. *American Journal of Psychiatry, 143*, 71–73.
5. Hagerman, R. J., Ono, M. Y., & Hagerman, P. J. (2005). Recent advances in fragile X: A model for autism and neurodegeneration. *Current Opinion Psychiatry, 18*, 490–496.
6. Hatton, D. D., Sideris, J., Skinner, M., Mankowski, J., Bailey Jr., D. B., Roberts, J., et al. (2006). Autistic behavior in children with fragile X syndrome: Prevalence, stability, and the impact of FMRP. *American Journal of Medical Genetics Part A, 140a*, 1804–1813.
7. Kaufmann, W. E., Cortell, R., Kau, A. S., Bukelis, I., Tierney, E., Gray, R. M., et al. (2004). Autism spectrum disorder in fragile X syndrome: Communication, social interaction, and specific behaviors. *American Journal of Medical Genetics Part A, 129*, 225–234.
8. Lathe, R. (2009). Fragile X and autism. *Autism: The International Journal of Research and Practice, 13*, 194–197.
9. Reiss, A. L., Feinstein, C., & Rosenbaum, K. N. (1986). Autism and genetic disorders. *Schizophrenia Bulletin, 12*, 724–738.
10. Pieretti, M., Zhang, F. P., Fu, Y. H., Warren, S. T., Oostra, B. A., Caskey, C. T., et al. (1991). Absence of expression of the FMR-1 gene in fragile X syndrome. *Cell, 66*, 817–822.
11. Verkerk, A. J., Pieretti, M., Sutcliffe, J. S., Fu, Y. H., Kuhl, D. P., Pizzuti, A., et al. (1991). Identification of a gene (FMR-1) containing a CGG repeat coincident with a breakpoint cluster region exhibiting length variation in fragile X syndrome. *Cell, 65*, 905–914.
12. Eberhart, D. E., Malter, H. E., Feng, Y., & Warren, S. T. (1996). The fragile X mental retardation protein is a ribonucleoprotein containing both nuclear localization and nuclear export signals. *Human Molecular Genetics, 5*, 1083–1091.
13. Liu-Yesucevitz, L., Bassell, G. J., Gitler, A. D., Hart, A. C., Klann, E., Richter, J. D., et al. (2011). Local RNA translation at the synapse and in disease. *Journal of Neuroscience, 31*, 16086–16093.
14. Sidorov, M. S., Auerbach, B. D., & Bear, M. F. (2013). Fragile X mental retardation protein and synaptic plasticity. *Molecular Brain, 6*, 15.
15. Li, Y., & Zhao, X. (2014). Concise review: Fragile X proteins in stem cell maintenance and differentiation. *Stem Cells, 32*, 1724–1733.
16. Hinton, V. J., Brown, W. T., Wisniewski, K., & Rudelli, R. D. (1991). Analysis of neocortex in three males with the fragile X syndrome. *American Journal of Medical Genetics, 41*, 289–294.
17. Irwin, S. A., Patel, B., Idupulapati, M., Harris, J. B., Crisostomo, R. A., Larsen, B. P., et al. (2001). Abnormal dendritic spine characteristics in the temporal and visual cortices of

patients with fragile-X syndrome: A quantitative examination. *American Journal of Medical Genetics, 98*, 161–167.
18. Wisniewski, K. E., Segan, S. M., Miezejeski, C. M., Sersen, E. A., & Rudelli, R. D. (1991). The Fra(X) syndrome: Neurological, electrophysiological, and neuropathological abnormalities. *American Journal of Medical Genetics, 38*, 476–480.
19. Chen, L. S., Tassone, F., Sahota, P., & Hagerman, P. J. (2003). The (CGG)n repeat element within the 5′ untranslated region of the FMR1 message provides both positive and negative cis effects on in vivo translation of a downstream reporter. *Human Molecular Genetics, 12*, 3067–3074.
20. Fu, Y. H., Kuhl, D. P., Pizzuti, A., Pieretti, M., Sutcliffe, J. S., Richards, S., et al. (1991). Variation of the CGG repeat at the fragile X site results in genetic instability: Resolution of the Sherman paradox. *Cell, 67*, 1047–1058.
21. Mailick, M. R., Hong, J., Rathouz, P., Baker, M. W., Greenberg, J. S., Smith, L., et al. (2014). Low-normal FMR1 CGG repeat length: Phenotypic associations. *Frontiers in Genetics, 5*, 309.
22. Jobe, E. M., McQuate, A. L., & Zhao, X. (2012). Crosstalk among epigenetic pathways regulates neurogenesis. *Frontiers in Neuroscience, 6*, 59.
23. Piccolo, F. M., & Fisher, A. G. (2014). Getting rid of DNA methylation. *Trends in Cell Biology, 24*, 136–143.
24. Bernstein, B. E., Meissner, A., & Lander, E. S. (2007). The mammalian epigenome. *Cell, 128*, 669–681.
25. Pastori, C., Peschansky, V. J., Barbouth, D., Mehta, A., Silva, J. P., & Wahlestedt, C. (2014). Comprehensive analysis of the transcriptional landscape of the human FMR1 gene reveals two new long noncoding RNAs differentially expressed in fragile X syndrome and fragile X-associated tremor/ataxia syndrome. *Human Genetics, 133*, 59–67.
26. Peschansky, V. J., Pastori, C., Zeier, Z., Motti, D., Wentzel, K., Velmeshev, D., et al. (2015). Changes in expression of the long non-coding RNA FMR4 associate with altered gene expression during differentiation of human neural precursor cells. *Frontiers in Genetics, 6*, 263.
27. Shaw, M. A., Chiurazzi, P., Romain, D. R., Neri, G., & Gecz, J. (2002). A novel gene, FAM11A, associated with the FRAXF CpG island is transcriptionally silent in FRAXF full mutation. *European Journal of Human Genetics, 10*, 767–772.
28. Ladd, P. D., Smith, L. E., Rabaia, N. A., Moore, J. M., Georges, S. A., Hansen, R. S., et al. (2007). An antisense transcript spanning the CGG repeat region of FMR1 is upregulated in premutation carriers but silenced in full mutation individuals. *Human Molecular Genetics, 16*, 3174–3187.
29. Stoger, R., Genereux, D. P., Hagerman, R. J., Hagerman, P. J., Tassone, F., & Laird, C. D. (2011). Testing the FMR1 promoter for mosaicism in DNA methylation among CpG sites, strands, and cells in FMR1-expressing males with fragile X syndrome. *PLoS One, 6*, e23648.
30. Hagerman, R. J., Hull, C. E., Safanda, J. F., Carpenter, I., Staley, L. W., O'Connor, R. A., et al. (1994). High functioning fragile X males: Demonstration of an unmethylated fully expanded FMR-1 mutation associated with protein expression. *American Journal of Medical Genetics, 51*, 298–308.
31. Loesch, D. Z., Huggins, R., Hay, D. A., Gedeon, A. K., Mulley, J. C., & Sutherland, G. R. (1993). Genotype-phenotype relationships in fragile X syndrome: A family study. *American Journal of Human Genetics, 53*, 1064–1073.
32. Loesch, D. Z., Huggins, R. M., & Hagerman, R. J. (2004). Phenotypic variation and FMRP levels in fragile X. *Mental Retardation and Developmental Disabilities Research Reviews, 10*, 31–41.
33. Loesch, D. Z., Sherwell, S., Kinsella, G., Tassone, F., Taylor, A., Amor, D., et al. (2012). Fragile X-associated tremor/ataxia phenotype in a male carrier of unmethylated full mutation in the FMR1 gene. *Clinical Genetics, 82*, 88–92.

34. Brouwer, J. R., Mientjes, E. J., Bakker, C. E., Nieuwenhuizen, I. M., Severijnen, L. A., Van der Linde, H. C., et al. (2007). Elevated Fmr1 mRNA levels and reduced protein expression in a mouse model with an unmethylated fragile X full mutation. *Experimental Cell Research, 313*, 244–253.
35. Clowry, G., Molnar, Z., & Rakic, P. (2010). Renewed focus on the developing human neocortex. *Journal of Anatomy, 217*, 276–288.
36. Dehay, C., & Kennedy, H. (2007). Cell-cycle control and cortical development. *Nature Reviews Neuroscience, 8*, 438–450.
37. Johnson, M. B., Kawasawa, Y. I., Mason, C. E., Krsnik, Z., Coppola, G., Bogdanovic, D., et al. (2009). Functional and evolutionary insights into human brain development through global transcriptome analysis. *Neuron, 62*, 494–509.
38. Kennedy, H., & Dehay, C. (2012). Self-organization and interareal networks in the primate cortex. *Progress in Brain Research, 195*, 341–360.
39. Rakic, P. (2009). Evolution of the neocortex: Perspective from developmental biology. *Nature Reviews Neuroscience, 10*, 724–735.
40. Molnár, Z., & Clowry, G. (2012). Chapter 3 - Cerebral cortical development in rodents and primates. In M. A. Hofman & D. Falk (Eds.), *Progress in brain research* (pp. 45–70). Amsterdam: Elsevier.
41. Dehay, C., Kennedy, H., & Kosik, K. S. (2015). The outer subventricular zone and primate-specific cortical complexification. *Neuron, 85*, 683–694.
42. Fish, J. L., Dehay, C., Kennedy, H., & Huttner, W. B. (2008). Making bigger brains-the evolution of neural-progenitor-cell division. *Journal of Cell Science, 121*, 2783–2793.
43. Hansen, D. V., Lui, J. H., Parker, P. R., & Kriegstein, A. R. (2010). Neurogenic radial glia in the outer subventricular zone of human neocortex. *Nature, 464*, 554–561.
44. Lui, J. H., Hansen, D. V., & Kriegstein, A. R. (2011). Development and evolution of the human neocortex. *Cell, 146*, 18–36.
45. Smart, I. H., Dehay, C., Giroud, P., Berland, M., & Kennedy, H. (2002). Unique morphological features of the proliferative zones and postmitotic compartments of the neural epithelium giving rise to striate and extrastriate cortex in the monkey. *Cerebral Cortex, 12*, 37–53.
46. Bakken, T. E., Miller, J. A., Ding, S. L., Sunkin, S. M., Smith, K. A., Ng, L., et al. (2016). A comprehensive transcriptional map of primate brain development. *Nature, 535*, 367–375.
47. Silbereis, J. C., Pochareddy, S., Zhu, Y., Li, M., & Sestan, N. (2016). The cellular and molecular landscapes of the developing human central nervous system. *Neuron, 89*, 248–268.
48. Sousa, A. M. M., Meyer, K. A., Santpere, G., Gulden, F. O., & Sestan, N. (2017). Evolution of the human nervous system function, structure, and development. *Cell, 170*, 226–247.
49. Florio, M., et al. (2015). Human-specific gene ARHGAP11B promotes basal progenitor amplification and neocortex expansion. *Science, 347*(6229), 1465–1470.
50. Florio, M., et al. (2018). Evolution and cell-type specificity of human-specific genes preferentially expressed in progenitors of fetal neocortex. Elife, 7.
51. Kalebic, N., Gilardi, C., Albert, M., Namba, T., Long, K. R., Kostic, M., et al. (2018). Human-specific ARHGAP11B induces hallmarks of neocortical expansion in developing ferret neocortex. *eLife, 7*, e41241.
52. Hansen, D. V., Lui, J. H., Flandin, P., Yoshikawa, K., Rubenstein, J. L., Alvarez-Buylla, A., et al. (2013). Non-epithelial stem cells and cortical interneuron production in the human ganglionic eminences. *Nature Neuroscience, 16*, 1576–1587.
53. LaMonica, B. E., Lui, J. H., Wang, X., & Kriegstein, A. R. (2012). OSVZ progenitors in the human cortex: An updated perspective on neurodevelopmental disease. *Current Opinion in Neurobiology, 22*, 747–753.
54. Marin, O. (2013). Human cortical interneurons take their time. *Cell Stem Cell, 12*, 497–499.
55. Tyson, J. A., & Anderson, S. A. (2013). The protracted maturation of human ESC-derived interneurons. *Cell Cycle, 12*, 3129–3130.
56. Zhao, X., & Bhattacharyya, A. (2018). Human models are needed for studying human neurodevelopmental disorders. *American Journal of Human Genetics, 103*, 829–857.

57. Thomson, J. A., Itskovitz-Eldor, J., Shapiro, S. S., Waknitz, M. A., Swiergiel, J. J., Marshall, V. S., et al. (1998). Embryonic stem cell lines derived from human blastocysts. *Science, 282*, 1145–1147.
58. Takahashi, K., Tanabe, K., Ohnuki, M., Narita, M., Ichisaka, T., Tomoda, K., et al. (2007). Induction of pluripotent stem cells from adult human fibroblasts by defined factors. *Cell, 131*, 861–872.
59. Yu, J., Vodyanik, M. A., Smuga-Otto, K., Antosiewicz-Bourget, J., Frane, J. L., Tian, S., et al. (2007). Induced pluripotent stem cell lines derived from human somatic cells. *Science, 318*, 1917–1920.
60. Verlinsky, Y., Strelchenko, N., Kukharenko, V., Rechitsky, S., Verlinsky, O., Galat, V., et al. (2005). Human embryonic stem cell lines with genetic disorders. *Reproductive Biomedicine Online, 10*, 105–110.
61. Ben-Yosef, D., Malcov, M., & Eiges, R. (2008). PGD-derived human embryonic stem cell lines as a powerful tool for the study of human genetic disorders. *Molecular and Cellular Endocrinology, 282*, 153–158.
62. Kuliev, A., Rechitsky, S., Tur-Kaspa, I., & Verlinsky, Y. (2005). Preimplantation genetics: Improving access to stem cell therapy. *Annals of the New York Academy of Sciences, 1054*, 223–227.
63. Pickering, S. J., Braude, P. R., Patel, M., Burns, C. J., Trussler, J., Bolton, V., et al. (2003). Preimplantation genetic diagnosis as a novel source of embryos for stem cell research. *Reproductive Biomedicine Online, 7*, 353–364.
64. Stephenson, E. L., Mason, C., & Braude, P. R. (2009). Preimplantation genetic diagnosis as a source of human embryonic stem cells for disease research and drug discovery. *BJOG: An International Journal of Obstetrics and Gynaecology, 116*, 158–165.
65. Avitzour, M., Mor-Shaked, H., Yanovsky-Dagan, S., Aharoni, S., Altarescu, G., Renbaum, P., et al. (2014). FMR1 epigenetic silencing commonly occurs in undifferentiated fragile X-affected embryonic stem cells. *Stem Cell Reports, 3*, 699–706.
66. Eiges, R., Urbach, A., Malcov, M., Frumkin, T., Schwartz, T., Amit, A., et al. (2007). Developmental study of fragile X syndrome using human embryonic stem cells derived from preimplantation genetically diagnosed embryos. *Cell Stem Cell, 1*, 568–577.
67. Gerhardt, J., Tomishima, M. J., Zaninovic, N., Colak, D., Yan, Z., Zhan, Q., et al. (2014). The DNA replication program is altered at the FMR1 locus in fragile X embryonic stem cells. *Molecular Cell, 53*, 19–31.
68. Mor-Shaked, H., & Eiges, R. (2016). Modeling fragile X syndrome using human pluripotent stem cells. *Genes, 7*, 77.
69. Bar-Nur, O., Caspi, I., & Benvenisty, N. (2012). Molecular analysis of FMR1 reactivation in fragile-X induced pluripotent stem cells and their neuronal derivatives. *Journal of Molecular Cell Biology, 4*, 180–183.
70. Bhattacharyya, A., & Zhao, X. (2016). Human pluripotent stem cell models of fragile X syndrome. *Molecular and Cellular Neurosciences, 73*, 43–51.
71. Brick, D. J., Nethercott, H. E., Montesano, S., Banuelos, M. G., Stover, A. E., Schutte, S. S., et al. (2014). The autism Spectrum disorders stem cell resource at Children's Hospital of Orange County: Implications for disease modeling and drug discovery. *Stem Cells Translational Medicine, 3*, 1275–1286.
72. de Esch, C. E., Ghazvini, M., Loos, F., Schelling-Kazaryan, N., Widagdo, W., Munshi, S. T., et al. (2014). Epigenetic characterization of the FMR1 promoter in induced pluripotent stem cells from human fibroblasts carrying an unmethylated full mutation. *Stem Cell Reports, 3*, 548–555.
73. Doers, M. E., Musser, M. T., Nichol, R., Berndt, E. R., Baker, M., Gomez, T. M., et al. (2014). iPSC-derived forebrain neurons from FXS individuals show defects in initial neurite outgrowth. *Stem Cells and Development, 23*, 1777–1787.

74. Halevy, T., Czech, C., & Benvenisty, N. (2015). Molecular mechanisms regulating the defects in fragile X syndrome neurons derived from human pluripotent stem cells. *Stem Cell Reports, 4*, 37–46.
75. Kaufmann, M., Schuffenhauer, A., Fruh, I., Klein, J., Thiemeyer, A., Rigo, P., et al. (2015). High-throughput screening using iPSC-derived neuronal progenitors to identify compounds counteracting epigenetic gene silencing in fragile X syndrome. *Journal of Biomolecular Screening, 20*, 1101.
76. Urbach, A., Bar-Nur, O., Daley, G. Q., & Benvenisty, N. (2010). Differential modeling of fragile X syndrome by human embryonic stem cells and induced pluripotent stem cells. *Cell Stem Cell, 6*, 407–411.
77. Brykczynska, U., Pecho-Vrieseling, E., Thiemeyer, A., Klein, J., Fruh, I., Doll, T., et al. (2016). CGG repeat-induced FMR1 silencing depends on the expansion size in human iPSCs and neurons carrying unmethylated full mutations. *Stem Cell Reports, 7*, 1059–1071.
78. Gafni, O., Weinberger, L., Mansour, A. A., Manor, Y. S., Chomsky, E., Ben-Yosef, D., et al. (2013). Derivation of novel human ground state naive pluripotent stem cells. *Nature, 504*, 282–286.
79. Bakker, C. E., de Diego Otero, Y., Bontekoe, C., Raghoe, P., Luteijn, T., Hoogeveen, A. T., et al. (2000). Immunocytochemical and biochemical characterization of FMRP, FXR1P, and FXR2P in the mouse. *Experimental Cell Research, 258*, 162–170.
80. Devys, D., Lutz, Y., Rouyer, N., Bellocq, J. P., & Mandel, J. L. (1993). The FMR-1 protein is cytoplasmic, most abundant in neurons and appears normal in carriers of a fragile X premutation. *Nature Genetics, 4*, 335–340.
81. Willemsen, R., Oostra, B. A., Bassell, G. J., & Dictenberg, J. (2004). The fragile X syndrome: From molecular genetics to neurobiology. *Mental Retardation and Developmental Disabilities Research Reviews, 10*, 60–67.
82. Arbab, T., Battaglia, F. P., Pennartz, C. M. A., & Bosman, C. A. (2018). Abnormal hippocampal theta and gamma hypersynchrony produces network and spike timing disturbances in the Fmr1-KO mouse model of fragile X syndrome. *Neurobiology of Disease, 114*, 65–73.
83. Cea-Del Rio, C. A., & Huntsman, M. M. (2014). The contribution of inhibitory interneurons to circuit dysfunction in fragile X syndrome. *Frontiers in Cellular Neuroscience, 8*, 245.
84. Goel, A., Cantu, D. A., Guilfoyle, J., Chaudhari, G. R., Newadkar, A., Todisco, B., et al. (2018). Impaired perceptual learning in a mouse model of fragile X syndrome is mediated by parvalbumin neuron dysfunction and is reversible. *Nature Neuroscience, 21*, 1404–1411.
85. Nomura, T., Musial, T. F., Marshall, J. J., Zhu, Y., Remmers, C. L., Xu, J., et al. (2017). Delayed maturation of fast-spiking interneurons is rectified by activation of the TrkB receptor in the mouse model of fragile X syndrome. *The Journal of Neuroscience, 37*, 11298–11310.
86. Wen, T. H., Afroz, S., Reinhard, S. M., Palacios, A. R., Tapia, K., Binder, D. K., et al. (2018). Genetic reduction of matrix metalloproteinase-9 promotes formation of Perineuronal nets around parvalbumin-expressing interneurons and normalizes auditory cortex responses in developing Fmr1 knock-out mice. *Cerebral Cortex, 28*, 3951–3964.
87. Yang, Y. M., Arsenault, J., Bah, A., Krzeminski, M., Fekete, A., Chao, O. Y., et al. (2018). Identification of a molecular locus for normalizing dysregulated GABA release from interneurons in the Fragile X brain. *Molecular Psychiatry*. https://doi.org/10.1038/s41380-018-0240-0
88. Cheng, C., Lau, S. K., & Doering, L. C. (2016). Astrocyte-secreted thrombospondin-1 modulates synapse and spine defects in the fragile X mouse model. *Molecular Brain, 9*, 74.
89. Gholizadeh, S., Halder, S. K., & Hampson, D. R. (2015). Expression of fragile X mental retardation protein in neurons and glia of the developing and adult mouse brain. *Brain Research, 1596*, 22–30.
90. Higashimori, H., Morel, L., Huth, J., Lindemann, L., Dulla, C., Taylor, A., et al. (2013). Astroglial FMRP-dependent translational down-regulation of mGluR5 underlies glutamate transporter GLT1 dysregulation in the fragile X mouse. *Human Molecular Genetics, 22*, 2041–2054.

91. Higashimori, H., Schin, C. S., Chiang, M. S., Morel, L., Shoneye, T. A., Nelson, D. L., et al. (2016). Selective deletion of astroglial FMRP dysregulates glutamate transporter GLT1 and contributes to fragile X syndrome phenotypes in vivo. *The Journal of Neuroscience, 36*, 7079–7094.
92. Hodges, J. L., Yu, X., Gilmore, A., Bennett, H., Tjia, M., Perna, J. F., et al. (2017). Astrocytic contributions to synaptic and learning abnormalities in a mouse model of fragile X syndrome. *Biological Psychiatry, 82*, 139–149.
93. Jacobs, S., Cheng, C., & Doering, L. C. (2012). Probing astrocyte function in fragile X syndrome. *Results and Problems in Cell Differentiation, 54*, 15–31.
94. Jacobs, S., Nathwani, M., & Doering, L. C. (2010). Fragile X astrocytes induce developmental delays in dendrite maturation and synaptic protein expression. *BMC Neuroscience, 11*, 132.
95. Pacey, L. K., & Doering, L. C. (2007). Developmental expression of FMRP in the astrocyte lineage: Implications for fragile X syndrome. *Glia, 55*, 1601–1609.
96. Wang, L., Wang, Y., Zhou, S., Yang, L., Shi, Q., Li, Y., et al. (2016). Imbalance between glutamate and GABA in Fmr1 knockout astrocytes influences neuronal development. *Genes (Basel), 7*, 45.
97. Giampetruzzi, A., Carson, J. H., & Barbarese, E. (2013). FMRP and myelin protein expression in oligodendrocytes. *Molecular and Cellular Neurosciences, 56*, 333–341.
98. Pacey, L. K., Xuan, I. C., Guan, S., Sussman, D., Henkelman, R. M., Chen, Y., et al. (2013). Delayed myelination in a mouse model of fragile X syndrome. *Human Molecular Genetics, 22*, 3920–3930.
99. Green, T., Barnea-Goraly, N., Raman, M., Hall, S. S., Lightbody, A. A., Bruno, J. L., et al. (2015). Specific effect of the fragile-X mental retardation-1 gene (FMR1) on white matter microstructure. *The British Journal of Psychiatry, 207*, 143.
100. Villalon-Reina, J., Jahanshad, N., Beaton, E., Toga, A. W., Thompson, P. M., & Simon, T. J. (2013). White matter microstructural abnormalities in girls with chromosome 22q11.2 deletion syndrome, fragile X or turner syndrome as evidenced by diffusion tensor imaging. *NeuroImage, 81*, 441–454.
101. Kim, D. S., et al. (2014). Optimizing neuronal differentiation from induced pluripotent stem cells to model ASD. *Front Cell Neurosci, 8*, 109.
102. Chambers, S. M., Fasano, C. A., Papapetrou, E. P., Tomishima, M., Sadelain, M., & Studer, L. (2009). Highly efficient neural conversion of human ES and iPS cells by dual inhibition of SMAD signaling. *Nature Biotechnology, 27*, 275–280.
103. Hu, B. Y., Du, Z. W., & Zhang, S. C. (2009). Differentiation of human oligodendrocytes from pluripotent stem cells. *Nature Protocols, 4*, 1614–1622.
104. Hu, B. Y., Weick, J. P., Yu, J., Ma, L. X., Zhang, X. Q., Thomson, J. A., et al. (2010). Neural differentiation of human induced pluripotent stem cells follows developmental principles but with variable potency. *Proceedings of the National Academy of Sciences of the United States of America, 107*, 4335–4340.
105. Krencik, R., Weick, J. P., Liu, Y., Zhang, Z. J., & Zhang, S. C. (2011). Specification of transplantable astroglial subtypes from human pluripotent stem cells. *National Biotechnology, 29*, 528–534.
106. Li, X. J., Du, Z. W., Zarnowska, E. D., Pankratz, M., Hansen, L. O., Pearce, R. A., et al. (2005). Specification of motoneurons from human embryonic stem cells. *Nature Biotechnology, 23*, 215–221.
107. Li, X. J., Hu, B. Y., Jones, S. A., Zhang, Y. S., Lavaute, T., Du, Z. W., et al. (2008). Directed differentiation of ventral spinal progenitors and motor neurons from human embryonic stem cells by small molecules. *Stem Cells, 26*, 886–893.
108. Li, X. J., & Zhang, S. C. (2006). In vitro differentiation of neural precursors from human embryonic stem cells. *Methods of Molecular Biology, 331*, 169–177.

109. Liu, Y., Liu, H., Sauvey, C., Yao, L., Zarnowska, E. D., & Zhang, S. C. (2013a). Directed differentiation of forebrain GABA interneurons from human pluripotent stem cells. *Nature Protocols, 8*, 1670–1679.
110. Liu, Y., Weick, J. P., Liu, H., Krencik, R., Zhang, X., Ma, L., et al. (2013b). Medial ganglionic eminence-like cells derived from human embryonic stem cells correct learning and memory deficits. *Nature Biotechnology, 31*, 440–447.
111. Maroof, A. M., Keros, S., Tyson, J. A., Ying, S. W., Ganat, Y. M., Merkle, F. T., et al. (2013). Directed differentiation and functional maturation of cortical interneurons from human embryonic stem cells. *Cell Stem Cell, 12*, 559–572.
112. Nicholas, C. R., Chen, J., Tang, Y., Southwell, D. G., Chalmers, N., Vogt, D., et al. (2013). Functional maturation of hPSC-derived forebrain interneurons requires an extended timeline and mimics human neural development. *Cell Stem Cell, 12*, 573–586.
113. Watanabe, K., Kamiya, D., Nishiyama, A., Katayama, T., Nozaki, S., Kawasaki, H., et al. (2005). Directed differentiation of telencephalic precursors from embryonic stem cells. *Nature Neuroscience, 8*, 288–296.
114. Yan, Y., Yang, D., Zarnowska, E. D., Du, Z., Werbel, B., Valliere, C., et al. (2005). Directed differentiation of dopaminergic neuronal subtypes from human embryonic stem cells. *Stem Cells, 23*, 781–790.
115. Yuan, F., Fang, K. H., Cao, S. Y., Qu, Z. Y., Li, Q., Krencik, R., et al. (2015). Efficient generation of region-specific forebrain neurons from human pluripotent stem cells under highly defined condition. *Scientific Reports, 5*, 18550.
116. Zhang, S. C., Wernig, M., Duncan, I. D., Brustle, O., & Thomson, J. A. (2001). In vitro differentiation of transplantable neural precursors from human embryonic stem cells. *Nature Biotechnology, 19*, 1129–1133.
117. Drouin-Ouellet, J., Lau, S., Brattas, P. L., Rylander Ottosson, D., Pircs, K., Grassi, D. A., et al. (2017). REST suppression mediates neural conversion of adult human fibroblasts via microRNA-dependent and -independent pathways. *EMBO Molecular Medicine, 9*, 1117–1131.
118. Drouin-Ouellet, J., Pircs, K., Barker, R. A., Jakobsson, J., & Parmar, M. (2017). Direct neuronal reprogramming for disease modeling studies using patient-derived neurons: What have we learned? *Frontiers in Neuroscience, 11*, 530.
119. Pang, Z. P., Yang, N., Vierbuchen, T., Ostermeier, A., Fuentes, D. R., Yang, T. Q., et al. (2011). Induction of human neuronal cells by defined transcription factors. *Nature, 476*, 220–223.
120. Vierbuchen, T., Ostermeier, A., Pang, Z. P., Kokubu, Y., Sudhof, T. C., & Wernig, M. (2010). Direct conversion of fibroblasts to functional neurons by defined factors. *Nature, 463*, 1035–1041.
121. Zhang, Y., Pak, C., Han, Y., Ahlenius, H., Zhang, Z., Chanda, S., et al. (2013). Rapid single-step induction of functional neurons from human pluripotent stem cells. *Neuron, 78*, 785–798.
122. Boland, M. J., Nazor, K. L., Tran, H. T., Szucs, A., Lynch, C. L., Paredes, R., et al. (2017). Molecular analyses of neurogenic defects in a human pluripotent stem cell model of fragile X syndrome. *Brain, 140*, 582–598.
123. Telias, M., Segal, M., & Ben-Yosef, D. (2013). Neural differentiation of fragile X human embryonic stem cells reveals abnormal patterns of development despite successful neurogenesis. *Developmental Biology, 374*, 32–45.
124. Sheridan, S. D., et al. (2011). Epigenetic characterization of the FMR1 gene and aberrant neurodevelopment in human induced pluripotent stem cell models of fragile X syndrome. *PLoS.One., 6*(10), e26203.
125. Guo, W., Allan, A. M., Zong, R., Zhang, L., Johnson, E. B., Schaller, E. G., et al. (2011). Ablation of Fmrp in adult neural stem cells disrupts hippocampus-dependent learning. *Nature Medicine, 17*, 559–565.
126. Guo, W., Murthy, A. C., Zhang, L., Johnson, E. B., Schaller, E. G., Allan, A. M., et al. (2012). Inhibition of GSK3beta improves hippocampus-dependent learning and rescues neurogenesis in a mouse model of fragile X syndrome. *Human Molecular Genetics, 21*, 681–691.

127. Guo, W., et al. (2015). Fragile X Proteins FMRP and FXR2P Control Synaptic GluA1 Expression and Neuronal Maturation via Distinct Mechanisms. *Cell Rep, 11*(10), 1651–1666.
128. Sunamura, N., Iwashita, S., Enomoto, K., Kadoshima, T., & Isono, F. (2018). Loss of the fragile X mental retardation protein causes aberrant differentiation in human neural progenitor cells. *Scientific Reports, 8*, 11585.
129. Telias, M., Kuznitsov-Yanovsky, L., Segal, M., & Ben-Yosef, D. (2015). Functional deficiencies in fragile X neurons derived from human embryonic stem cells. *The Journal of Neuroscience, 35*, 15295–15306.
130. Telias, M., Segal, M., & Ben-Yosef, D. (2016). Immature responses to GABA in fragile X neurons derived from human embryonic stem cells. *Frontiers in Cellular Neuroscience, 10*, 121.
131. Zhang, Z., Marro, S. G., Zhang, Y., Arendt, K. L., Patzke, C., Zhou, B., et al. (2018). The fragile X mutation impairs homeostatic plasticity in human neurons by blocking synaptic retinoic acid signaling. *Science Translational Medicine, 10, eaar4338.*
132. Niedringhaus, M., Dumitru, R., Mabb, A. M., Wang, Y., Philpot, B. D., Allbritton, N. L., et al. (2015). Transferable neuronal mini-cultures to accelerate screening in primary and induced pluripotent stem cell-derived neurons. *Scientific Reports, 5*, 8353.
133. Drouin, R., Angers, M., Dallaire, N., Rose, T. M., Khandjian, E. W., & Rousseau, F. (1997). Structural and functional characterization of the human FMR1 promoter reveals similarities with the hnRNP-A2 promoter region. *Human Molecular Genetics, 6*, 2051–2060.
134. Schwemmle, S. (1999). In vivo footprinting analysis of the FMR1 gene: Proposals concerning gene regulation in high-functioning males. *American Journal of Medical Genetics, 84*, 266–267.
135. Schwemmle, S., de Graaff, E., Deissler, H., Glaser, D., Wohrle, D., Kennerknecht, I., et al. (1997). Characterization of FMR1 promoter elements by in vivo-footprinting analysis. *American Journal of Human Genetics, 60*, 1354–1362.
136. Gheldof, N., Tabuchi, T. M., & Dekker, J. (2006). The active FMR1 promoter is associated with a large domain of altered chromatin conformation with embedded local histone modifications. *Proceedings of the National Academy of Sciences of the United States of America, 103*, 12463–12468.
137. Coffee, B., Zhang, F., Warren, S. T., & Reines, D. (1999). Acetylated histones are associated with FMR1 in normal but not fragile X-syndrome cells. *Nature Genetics, 22*, 98–101.
138. Coffee, B., Zhang, F., Ceman, S., Warren, S. T., & Reines, D. (2002). Histone modifications depict an aberrantly heterochromatinized FMR1 gene in fragile X syndrome. *American Journal of Human Genetics, 71*, 923–932.
139. Pietrobono, R., Tabolacci, E., Zalfa, F., Zito, I., Terracciano, A., Moscato, U., et al. (2005). Molecular dissection of the events leading to inactivation of the FMR1 gene. *Human Molecular Genetics, 14*, 267–277.
140. Tabolacci, E., Moscato, U., Zalfa, F., Bagni, C., Chiurazzi, P., & Neri, G. (2008b). Epigenetic analysis reveals a euchromatic configuration in the FMR1 unmethylated full mutations. *European Journal of Human Genetics, 16*, 1487–1498.
141. Kumari, D., & Usdin, K. (2014). Polycomb group complexes are recruited to reactivated FMR1 alleles in fragile X syndrome in response to FMR1 transcription. *Human Molecular Genetics, 23*, 6575.
142. Colak, D., Zaninovic, N., Cohen, M. S., Rosenwaks, Z., Yang, W. Y., Gerhardt, J., et al. (2014). Promoter-bound trinucleotide repeat mRNA drives epigenetic silencing in fragile X syndrome. *Science, 343*, 1002–1005.
143. Smeets, H. J., Smits, A. P., Verheij, C. E., Theelen, J. P., Willemsen, R., van de Burgt, I., et al. (1995). Normal phenotype in two brothers with a full FMR1 mutation. *Human Molecular Genetics, 4*, 2103–2108.
144. Gholizadeh, S., Arsenault, J., Xuan, I. C., Pacey, L. K., & Hampson, D. R. (2014). Reduced phenotypic severity following adeno-associated virus-mediated Fmr1 gene delivery in fragile X mice. *Neuropsychopharmacology, 39*, 3100–3111.

145. Zeier, Z., Kumar, A., Bodhinathan, K., Feller, J. A., Foster, T. C., & Bloom, D. C. (2009). Fragile X mental retardation protein replacement restores hippocampal synaptic function in a mouse model of fragile X syndrome. *Gene Therapy, 16*, 1122–1129.
146. Chiurazzi, P., Pomponi, M. G., Willemsen, R., Oostra, B. A., & Neri, G. (1998). In vitro reactivation of the FMR1 gene involved in fragile X syndrome. *Human Molecular Genetics, 7*, 109–113.
147. Pietrobono, R., Pomponi, M. G., Tabolacci, E., Oostra, B., Chiurazzi, P., & Neri, G. (2002). Quantitative analysis of DNA demethylation and transcriptional reactivation of the FMR1 gene in fragile X cells treated with 5-azadeoxycytidine. *Nucleic Acids Research, 30*, 3278–3285.
148. Brendel, C., Mielke, B., Hillebrand, M., Gartner, J., & Huppke, P. (2013). Methotrexate treatment of FraX fibroblasts results in FMR1 transcription but not in detectable FMR1 protein levels. *Journal of Neurodevelopmental Disorders, 5*, 23.
149. Tabolacci, E., De Pascalis, I., Accadia, M., Terracciano, A., Moscato, U., Chiurazzi, P., et al. (2008a). Modest reactivation of the mutant FMR1 gene by valproic acid is accompanied by histone modifications but not DNA demethylation. *Pharmacogenetics and Genomics, 18*, 738–741.
150. Chiurazzi, P., Pomponi, M. G., Pietrobono, R., Bakker, C. E., Neri, G., & Oostra, B. A. (1999). Synergistic effect of histone hyperacetylation and DNA demethylation in the reactivation of the FMR1 gene. *Human Molecular Genetics, 8*, 2317–2323.
151. Biacsi, R., Kumari, D., & Usdin, K. (2008). SIRT1 inhibition alleviates gene silencing in fragile X mental retardation syndrome. *PLoS Genetics, 4*, e1000017.
152. Kumari, D., Swaroop, M., Southall, N., Huang, W., Zheng, W., & Usdin, K. (2015). High-throughput screening to identify compounds that increase fragile X mental retardation protein expression in neural stem cells differentiated from fragile X syndrome patient-derived induced pluripotent stem cells. *Stem Cells Translational Medicine, 4*, 800–808.
153. Jang, S. W., Lopez-Anido, C., MacArthur, R., Svaren, J., & Inglese, J. (2012). Identification of drug modulators targeting gene-dosage disease CMT1A. *ACS Chemical Biology, 7*, 1205–1213.
154. Hunt, J. F. V. S., Li, M., Zhao, X., & Bhattacharyya, A. (2019). Using human neural progenitor cell models to conduct large-scale drug screens for neurological and psychiatric diseases. In D. Ben-Yosef & Y. Mayshar (Eds.), *Fragile-X syndrome: Methods and protocols, methods in molecular biology*. New York): Springer Science+Business Media, LLC, part of Springer Nature.
155. Li, M., Hunt, J. F. V. S., Bhattacharyya, A., & Zhao, X. (2019). One-step generation of seamless luciferase gene knockin using CRISPR/Cas9 genome editing in human pluripotent stem cells. In D. Ben-Yosef & Y. Mayshar (Eds.), *Fragile-X syndrome: Methods and protocols, methods in molecular biology*. New York: Springer Science+Business Media, LLC, part of Springer Nature.
156. Li, M., Zhao, H., Ananiev, G. E., Musser, M. T., Ness, K. H., Maglaque, D. L., et al. (2017). Establishment of reporter lines for detecting fragile X mental retardation (FMR1) gene reactivation in human neural cells. *Stem Cells, 35*, 158–169.
157. Hsu, P. D., Lander, E. S., & Zhang, F. (2014). Development and applications of CRISPR-Cas9 for genome engineering. *Cell, 157*, 1262–1278.
158. Ma, K., Wang, J., Shen, B., Qiu, L., Huang, X., & Li, Z. (2014). Efficient targeting of FATS at a common fragile site in mice through TALEN-mediated double-hit genome modification. *Biotechnology Letters, 36*, 471–479.
159. Zhang, F., Wen, Y., & Guo, X. (2014). CRISPR/Cas9 for genome editing: Progress, implications and challenges. *Human Molecular Genetics, 23*, R40–R46.
160. Park, C. Y., Halevy, T., Lee, D. R., Sung, J. J., Lee, J. S., Yanuka, O., et al. (2015). Reversion of FMR1 methylation and silencing by editing the triplet repeats in fragile X iPSC-derived neurons. *Cell Reports, 13*, 234–241.

161. Xie, N., Gong, H., Suhl, J. A., Chopra, P., Wang, T., & Warren, S. T. (2016). Reactivation of FMR1 by CRISPR/Cas9-mediated deletion of the expanded CGG-repeat of the fragile X chromosome. *PLoS One, 11*, e0165499.
162. Qi, L. S., Larson, M. H., Gilbert, L. A., Doudna, J. A., Weissman, J. S., Arkin, A. P., et al. (2013). Repurposing CRISPR as an RNA-guided platform for sequence-specific control of gene expression. *Cell, 152*, 1173–1183.
163. Adli, M. (2018). The CRISPR tool kit for genome editing and beyond. *Nature Communications, 9*, 1911.
164. Haenfler, J. M., Skariah, G., Rodriguez, C. M., Monteiro da Rocha, A., Parent, J. M., Smith, G. D., et al. (2018). Targeted reactivation of FMR1 transcription in fragile X syndrome embryonic stem cells. *Frontiers in Molecular Neuroscience, 11*, 282.
165. Balboa, D., Weltner, J., Eurola, S., Trokovic, R., Wartiovaara, K., & Otonkoski, T. (2015). Conditionally stabilized dCas9 activator for controlling gene expression in human cell reprogramming and differentiation. *Stem Cell Reports, 5*, 448–459.
166. Liu, X. S., Wu, H., Krzisch, M., Wu, X., Graef, J., Muffat, J., et al. (2018). Rescue of Fragile X syndrome neurons by DNA methylation editing of the FMR1 gene. *Cell, 172*, 979–992. e976.
167. Lee, B., Lee, K., Panda, S., Gonzales-Rojas, R., Chong, A., Bugay, V., et al. (2018). Nanoparticle delivery of CRISPR into the brain rescues a mouse model of fragile X syndrome from exaggerated repetitive behaviours. *Nature Biomedical Engineering, 2*, 497–507.
168. Wagner, D. L., Amini, L., Wendering, D. J., Burkhardt, L.-M., Akyüz, L., Reinke, P., et al. (2018). High prevalence of Streptococcus pyogenes Cas9-reactive T cells within the adult human population. *Nature Medicine, 25*(2), 242–248.

IPSC Models of Chromosome 15Q Imprinting Disorders: From Disease Modeling to Therapeutic Strategies

Noelle D. Germain, Eric S. Levine, and Stormy J. Chamberlain

The chromosome 15q11-q13 region of the human genome is regulated by genomic imprinting, an epigenetic phenomenon in which genes are expressed exclusively from one parental allele. Several genes within the 15q11-q13 region are expressed exclusively from the paternally inherited chromosome 15. At least one gene *UBE3A*, shows exclusive expression of the maternal allele, but this allele-specific expression is restricted to neurons.

The appropriate regulation of imprinted gene expression across chromosome 15q11-q13 has important implications for human disease. Three different neurodevelopmental disorders result from aberrant expression of imprinted genes in this region: Prader–Willi syndrome (PWS), Angelman syndrome (AS), and 15q duplication syndrome. These three disorders each occur at an estimated frequency of approximately 1/15,000–1/30,000 live births [1, 2]. The Prader–Willi and Angelman syndromes most commonly result from large, 5–7 Mb deletions that include both imprinted and non-imprinted genes. These deletions occur during meiosis and are mediated by local repetitive sequences [3–5]. 15q duplication syndrome results when the same repetitive sequences mediate duplication, rather than deletion, by unequal homologous recombination [6]. In addition to interstitial duplication, an extra isodicentric chromosome 15q can also result, yielding two extra copies of the chromosome 15q11-q13 region [7]. Induced pluripotent stem cells (iPSCs) have been generated from individuals with each of these disorders by multiple groups.

A map of the chromosome 15q11-q13 region is shown in Fig. 1. *MKRN3, MAGEL2, NDN, C15ORF2, SNURF-SNRPN* (heretofore called *SNRPN*),

N. D. Germain · S. J. Chamberlain
Department of Genetics and Genome Sciences, University of Connecticut School of Medicine, Farmington, CT, USA

E. S. Levine (✉)
Department of Neuroscience, University of Connecticut School of Medicine, Farmington, CT, USA
e-mail: eslevine@uchc.edu

E. DiCicco-Bloom, J. H. Millonig (eds.), *Neurodevelopmental Disorders*, Advances in Neurobiology 25, https://doi.org/10.1007/978-3-030-45493-7_3

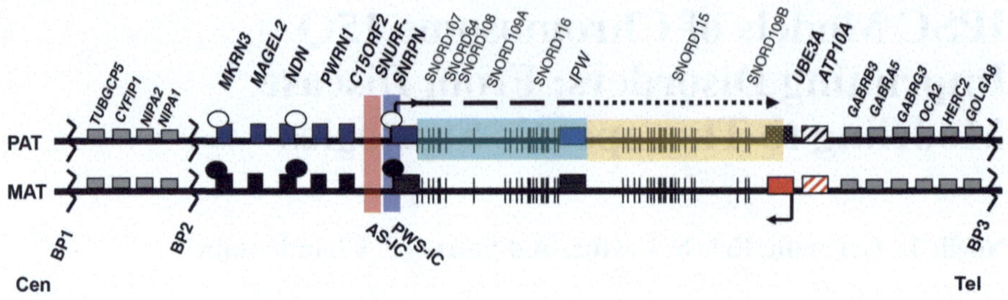

Fig. 1 Diagram of the human 15q11-q13 genetic region. Individual genes are indicated by rectangles. Paternally expressed imprinted genes are denoted by blue rectangles, the maternally expressed imprinted gene is denoted by a red rectangle. Black rectangles represent silences alleles. Gray rectangles represent genes expressed from both parental alleles. *BP1*, BP2, and BP3 indicate common deletion/duplication break points. Dashed line represents the *SNHG14* non-coding RNA. The large dashes represent *SNHG14* expressed in all cell types, while the short dashes represent neuron-specific *SNHG14/UBE3A-ATS* expression. The PWS critical region is highlighted in blue. The neuron-specific portion of *SNHG14* is highlighted in pale orange

SNORD107, SNORD64, SNORD108, SNORD109A, SNORD116, IPW, SNORD115, and *SNORD109B* are exclusively expressed from the paternally inherited allele. *UBE3A* is expressed from the maternally inherited allele, and shows brain-specific imprinted expression. ATP10A is similarly imprinted and expressed from the maternally inherited allele in brain, although not in all individuals [8]. Several genes in the 15q11-q13 region lie within the region deleted or duplicated in PWS, AS, or Dup15q syndrome but are not imprinted. Proximal to the imprinted domain are *GOLGA8E, TUBGCP5, CYFIP1,* and *NIPA2* [9]. Distal to the imprinted domain, lie a cluster of three non-imprinted $GABA_A$ receptor subunit genes, *GABRB3, GABRA5,* and *GABRG3*, and the genes *OCA2, HERC2,* and *GOLGA8G* [10]. Each of the three disorders is described more fully below.

1 Prader–Willi Syndrome

Prader–Willi syndrome (PWS) is a multifaceted disorder that first presents as hypotonia and failure to thrive in affected infants, which gives way to hyperphagia and morbid obesity. Phenotypic hallmarks also include small stature with small hands and feet, mild to moderate cognitive deficit, and behavioral problems, including obsessive-compulsive disorder [1].

PWS is caused by the loss of expression from the paternal allele of 15q11-q13. Approximately 70% of PWS patients have a ~ 5–7 Mb deletion of the entire region; approximately 25% of PWS patients have inherited both copies of chromosome 15 from their mother, a condition called maternal uniparental disomy (UPD); and approximately 5% of PWS can be attributed to an imprinting defect whereby the individual has inherited one chromosome 15 from each parent, however, the paternally inherited allele of 15q11-q13 behaves as if it were the maternally inherited

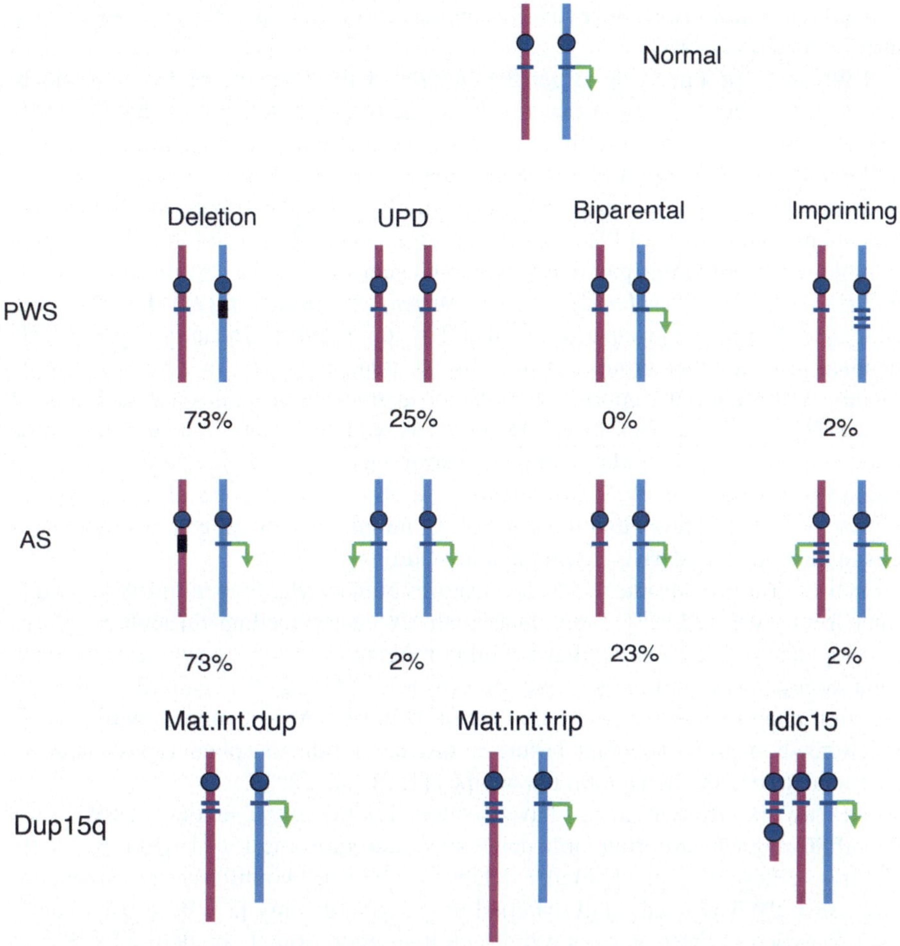

Fig. 2 Genetic etiologies of Prader–Willi, Angelman, and Dup15q syndromes. Maternal and paternal chromosomes 15 are depicted in pink and blue, respectively. Circle represents the centromeres, and the horizontal lines represent the 15q11-q13 region. The green arrow represents the active paternal PWS-IC and ensuing transcription. Approximate percentages of each genetic etiology is reported for PWS and AS. *UPD* uniparental disomy, *Imprinting* imprinting defect, *Mat. int. dup.* maternal interstitial duplication, *Mat.int.trip* maternal interstitial triplication, and *idic15* isodicentric chromosome 15

allele (Fig. 2). Originally thought to be a true contiguous gene syndrome because no single gene mutation was found to lead to the disorder, recently identified patients have shown that PWS may result from the loss of one of the snoRNA clusters that reside in the region. Individuals lacking the *SNORD116* snoRNA cluster and *IPW* suffer the same failure to thrive, hypotonia, and hyperphagia that is observed in patients with larger deletions [11, 12]. Therefore, this region may encompass the PWS critical region. Despite the further refinement of the PWS critical region to a defined set of transcripts, their function and contribution to the PWS disease mechanism remain a mystery.

PWS mice produced by targeted mutation of the PWS-IC or by a paternally inherited transgene insertion that results in the physical deletion of the paternally expressed genes, recapitulate the early failure-to-thrive phenotype and show perinatal lethality, depending on their genetic background [13–15]. Although the regulation of imprinted gene expression seems similarly regulated between mouse and human, murine models of PWS have yet to show hyperphagia and obesity. Deletion or mutation of individual paternally expressed genes within the chromosome 15q11-q13 (i.e. *Ndn* [16, 17], *Magel2* [18, 19], *Snrpn* [13], *Snord116* [20]) has also been generated. While one group reported that deletion of *Snord116* caused hyperphagia in mice [20], another group did not observe that phenotype in a similar murine model. A third group reports hyperphagia in mice with adult-onset deletion of *Snord116* in the mediobasal hypothalamus, as well as obesity in a subset of those mice [21]. None of the individual gene disruptions has recapitulated the fully penetrant obesity seen in PWS individuals. The absence of an obesity phenotype in murine models has propelled the scientific community to investigate neuronal deficits in human iPSC-derived hypothalamic neurons.

Schaaf–Yang syndrome (SYS) has features both overlapping with PWS and distinct from PWS [22, 23]. Individuals with SYS have feeding difficulties, global developmental delay, intellectual disability ranging from mild to severe, hypotonia, joint contractures, and autism spectrum disorder [23]. SYS is caused by loss-of-function mutations in the paternal allele of *MAGEL2*. Mouse models with *Magel2* deletions show growth deficits, failure-to-thrive, reproductive phenotypes, circadian rhythm deficits, and behavioral phenotypes [18, 19, 24, 25].

At least six different groups have developed human iPSC models of PWS [26–31]. iPSC models harboring large deletions of chromosome 15q11-q13 [26, 28, 30, 31], two different atypical smaller deletions [29, 32], two different translocations that cause PWS [27, 28], and maternal uniparental disomy [33] have been generated. For each of these cases, multiple cell lines appropriately modeling PWS gene expression and epigenotype have been generated. PWS is a disorder primarily affecting hypothalamic function. Generation of highly enriched populations of mature hypothalamic neurons from human pluripotent stem cells is challenging and is a limiting factor for PWS iPSC research.

2 Angelman Syndrome

The human neurogenetic disorder, Angelman Syndrome (AS), is caused by loss of function from the maternally inherited allele of *UBE3A*. *UBE3A* is an E3 ubiquitin ligase that is expressed from both parental alleles in most tissues. However, *UBE3A* is expressed exclusively from the maternally inherited allele in the brain, so mutation or deletion of the maternal copy of the gene results in complete loss of the ubiquitin ligase in this tissue [34].

Loss of *UBE3A* expression in AS patients occurs due to one of five classes of genetic abnormalities (see Fig. 2). 70% of patients suffer from a ~ 5–7 Mb deletion of the maternal allele of chromosome 15q11-q13 and 10% of patients harbor a loss-of-function mutation in the *UBE3A* gene. These two classes comprise the most common genetic etiologies of AS. The remainder of patients attribute AS to either paternal uniparental disomy or an imprinting defect in which the maternal allele behaves as if it were paternally inherited [35]. A small but significant portion of individuals with a clinical diagnosis of AS has none of the aforementioned genetic etiologies. In some of these cases, mutations in other genes such as *SLC9A6* or *TCF4* have been identified. Indeed, there are a handful of rare neurodevelopmental disorders that can present as Angelman-like [36].

Phenotypic hallmarks of AS include motor dysfunction leading to an ataxic gait, frequent seizures that can be severe, profound learning disability coupled with a short attention span, absent speech, and characteristic happy demeanor [35]. Mouse models of AS have been generated by targeted mutation of the murine *Ube3a* gene [37, 38]. AS mice have increased incidence of seizures, poor performance on rotarod assays, and defects in long-term potentiation (LTP), a measure of learning and memory [37–40]. These phenotypes, while similar to human AS symptoms, are less severe in mice and some are dependent on particular inbred genetic backgrounds [38, 39]. The ubiquitylation targets of UBE3A that cause the neuronal deficits in the AS mouse model and whether UBE3A ubiquitylates the same proteins in human neurons is not known.

At least four different groups have generated iPSC models of AS [26, 41–43]. Genetic subtypes of the iPSCs generated include large deletion of chromosome 15q11-q13 [26, 43], non-sense mutations in UBE3A [44], missense mutation in UBE3A [41], and paternal uniparental disomy [42]. In each of these instances, multiple clonal AS iPSC lines have been generated with the appropriate epigenotype and have been successfully differentiated into neurons. Many different neuronal subtypes are thought to be affected in AS, since *UBE3A* is imprinted in nearly all neuronal subtypes investigated. iPSC models of AS have yet to delve into the effect loss of *UBE3A* has on different neuronal subtypes.

3 15q11-q13 Duplication

Dup15q syndrome results from either interstitial duplication or triplication of 15q11-q13 or an extra, isodicentric chromosome 15 [2] (see Fig. 2). Individuals with Dup15q syndrome commonly present with hypotonia, delay in motor skills and language development, cognitive and learning disabilities, epilepsy, and characteristic facial features. Affected individuals also usually meet the diagnostic criteria for autism. In fact, one study of autistic children identified 15q aberrations as the single most common cause, and estimates place the prevalence of 15q duplication in autism at 1–3% [45, 46]. Some individuals also present with anxiety, hyperactivity, and short stature [2]. Dup15q syndrome is more severe in individuals with idic15, presumably because they have three copies of maternal 15q11-q13, compared with two copies in interstitial duplication patients [7].

The phenotypic 15q11-q13 duplications are often of maternal origin. In contrast, individuals who inherit a paternal chromosome with an interstitial 15q11-q13 duplication usually display a normal phenotype [47, 48], although there have been reports of individuals with a paternally inherited duplication of 15q and developmental delay and/or intellectual disability [49]. It is possible that individuals with paternally inherited interstitial duplications have a sub-clinical phenotype or present with other neuropsychiatric disorders. Idic15 is almost exclusively of maternal origin, likely due to the mechanisms underlying the construction of this chromosomal aberration, thus the pathology of individuals with an analogous supernumerary chromosome 15 of paternal origin is unclear. Since the phenotypic duplications are usually of maternal origin, studies suggest that maternal duplication of *UBE3A* underlies many of the phenotypes associated with 15q11-q13 duplications [45, 50, 51], although no mechanism has been proposed, and more than 20 genes have been identified within the duplication region.

A sophisticated mouse model of Dup15q syndrome has been used to investigate the roles of maternal versus paternal duplications [52]. This mouse model used chromosomal engineering to generate an interstitial duplication of a 6.3 Mb region encompassing the entire imprinted domain, as well as several non-imprinted genes hypothesized to play a role in Dup15q syndrome. The duplicated allele assumes the appropriate epigenotype when transmitted either maternally- or paternally; maternal duplication resulted in increased expression of *Ube3a* and paternal duplication results in increased expression of *Snord116* and *Snord115* in brain. Unexpectedly, the presence of the duplicated allele in the mice led to autistic-like behaviors, such as reduced social interactions and vocalizations, when paternally transmitted. There was no observable phenotype when maternally transmitted. Other mouse models of Dup15q syndrome are focused on overexpression of *Ube3a*. At least two different transgenic lines overexpressing *Ube3a* have been reported. The large duplication mouse model has excellent construct validity, but tenuous face validity, while the Ube3a overexpression models show some phenotypes that may be analogous to human Dup15q phenotypes, but do not appropriately model the genetics of the disorder. Human iPSC models of Dup15q syndrome may help determine the relative roles of individual genes in Dup15q syndrome and contribute to the development of optimal mouse models.

Only one group has published iPSC models of Dup15q syndrome [53]. iPSCs from individuals with maternal and paternal interstitial duplications as well as multiple individuals with idic15 were generated. The iPSCs reported by this group also maintained the expected epigenotype in the iPSCs compared to the somatic cells sampled from the patients. Forebrain cortical neurons were also successfully generated from these iPSCs, although the disorder likely affects other neuronal subtypes.

4 Special Considerations for Disease Models of Imprinting Disorders

Appropriate parent-of-origin specific gene expression is essential for faithful modeling of chromosome 15q imprinting disorders. Germline imprints are erased and re-established in the germline, while somatic imprints are often established post-zygotically, and in some cases can be tissue-specific. Since cells do not transit through a germ cell-like state during the reprogramming process, germline imprints are not frequently disrupted during reprogramming. On the other hand, somatic imprints can be disrupted by simple maintenance in culture. Since epigenetic reprogramming disrupts both DNA methylation and histone modifications, attention to both germline and somatic imprints would be optimal, although somatic imprints can depend on the particular differentiated tissue.

The germline imprint for the chromosome 15q11-q13 locus is comprised of a region of differential DNA methylation between the parental alleles termed the PWS imprinting center [54]. The PWS-IC encompasses exon 1 of the SNRPN gene and extends into the first intron. The CpG island on the paternal allele is unmethylated, while the same region on the maternally inherited allele is methylated (Fig. 1). The PWS-IC is a promoter for the paternally expressed coding SNRPN and non-coding SNHG14 RNAs [54, 55]. The latter includes all of the snoRNAs and non-coding RNA species mentioned above (SNORD107, SNORD64, SNORD108, SNORD109A, SNORD116, IPW, SNORD115, and SNORD109B) as well as UBE3A-ATS. How the PWS-IC influences expression of the proximal cluster of genes, including MKRN3, MAGEL2, and NDN, more than 1 Mb proximal to it, is not understood. The PWS-IC is functionally conserved between mouse and human. The DNA methylation imprint at the PWS-IC is highly stable in murine and human cell culture [56–58] and even in mouse iPSCs and cloned embryos [59, 60].

Another, lesser-known regulatory element in the region is the Angelman syndrome imprinting center (AS-IC). It represses the PWS-IC on the maternal allele in the maternal germline [54] by driving transcription through the PWS-IC in oocytes. The expression leads to gene body methylation at the PWS-IC, which later is interpreted as repressive promoter methylation when the PWS-IC becomes the major SNRPN promoter in the zygote. Disruption of the AS-IC leads to a paternal epigenotype (loss of DNA methylation), and maternal transmission of this epigenetic abnormality leads to AS [61–63] by epigenetic silencing of maternal UBE3A. The human AS-IC has been mapped to a small region approximately 35 kb upstream of SNRPN. Although deletions of DNA analogous to the AS-IC fail to cause AS in the mouse [64], the function of the AS-IC is conserved between human and mouse [65]. In mouse, multiple upstream SNRPN exons promote expression through the PWS-IC. In humans, the AS-IC is most likely a single oocyte-specific promoter, but this hypothesis will require testing in human cell types. While the AS-IC acts in the maternal germline, oocytes are not finished developing until after fertilization [66]. Therefore, the activity of this germline imprint may not be fully realized in all early zygotes. This may contribute to the prevalence of individuals with AS who are mosaic for an imprinting center defect.

Somatic imprints are not well understood. Within the chromosome 15q11-q13 region, the best known somatic imprint is the repression of paternal *UBE3A* in neurons. Unlike most imprints, the *UBE3A* imprint does not involve a repressive DNA or histone modification, to our current understanding. Rather, paternal *UBE3A* is repressed by transcription of a paternally expressed transcript, *UBE3A-ATS*, in the antisense direction. This somatic imprint has been shown to be intact in human iPSCs upon neuronal differentiation [26, 41, 53]. The somatic imprints governing *NDN, MAGEL2*, and *MKRN3* involve two differentially methylated regions at *NDN* and *MKRN3*, which are thought to be downstream of the methylation at the PWS-IC. A recent paper suggested that these somatic imprints may either be unfaithfully re-established during the reprogramming process or not stably maintained during iPSC culture [43]. Another recent paper revealed an unexpected somatic imprint that represses maternal *SNRPN* in human neurons. Knockdown of ZNF274 was shown to de-repress coding and non-coding transcripts of *SNRPN* in neural derivatives of PWS iPSCs. This de-repression involved activation of upstream *SNRPN* exons and did not change the methylation imprint at the PWS-IC. Together, these data suggest that we do not fully understand the somatic imprints in the chromosome 15q11-q13 region and how they are regulated in different tissues.

5 Physiological and Morphological Phenotypes

In order to understand the functional consequences of the genetic disruptions in AS, PWS, and Dup15q, it is necessary to examine the intrinsic excitability and network activity of iPSC-derived neurons. Such studies are important, as understanding the functional consequence of genetic disruptions on neuronal communication will ultimately better inform our understanding of neural pathophysiology. Characterizing the spectrum of functional deficits and unraveling primary phenotypes are critical for identifying pathways and molecules for therapeutic targets. Patient-specific iPSC-derived neurons have been used to characterize changes in electrophysiological properties, network activity, and neuronal morphology. Below we summarize results using AS and Dup15q patient-derived cell lines; however, similar studies have not yet been carried out on PWS-derived neurons.

5.1 Cellular Phenotypes in Angelman Syndrome-Derived Neurons

The functional maturation of neurons from AS patients has been characterized in a recent study using multiple control and patient-derived lines [67]. Electrophysiological analysis of control iPSC-derived neurons over 20 weeks in culture revealed maturation of physiological properties including hyperpolarization of the resting membrane potential (RMP), increases in action potential (AP) firing pattern, decreases in AP duration, increases in AP amplitude and transient potassium current density, and increases in the frequency of synaptic activity as well as the percent of synaptically active cells. Spontaneous AP activity as assayed by calcium imaging was also present throughout in vitro development in control neurons. Thus, these data provide a robust timeline for electrophysiological maturation of iPSC-derived neurons.

Compared to control neurons, AS-derived neurons failed to show the same degree of maturation across 20 weeks in culture. AS neurons showed immature resting membrane potentials and AP firing, as well as significantly fewer spontaneous AP-dependent calcium transients. Levels of spontaneous synaptic activity were also decreased in AS neurons, along with decreased dendritic branching and decreased density of dendritic spine-like protrusions. These cellular deficits resulted specifically from the loss of *UBE3A*. The same cellular phenotype that was seen in cells derived from AS patients with a large deletion was also seen in a line derived from an AS patient with a specific *UBE3A* mutation. To further confirm the critical role of *UBE3A* loss, knocking out *UBE3A* in an isogenic CRISPR-Cas9 gene-edited cell line completely replicated the AS electrophysiological phenotype. Moreover, this phenotype could be rescued by unsilencing *UBE3A* expression from the paternal allele with the topoisomerase inhibitor topotecan [67].

It is known from mouse models of AS that there is a critical period (early in development) in which rescue of *UBE3A* can also rescue behavioral phenotypes [40]. To determine whether acute loss of *UBE3A* causes similar phenotypes, *UBE3A* was knocked down using antisense oligonucleotides against *UBE3A* at early (6–9 weeks in culture) and late (18–20 weeks in culture), which reduced *UBE3A* message and protein ~50%. While both early and late *UBE3A* knockdown in control neurons could recapitulate the depolarization of RMP, only early knockdown altered AP firing and synaptic activity. This suggests the possibility that *UBE3A* loss results in a primary change to RMP that then drives changes in firing and synaptic activity, as changes in resting potential can disrupt neuronal excitability and function and thus trigger compensatory changes. To test this, control neurons were depolarized (~10 mV) with potassium chloride for the duration of their development followed by electrophysiological analysis. As with AS and *UBE3A* KD/KO neurons, potassium-treated control neurons displayed the entire spectrum of AS phenotypes, despite the normal expression of *UBE3A*.

These phenotypes are not all observed in mouse models of AS, though there has been a report of a hyperpolarized RMP later in development [68]. Such differences might be due to the differences in the developmental time periods being studied in mouse models vs iPSC-derived neuronal cultures. Changes in synaptic activity and plasticity commonly observed at mature ages in mouse models of AS may be a consequence of earlier changes in RMP, as suggested in our study. Relevant targets of UBE3A that may relate to changes in synaptic activity and plasticity have been hard to identify. It is known that ubiquitin ligases are expressed in different cell types and change their targets in a brain region- and developmental stage-specific manner. Thus, relevant targets of UBE3A may need to be identified during critical developmental periods.

Because of the relevance of neuronal and synaptic plasticity to the behavioral phenotypes in AS, Dup15q, and other neurodevelopmental disorders, one of the critical challenges to the use of human stem cell models is generating neurons with the capacity to undergo activity-dependent plasticity. Pharmacologically induced plasticity protocols, which have successfully been used in mouse brain slices and cultured neurons to elicit LTP [69], have also been used in studies of AS iPSC-derived neurons [67]. For example, plasticity induction via increased cAMP levels and enhanced activation of NMDA receptors caused a long-term elevation of spontaneous synaptic event frequency in control neurons which was absent in AS-derived neurons, paralleling LTP deficits in AS mouse models. The deficits in plasticity of synaptic transmission were paralleled by deficits in plasticity of AP firing as revealed by calcium imaging.

5.2 Cellular Phenotypes in Dup15q Syndrome-Derived Neurons

iPSC-derived neurons have also been used to characterize functional development and morphology in neurons derived from Dup15q patients. In contrast to AS-derived neurons, development of a mature RMP in Dup15q neurons was similar to controls over 20 weeks of in vitro development, shifting to more hyperpolarized potentials over time in culture. Development of AP firing was delayed in Dup15q neurons compared to controls, but was not as strongly impaired as observed in AS neurons. Dup15q neurons also showed significant increases in synaptic event frequency and amplitude compared to controls, which is maintained over 20 weeks of in vitro development, but no differences in dendritic branching. Similar increases in synaptic event amplitude have been observed in AS-derived cultures, though these cells show significant decreases in synaptic frequency [70]. Neurons derived from a patient with a maternal interstitial triplication showed a phenotype similar to isodicentric-derived Dup15q neurons, and neurons from a paternal interstitial duplication individual were similar to unaffected controls.

Synaptic plasticity deficits were also seen in the Dup15q neurons. Whereas control neurons display a long-term increase in the frequency and amplitude of synaptic events in response to pharmacologically induced plasticity, Dup15q neurons fail to show similar long-term plasticity. Another common form of plasticity is induced by long-term changes in network activity. Blocking AP firing (with the sodium channel blocker tetrodotoxin) or increasing AP firing (via GABA receptor blockade) results in changes to AMPA receptor expression. Such plasticity, termed homeostatic synaptic scaling, has been tied to a variety of neurodevelopmental syndromes in mouse models. In a study of Dup15q using human iPSC-derived neurons, control cells displayed the expected increases and decreases in AMPA event amplitude in response to manipulations in network activity as measured by both immunocytochemistry and electrophysiology. Interestingly, Dup15q neurons were not able to scale AMPA events either up or down, which could contribute to a hyperexcitable phenotype in these cells [70]. These differences were reflected in both synaptic activity and spontaneous AP firing. The goal of homeostatic scaling and plasticity is to establish a mechanism to avoid positive-feedback loops that continually increase AP firing in response to long-term potentiation. Therefore, an impaired ability of Dup15q neurons to decrease synaptic amplitudes in response to changes in network activity represents a potential mechanism for hyperexcitability in these neurons [70].

Given the seizure phenotype commonly associated with Dup15q, spontaneous AP firing in control, AS, and Dup15q neurons was also monitored. Control and AS neurons showed low levels of baseline spontaneous firing; however, Dup15q neurons fired at triple the rate of controls. Increased firing rate in Dup15q neurons was confirmed with population calcium imaging. Together these data suggest an additional mechanism of hyperexcitability specific to Dup15q neurons. Pharmacological manipulations suggest that KCNQ2 channels may be impaired in Dup15q neurons. KCNQ2 channels are potassium channels that are modulated by muscarinic receptor activation and act at subthreshold voltages as a brake on repetitive AP firing. Moreover, mutations in KCNQ2 result in benign familial neonatal convulsions, a genetic form of epilepsy. Interestingly, Dup15q neurons failed to respond to either pharmacological activation or blockade of KCNQ2 channels, suggesting impaired function and/or a reduction of these channels in Dup15q neurons. Interestingly, blockade of KCNQ2 in control neurons resulted in firing rates that were similar to the baseline firing of Dup15q neurons. In line with these data, KCNQ2 expression was significantly diminished in neurons from Dup15q patients as measured by immunostaining and flow cytometry [70].

These results establish three mechanisms of hyperexcitability in Dup15q neurons: increased synaptic event frequency and amplitude, impaired synaptic scaling, and increased AP firing due to KCNQ2 disruptions. Although primary vs. secondary targets have not been established, it is interesting to speculate that the impaired down-scaling in Dup15q neurons may contribute to the development of excessive firing and synchronous activity. Likewise, impaired up-scaling may be occluded by excessive baseline firing. Deficits in KCNQ2 channels may also contribute to synaptic scaling deficits.

5.3 *Cellular Phenotypes in Prader–Willi Syndrome-Derived Neurons*

Unlike Angelman and Dup15q syndromes, where many different brain regions likely have deficits that contribute to the phenotypic presentation of the disorder, PWS is thought to be caused almost entirely by deficits in the hypothalamus. Although deficits in PWS-derived cortical neurons cannot be entirely ruled out, disease relevant phenotypes are most likely confined to hypothalamic neurons. Wang et al. reported the differentiation of human pluripotent stem cells into hypothalamic-like neurons [71]. Recently, Rajamani et al., reported an improved protocol to generate hypothalamic neurons that respond to hormones like ghrelin and leptin [72]. With these improvements in hypothalamic neuron differentiation, assessment of function in this neuronal subtype should be possible for PWS-derived neurons.

6 Regulation of 15Q11-Q13 Gene Expression and Therapeutic Strategies

While many pharmaceutical therapies currently exist or are in the pipeline to treat the various symptoms of AS, PWS, and Dup15q (reviewed elsewhere), in vitro neuronal models are often not appropriate for testing their efficacy given the complex nature of the systems the drugs target. In this chapter, we instead focus on the use of iPSC models to advance therapeutic approaches aimed at targeting the genetic and epigenetic causes of these disorders.

6.1 Angelman Syndrome

In non-neuronal tissues, *UBE3A* is expressed from both parental alleles. However, in neurons, a long non-coding antisense transcript (*UBE3A-ATS*) originating from the *SNURF-SNRPN* promoter silences paternally inherited *UBE3A* [73, 74]. Therefore, loss of maternal *UBE3A* in AS results in complete absence of the protein in neurons. Gene replacement therapy has been proposed to introduce a functional copy of *UBE3A* using an adeno-associated virus (AAV) vector into AS patients and is currently under development independently by two different groups, PTC therapeutics and the Gene Therapy Center at the University of Pennsylvania. However, the intact, but silenced, paternal copy of *UBE3A* that already exists in AS neurons is an attractive therapeutic target since reactivation of this copy could potentially restore proper levels of *UBE3A* expression and function.

Studies in the AS mouse model have shed light on the mechanism whereby *UBE3A-ATS* silences paternal *UBE3A*. The Beaudet lab first showed that by engineering deletions of the *Snrpn* promoter, *Ube3a-ATS* levels were significantly reduced and paternal *Ube3a* expression was increased [75]. This data supported the hypothesis that *Ube3a-ATS* expression in *cis* is required for silencing paternal *Ube3a*. Further experiments determined that reduction specifically in *Ube3a-ATS*—and not in the several other transcripts that are processed from the long non-coding RNA initiating at the *Snrpn* promoter—is responsible for *Ube3a* unsilencing. By inserting a transcriptional stop cassette between *Snord115* and *Ube3a* on the paternal chromosome, the Beaudet lab prematurely truncated the transcript and reduced expression of *Ube3a-ATS* [76]. This resulted in the complete unsilencing of paternal *Ube3a* and rescue of some AS behavioral phenotypes in the mouse model. These studies set the groundwork for developing therapeutic approaches centered on the hypothesis that reducing *UBE3A-ATS* activity could restore paternal *UBE3A* expression.

Imprinting of paternal *UBE3A* by the *UBE3A-ATS* transcript is hypothesized to be the result of transcriptional interference. The transcriptional interference model for imprinting of paternal *UBE3A* purports that active transcription of *UBE3A-ATS* on the plus strand prevents full transcription of *UBE3A* from the minus strand. This competition between transcripts should be amenable to manipulation by adjusting the level of transcription from either strand. In fact, preliminary work from the Chamberlain lab shows that drastically increasing the level of *UBE3A-ATS* transcription in AS iPSCs can completely silence paternal *UBE3A* in non-neuronal cells [77]. Using CRISPR-cas9 technology to delete the region between *SNRPN* intron 1 and *SNORD115-47*, directly placing *UBE3A-ATS* under control of the *SNURF/SNRPN* promoter, they were able to significantly increase *UBE3A-ATS* expression and silence paternal *UBE3A* in iPSCs. By comparing these iPSCs to other iPSC lines in which they had engineered smaller deletions to remove boundary elements regulating *UBE3A-ATS* expression, they determined that increasing *UBE3A-ATS* levels past a certain threshold resulted in complete silencing of the *UBE3A* sense transcript. This evidence strongly implicates a competition between the levels of *UBE3A-ATS* and *UBE3A* sense transcription in establishing the *UBE3A* imprint.

Two main therapeutic paradigms for unsilencing the imprinted paternal *UBE3A* allele in AS have been under development over the past several years. High-throughput screening experiments have identified small molecules, such as topoisomerase inhibitors, that are capable of restoring *Ube3a* expression. The other approach involves targeting cleavage of the *Ube3a-ATS* transcript using RNA-based therapeutics.

The Philpot lab used primary cortical neurons derived from mice with a *Ube3a*-GFP reporter knocked in to the paternal allele in high-throughput screens to identify small molecules capable of inducing expression of paternal *Ube3a* [78]. These studies identified irinotecan and topotecan, two FDA approved camptothecin-derived topoisomerase inhibitors, as well as 14 other compounds that inhibit either topoisomerase I or II, as drugs that could induce paternal *Ube3a* expression. Topotecan treatment restored paternal *Ube3a* expression in various neuronal populations in the mouse brain and spinal cord as well as in human AS iPSC-derived neurons. Because of the limited bioavailability of topotecan to the central nervous system, the Philpot group subsequently identified 13 indenoisoquinoline-derived topoisomerase inhibitors that all restored paternal *Ube3a* to levels similar to topotecan [79]. They suggest that one compound, indotecan, may be more efficient at reactivating *Ube3a*; however, whether indotecan has improved bioavailability over topotecan remains to be determined. Mechanistic studies determined that topotecan likely works to inhibit *Ube3a-ATS* by a mechanism independent of the enzymatic function of topoisomerase I [80]. Instead, topotecan works by forming stable complexes between topoisomerase I and the DNA to which it is bound (called the TOPI cleavage complex) thereby inhibiting transcription elongation.

Given that these candidate topoisomerase inhibitors are already FDA approved and widely used as cancer therapeutics, they represent an encouraging therapeutic avenue for AS. However, there are also some weaknesses to their application for AS. For example, topoisomerase inhibition was shown to not only reduce expression of *Ube3a-ATS* but of many other genes, including several autism candidate genes, which are important for neuronal development [81]. Additionally, inhibition of topoisomerase I was shown to impair synapse formation and function in cultured primary cortical neurons [82]. Topotecan and indotecan also have high cellular toxicity. Ongoing efforts to identify and screen additional topoisomerase inhibitors may identify candidates with safer toxicity profiles.

RNA-based therapeutics are under development or are already in clinical trials to treat a variety of genetic disorders. These approaches aim to reduce expression of target genes by reducing target mRNA levels or inhibiting protein translation. RNA-therapeutics can also be used to correct aberrant mRNA splicing caused by genetic mutations. Antisense oligonucleotides (ASOs) are one such RNA-based therapeutic currently under development for AS by several pharmaceutical companies including Ionis Pharmaceuticals, Biogen, and Roche. ASOs are single-stranded oligonucleotide sequences that can bind to and target specific RNAs for degradation [83]. ASOs reduce RNA levels by sterically inhibiting transcription or translation, by triggering RNase H-mediated degradation of the RNA in the DNA/RNA hybrid formed by the ASO and its target, or by inhibiting RNA splicing [83]. Ease of delivery to the CNS by intrathecal injection into the spinal fluid and high bioavailability make ASOs a promising approach [84]. In fact, several ASOs are currently FDA approved to treat a host of diseases including neurodegenerative disorders [85].

Building upon previous findings that reduction of *Ube3a-ATS* could unsilence paternal *Ube3a*, the Beaudet lab collaborated with Ionis Pharmaceuticals to develop ASOs to knock down *Ube3a-ATS* [86]. Using cultured cortical neurons from mice with YFP knocked in to the paternal allele of *Ube3a*, ASOs targeting the region downstream of *Snord115* were screened for their ability to activate paternal *Ube3a* expression. Select ASOs reduced *Ube3a-ATS* up to 90% and increased *Ube3a* protein levels in primary neurons from maternal *Ube3a* knock-out mice up to 90% of wild type levels. When injected into the lateral ventricle of adult paternal *Ube3a*-YFP mice, ASOs produced a two- to fivefold increase in *Ube3a*-YFP RNA. *Ube3a* protein levels were also significantly increased throughout the brain following ASO injection; however, these levels did not reach that of normal maternal allelic expression. Encouragingly, a single ASO injection was capable of sustaining increased paternal *Ube3a* expression for up to 4 months. When ASOs were administered to maternal *Ube3a*-deficient mice, moderate restoration of paternal *Ube3a* protein was able to rescue two AS phenotypes—the contextual fear conditioning deficit and obesity.

One important observation from the mouse ASO studies is that only *Ube3a-ATS*-targeting ASOs which trigger an RNase H response were effective at restoring paternal *Ube3a* expression [86]. Recently, two separate studies demonstrated that the $5'$- > $3'$ exonuclease XRN2 is required for the depletion of RNase H cleavage products generated after ASO binding to pre-mRNAs and nuclear retained RNAs [87, 88]. XRN2 is implicated in the torpedo model of transcription termination where it degrades remaining RNA cleavage products and disengages RNA Pol II to terminate transcription [89]. Together these findings suggest that ASOs targeting *UBE3A*-ATS may work by disengaging RNA Pol II. Other RNA-based therapeutics that terminate *UBE3A-ATS* transcription by this mechanism are also likely to be effective approaches for restoring paternal *UBE3A* expression.

It is important to note that with any of the approaches to restore *UBE3A* expression in AS neurons, care must be taken to regulate the amount of *UBE3A* expressed since overexpression of *UBE3A* is associated with Dup15q syndrome. While overexpression is likely to be a concern with proposed gene replacement approaches, it is less likely to be a problem with approaches that aim to unsilence the paternal allele. In addition, approaches that unsilence the paternal allele by reducing levels of the *UBE3A-ATS* must not also reduce levels of the upstream transcripts including *SNORD116* sufficient to cause PWS.

6.2 15q Duplication Syndrome

While several studies support the hypothesis that maternal duplication of *UBE3A* underlies the majority of the Dup15q symptoms, including the autism phenotype [45, 50, 90, 91], the duplicated region also contains several other biallelically expressed genes, which may contribute to the pathophysiology of Dup15q (Fig. 1). Several non-imprinted genes with known neuronal functions are duplicated in some int. dup15q individuals and all cases of Idic15. These include a cluster of GABA receptor genes; *CYFIP1*, a gene whose protein product binds to and antagonizes FMRP, the gene disrupted in Fragile X syndrome [92]; *TUBGCP5*, a gene encoding a member of the gamma tubulin complex [93]; *HERC2*, another E3A ubiquitin ligase and binding partner of UBE3A [94, 95]; and *NIPA1* and *NIPA2*, genes encoding putative magnesium transporters involved in seizures, schizophrenia, and hereditary spastic paraplegia [96, 97].

In contrast to AS, which is largely a single gene disorder, less is known about the individual contributions of the many duplicated genes to the various Dup15q symptoms. Furthermore, combinations of duplicated genes may interact to produce the overall phenotype. At the gene regulation level, sets of genes such as *UBE3A* and the *GABA* receptor cluster may be co-regulated, making targeted modulation of any individual gene difficult. Due to this complexity, far fewer therapeutic approaches have been investigated to treat Dup15q at the genetic or epigenetic level. This is expected to improve as we gain understanding of the various gene contributions through modeling of Dup15q with human iPSC-derived neurons. Theoretically, approaches to reduce the expression levels of individual duplicated genes or combinations of genes to those of normal neurons—for example, using ASOs—could be a therapeutic strategy for Dup15q.

6.3 Prader–Willi Syndrome

PWS is caused by loss of a group of paternally imprinted genes, minimally the *SNORD116* host gene, which generates a set of C/D box snoRNAs, and *IPW* [11, 12]. Similar to the case for AS, reactivation of the silent second allele—in PWS this is the maternal allele—represents a promising epigenetic therapeutic approach.

While the germline imprint of the PWS-IC has been well studied (discussed earlier), recent work from the Lalande and Chamberlain labs suggests that a repressive complex comprised of the zinc finger protein ZNF274 and the histone methyltransferase SETDB1 represses expression of *SNORD116* on the maternal allele [33, 98]. Using PWS iPSCs and iPSC-derived neurons, they showed that ZNF274 and SETDB1 preferentially bind the maternal allele at a set of six binding sites within *SNORD116*. This binding was associated with maternal allele-specific deposition of the repressive H3K9me3 chromatin mark. Knockdown of SETDB1 and ZNF274 by shRNA or deletion of *ZNF274* using CRISPR/Cas9 technology in PWS iPSCs resulted in reduced H3K9me3 and concomitant unsilencing of maternal *SNORD116*.

Importantly, ZNF274-knockout-induced transcription of *SNORD116* was shown to be initiated from the upstream exon promoters of *SNRPN* and not at the *SNRPN* exon 1 promoter that is regulated by the PWS-IC. Consistent with findings that transcription of the SNRPN long non-coding RNA is driven predominantly from these upstream promoters in the brain, activation of *SNORD116* was even more robust following neuronal differentiation of PWS iPSCs lacking ZNF274 [33]. Ongoing work aims to further test the therapeutic capacity of reducing ZNF274 levels or disrupting binding of the ZNF274/SETDB1 repressive complex to the maternal allele in order to treat PWS.

Studies in mouse ESCs determined that an additional histone methyltransferase G9a is required to maintain maternal-specific H3K9 and CpG island methylation at the PWS-IC. Xin et al. [99] showed that homozygous deletion of *Ehmt2/G9a* in mouse ESCs resulted in 70% reduction of H3K9 methylation and complete loss of CpG island methylation at the PWS-IC. *G9a*-null ESCs also showed biallelic expression of *Snrpn*. Restoration of G9a function in these ESCs, through expression of a transgene, was able to restore H3K9 methylation but did not result in re-methylation of CpG islands at the PWS-IC or re-silencing of maternal *Snprn*. This data suggested that G9a is necessary for maintaining the already established imprint of *Snrpn* but is not sufficient to set up the initial imprinting.

A more recent study used mouse embryonic fibroblast cultures, which carried a maternally inherited *Snrpn*-GFP reporter, in a high-throughput screen to identify small molecules capable of inducing transcription of *Snrpn* from the maternal allele [100]. A screen of over 9000 molecules identified four compounds, which are catalytic inhibitors of G9A, that were capable of activating maternal *Snprn*-GFP expression. These compounds were confirmed to induce maternal expression of *SNRPN* mRNA as well as the snoRNA host transcripts from the *SNPRN* long non-coding RNA (*SNHG14*), including *SNORD116* and *SNORD115*, in human PWS fibroblast cultures. When tested in a mouse model of PWS, which have a paternal deletion from *Snrpn* to *Ube3a*, G9A inhibition induced long lasting expression (12 weeks following last drug injection) of maternal *Snrpn* and *Snord116* to almost half of normal levels in brain and liver. Drug treatment was also able to partially rescue a growth deficit and premature death phenotype of the PWS mouse. In contrast to the G9A deletion studies, activation of the maternal allele by G9A inhibitors was associated only with reduced H3K9 methylation at the PWS-IC and not a reduction in CpG island methylation. The authors suggest a model whereby G9A inhibition reduces H3K9 methylation at the PWS-IC opening up the chromatin structure to allow transcription from the normally repressed maternal allele [100].

One important consideration with epigenetic therapies for PWS is that since the distal portion of *SNHG14* (the *UBE3A-ATS* transcript) represses paternal *UBE3A* in *cis*, any approach to induce expression of maternal *SNORD116* must not also increase *UBE3A-ATS* levels sufficient to silence maternal *UBE3A*. Encouragingly, data from both the ZNF274 and G9A targeting studies do not suggest any detrimental effect on maternal *UBE3A* expression.

7 Synopsis

Using iPSCs to model the genomic imprinting disorders on chromosome 15q11-q13 presents unique opportunities and challenges. Three distinct neurodevelopmental disorders result from copy number variation at this locus. This necessitates diverse understanding of neuronal function in different regions of the brain. Angelman and Dup15q syndrome likely affect multiple regions of the brain. It remains to be determined whether the presentations of these syndromes result predominantly from a single brain region, or whether they are the result of an amalgamation of deficits affecting multiple brain regions. Prader–Willi syndrome, on the other hand is primarily a disorder of the hypothalamus. These disorders also necessitate broad understanding of gene regulation and gene function. Angelman syndrome is caused by loss of function from a single protein-coding gene. Prader–Willi syndrome is principally caused by the loss of a cluster of non-coding RNAs. The precise gene(s) underlying Dup15q syndrome has not yet been determined. The genetic aberrations associated with Dup15q are complex and difficult to replicate in mice. Thus, iPSC models of Dup15 syndrome will help inform development of appropriate mouse models. Faithful models of all three disorders depend heavily on maintenance of genomic imprints (both germline and somatic) during the reprogramming, maintenance, and differentiation processes. Finally, each of these disorders present a unique opportunity for therapeutic development. Individuals with Angelman and Prader–Willi syndromes may soon benefit from activation of an epigenetically silenced version of the gene involved in their respective disorders. However, careful consideration for all three disorders is essential to ensure that therapeutic interventions do not unintentionally disrupt SNORD116 or UBE3A, or lead to too much UBE3A and cause deficits associated with another neurodevelopmental disorder.

Acknowledgements The work was supported by the following funding sources: NIH HD068730, NIH HD091823-01, Foundation for Prader–Willi Research, and Connecticut DPH Stem Cell Research Program (12SCBUCHC) to SJC, the Joseph Wagstaff Postdoctoral Fellowship to NDG, NIH MH094896 to ESL, and the Angelman Syndrome Foundation and Connecticut DPH Stem Cell Research Program (14-SCDIS) to both SJC and ESL.

References

1. Cassidy, S.B. and D.J. Driscoll, (2009). Prader-Willi syndrome. *European Journal of Human Genetics, 17*(1), 3–13.
2. Battaglia, A. (2005). The inv dup(15) or idic(15) syndrome: A clinically recognisable neurogenetic disorder. *Brain Dev, 27*(5), 365–369.
3. Christian, S.L., N.K. Bhatt, S.A. Martin, J.S. Sutcliffe, T. Kubota, B. Huang, et al., (1998). Integrated YAC contig map of the Prader-Willi/Angelman region on chromosome 15q11-q13 with average STS spacing of 35 kb. *Genome Research, 8*(2), 146–157.
4. Amos-Landgraf, J.M., Y. Ji, W. Gottlieb, T. Depinet, A.E. Wandstrat, S.B. Cassidy, et al., (1999). Chromosome breakage in the Prader-Willi and Angelman syndromes involves recombination between large, transcribed repeats at proximal and distal breakpoints. *American Journal of Human Genetics, 65*(2), 370–386.
5. Christian, S.L., J.A. Fantes, S.K. Mewborn, B. Huang, and D.H. Ledbetter (1999). Large genomic duplicons map to sites of instability in the Prader-Willi/Angelman syndrome chromosome region (15q11-q13). *Human Molecular Genetics, 8*(6), 1025–1037.
6. Wang, N.J., D. Liu, A.S. Parokonny, and N.C. Schanen (2004). High-resolution molecular characterization of 15q11-q13 rearrangements by array comparative genomic hybridization (array CGH) with detection of gene dosage. *American Journal of Human Genetics, 75*(2), 267–281.
7. Battaglia, A. (2008). The inv dup (15) or idic (15) syndrome (Tetrasomy 15q). *Orphanet Journal of Rare Diseases, 3*, 30.
8. Hogart, A., K.A. Patzel, and J.M. LaSalle, (2008). Gender influences monoallelic expression of ATP10A in human brain. *Human Genetics, 124*(3), 235–242.
9. Chai, J.H., D.P. Locke, J.M. Greally, J.H. Knoll, T. Ohta, J. Dunai, et al., (2003). Identification of four highly conserved genes between breakpoint hotspots BP1 and BP2 of the Prader-Willi/Angelman syndromes deletion region that have undergone evolutionary transposition mediated by flanking duplicons. *American Journal of Human Genetics, 73*(4), 898–925.
10. Ji, Y., E.E. Eichler, S. Schwartz, and R.D. Nicholls, (2000). Structure of chromosomal duplicons and their role in mediating human genomic disorders. *Genome Research, 10*(5), 597–610.
11. Sahoo, T., D. del Gaudio, J.R. German, M. Shinawi, S.U. Peters, R.E. Person, et al., (2008). Prader-Willi phenotype caused by paternal deficiency for the HBII-85 C/D box small nucleolar RNA cluster. *Nature Genetics, 40*(6), 719–721.
12. dde Smith, A.J., C. Purmann, R.G. Walters, R.J. Ellis, S.E. Holder, M.M. Van Haelst, et al., (2009). A deletion of the HBII-85 class of small nucleolar RNAs (snoRNAs) is associated with hyperphagia, obesity and hypogonadism. *Human Molecular Genetics, 18*(17), 3257–3265.
13. Yang, T., T.E. Adamson, J.L. Resnick, S. Leff, R. Wevrick, U. Francke, et al., (1998). A mouse model for Prader-Willi syndrome imprinting-Centre mutations. *Nature Genetics, 19*(1), 25–31.
14. Chamberlain, S.J., K.A. Johnstone, A.J. DuBose, T.A. Simon, M.S. Bartolomei, J.L. Resnick, et al., (2004). Evidence for genetic modifiers of postnatal lethality in PWS-IC deletion mice. *Human Molecular Genetics, 13*(23), 2971–2977.

15. Gabriel, J.M., M. Merchant, T. Ohta, Y. Ji, R.G. Caldwell, M.J. Ramsey, et al., (1999). A transgene insertion creating a heritable chromosome deletion mouse model of Prader-Willi and Angelman syndromes. *Proceedings of the National Academy of Sciences of the United States of America, 96*(16), 9258–9263.
16. Gerard, M., L. Hernandez, R. Wevrick, and C.L. Stewart, (1999). Disruption of the mouse necdin gene results in early post-natal lethality. *Nature Genetics, 23*(2), 199–202.
17. Muscatelli, F., D.N. Abrous, A. Massacrier, I. Boccaccio, M. Le Moal, P. Cau, et al., (2000). Disruption of the mouse Necdin gene results in hypothalamic and behavioral alterations reminiscent of the human Prader-Willi syndrome. *Human Molecular Genetics, 9*(20), 3101–3110.
18. Mercer, R.E. and R. Wevrick, (2009). Loss of magel2, a candidate gene for features of Prader-Willi syndrome, impairs reproductive function in mice. *PLoS One, 4*(1), e4291.
19. Bischof, J.M., C.L. Stewart, and R. Wevrick (2007). Inactivation of the mouse Magel2 gene results in growth abnormalities similar to Prader-Willi syndrome. *Human Molecular Genetics, 16*(22), 2713–2719.
20. Ding, F., H.H. Li, S. Zhang, N.M. Solomon, S.A. Camper, P. Cohen, et al., (2008). SnoRNA Snord116 (Pwcr1/MBII-85) deletion causes growth deficiency and hyperphagia in mice. *PLoS One, 3*(3), e1709.
21. Polex-Wolf, J., B.Y. Lam, R. Larder, J. Tadross, D. Rimmington, F. Bosch, et al., (2018). Hypothalamic loss of Snord116 recapitulates the hyperphagia of Prader-Willi syndrome. *The Journal of Clinical Investigation, 128*(3), 960–969.
22. Schaaf, C.P., M.L. Gonzalez-Garay, F. Xia, L. Potocki, K.W. Gripp, B. Zhang, et al., (2013). Truncating mutations of MAGEL2 cause Prader-Willi phenotypes and autism. *Nature Genetics, 45*(11), 1405–1408.
23. Fountain, M.D., E. Aten, M.T. Cho, J. Juusola, M.A. Walkiewicz, J.W. Ray, et al., (2017). The phenotypic spectrum of Schaaf-Yang syndrome: 18 new affected individuals from 14 families. *Genetics in Medicine, 19*(1), 45–52.
24. Fountain, M.D., H. Tao, C.A. Chen, J. Yin, and C.P. Schaaf, (2017). Magel2 knockout mice manifest altered social phenotypes and a deficit in preference for social novelty. *Genes, Brain, and Behavior, 16*(6), 592–600.
25. Kozlov, S.V., J.W. Bogenpohl, M.P. Howell, R. Wevrick, S. Panda, J.B. Hogenesch, et al., (2007). The imprinted gene Magel2 regulates normal circadian output. *Nature Genetics, 39*(10), 1266–1272.
26. Chamberlain, S.J., P.F. Chen, K.Y. Ng, F. Bourgois-Rocha, F. Lemtiri-Chlieh, E.S. Levine, et al., (2010). Induced pluripotent stem cell models of the genomic imprinting disorders Angelman and Prader-Willi syndromes. *Proceedings of the National Academy of Sciences of the United States of America, 107*(41), 17668–17673.
27. Yang, J., J. Cai, Y. Zhang, X. Wang, W. Li, J. Xu, et al., (2010). Induced pluripotent stem cells can be used to model the genomic imprinting disorder Prader-Willi syndrome. *The Journal of Biological Chemistry, 285*(51), 40303–40311.
28. Stelzer, Y., I. Sagi, O. Yanuka, R. Eiges, and N. Benvenisty, (2014). The noncoding RNA IPW regulates the imprinted DLK1-DIO3 locus in an induced pluripotent stem cell model of Prader-Willi syndrome. *Nature Genetics, 46*(6), 551–557.
29. Burnett, L.C., C.A. LeDuc, C.R. Sulsona, D. Paull, S. Eddiry, B. Levy, et al., (2016). Induced pluripotent stem cells (iPSC) created from skin fibroblasts of patients with Prader-Willi syndrome (PWS) retain the molecular signature of PWS. *Stem Cell Research, 17*(3), 526–530.
30. Okuno, H., K. Nakabayashi, K. Abe, T. Ando, T. Sanosaka, J. Kohyama, et al., (2017). *Changeability of the fully methylated status of the 15q11.2 region in induced pluripotent stem cells derived from a patient with Prader-Willi syndrome. Congenit Anom (Kyoto),* ***57***(4), 96–103.
31. Eldar-Geva, T., V. Gross-Tsur, H.J. Hirsch, G. Altarescu, R. Segal, S. Zeligson, et al., (2018). Incomplete methylation of a germ cell tumor (seminoma) in a Prader-Willi male. *Molecular Genetics & Genomic Medicine, 6*, 811.

32. Martins-Taylor, K., J.S. Hsiao, P.F. Chen, H. Glatt-Deeley, A.J. De Smith, A.I. Blakemore, et al., (2014). Imprinted expression of UBE3A in non-neuronal cells from a Prader-Willi syndrome patient with an atypical deletion. *Human Molecular Genetics, 23*(9), 2364–2373.
33. Langouet, M., H.R. Glatt-Deeley, M.S. Chung, C.M. Dupont-Thibert, E. Mathieux, E.C. Banda, et al., (2018). Zinc finger protein 274 regulates imprinted expression of transcripts in Prader-Willi syndrome neurons. *Human Molecular Genetics, 27*(3), 505–515.
34. Yamasaki, K., K. Joh, T. Ohta, H. Masuzaki, T. Ishimaru, T. Mukai, et al., (2003). Neurons but not glial cells show reciprocal imprinting of sense and antisense transcripts of Ube3a. *Human Molecular Genetics, 12*(8), 837–847.
35. Lossie, A.C., M.M. Whitney, D. Amidon, H.J. Dong, P. Chen, D. Theriaque, et al., (2001). Distinct phenotypes distinguish the molecular classes of Angelman syndrome. *Journal of Medical Genetics, 38*(12), 834–845.
36. Tan, W.H., L.M. Bird, R.L. Thibert, and C.A. Williams, (2014). If not Angelman, what is it? A review of Angelman-like syndromes. *American Journal of Medical Genetics. Part A, 164A*(4), 975–992.
37. Jiang, Y.H., D. Armstrong, U. Albrecht, C.M. Atkins, J.L. Noebels, G. Eichele, et al., (1998). Mutation of the Angelman ubiquitin ligase in mice causes increased cytoplasmic p53 and deficits of contextual learning and long-term potentiation. *Neuron, 21*(4), 799–811.
38. Miura, K., T. Kishino, E. Li, H. Webber, P. Dikkes, G.L. Holmes, et al., (2002). Neurobehavioral and electroencephalographic abnormalities in Ube3a maternal-deficient mice. *Neurobiology of Disease, 9*(2), 149–159.
39. Huang, H.S., A.J. Burns, R.J. Nonneman, L.K. Baker, N.V. Riddick, V.D. Nikolova, et al., (2013). Behavioral deficits in an Angelman syndrome model: Effects of genetic background and age. *Behavioural Brain Research, 243*, 79–90.
40. Silva-Santos, S., G.M. van Woerden, C.F. Bruinsma, E. Mientjes, M.A. Jolfaei, B. Distel, et al., (2015). Ube3a reinstatement identifies distinct developmental windows in a murine Angelman syndrome model. *The Journal of Clinical Investigation, 125*(5), 2069–2076.
41. Stanurova, J., A. Neureiter, M. Hiber, H. de Oliveira Kessler, K. Stolp, R. Goetzke, et al., (2016). Angelman syndrome-derived neurons display late onset of paternal UBE3A silencing. *Scientific Reports, 6*, 30792.
42. Takahashi, Y., J. Wu, K. Suzuki, P. Martinez-Redondo, M. Li, H.K. Liao, et al., (2017). Integration of CpG-free DNA induces de novo methylation of CpG islands in pluripotent stem cells. *Science, 356*(6337), 503–508.
43. Polvora-Brandao, D., M. Joaquim, I. Godinho, D. Aprile, A.R. Alvaro, I. Onofre, et al., (2018). Loss of hierarchical imprinting regulation at the Prader-Willi/Angelman syndrome locus in human iPSCs. *Human Molecular Genetics, 27*(23), 3999–4011.
44. Fink, J.J., Robinson, T.M., Germain, N.D., Sirois, C.L., Bolduc, K.A., Ward, A.J., Rigo, F., Chamberlain, S.J., and Levine, E.S., (2017). Disrupted neuronal maturation in Angelman syndrome-derived induced pluripotent stem cells. *Nature Communications, 8*, 15038.
45. Schroer, R.J., M.C. Phelan, R.C. Michaelis, E.C. Crawford, S.A. Skinner, M. Cuccaro, et al., (1998). Autism and maternally derived aberrations of chromosome 15q. *American Journal of Medical Genetics, 76*(4), 327–336.
46. Cook, E.H., Jr., R.Y. Courchesne, N.J. Cox, C. Lord, D. Gonen, S.J. Guter, et al., (1998). Linkage-disequilibrium mapping of autistic disorder, with 15q11-13 markers. *American Journal of Human Genetics, 62*(5), 1077–1083.
47. Cook, E.H., Jr., V. Lindgren, B.L. Leventhal, R. Courchesne, A. Lincoln, C. Shulman, et al., (1997). Autism or atypical autism in maternally but not paternally derived proximal 15q duplication. *American Journal of Human Genetics, 60*(4), 928–934.
48. Reiter, L., Cleary, J., Brewer, V., Jabbour, J.T., Schanen, N.C., and Urraca, N., (2009). Maternal, but not paternal, interstitial duplications of chromosome 15q11.2-q13 are associated with ASD in 7 individuals. In *Presented at the 59th annual meeting of the American society for human genetics, October 23, 2009, Honolulu, Hawaii.*

49. Mohandas, T.K., J.P. Park, R.A. Spellman, J.J. Filiano, A.C. Mamourian, A.B. Hawk, et al., (1999). Paternally derived de novo interstitial duplication of proximal 15q in a patient with developmental delay. *American Journal of Medical Genetics, 82*(4), 294–300.
50. Marshall, C.R., A. Noor, J.B. Vincent, A.C. Lionel, L. Feuk, J. Skaug, et al., (2008). Structural variation of chromosomes in autism spectrum disorder. *American Journal of Human Genetics, 82*(2), 477–488.
51. Szatmari, P., A.D. Paterson, L. Zwaigenbaum, W. Roberts, J. Brian, X.Q. Liu, et al., (2007). Mapping autism risk loci using genetic linkage and chromosomal rearrangements. *Nature Genetics, 39*(3), 319–328.
52. Nakatani, J., K. Tamada, F. Hatanaka, S. Ise, H. Ohta, K. Inoue, et al., (2009). Abnormal behavior in a chromosome-engineered mouse model for human 15q11-13 duplication seen in autism. *Cell, 137*(7), 1235–1246.
53. Germain ND, C.P., Plocik AM, Glatt-Deeley H, Brown J, Fink JJ, Bolduc KA, Robinson TM, Levine ES, Reiter LT, Graveley BR, Lalande M, Chamberlain SJ., (2014). *Gene expression analysis of human induced pluripotent stem cell-derived neurons carrying copy number variants of chromosome 15q11-q13.1. Molecular Autism, 5*, 44.
54. Brannan, C.I. and M.S. Bartolomei, (1999). Mechanisms of genomic imprinting. *Current Opinion in Genetics & Development, 9*(2), 164–170.
55. Johnstone, K.A., A.J. DuBose, C.R. Futtner, M.D. Elmore, C.I. Brannan, and J.L. Resnick, (2006). A human imprinting Centre demonstrates conserved acquisition but diverged maintenance of imprinting in a mouse model for Angelman syndrome imprinting defects. *Human Molecular Genetics, 15*(3), 393–404.
56. Schumacher, A. and W. Doerfler, (2004). Influence of in vitro manipulation on the stability of methylation patterns in the Snurf/Snrpn-imprinting region in mouse embryonic stem cells. *Nucleic Acids Research, 32*(4), 1566–1576.
57. Kim, K.P., A. Thurston, C. Mummery, D. Ward-van Oostwaard, H. Priddle, C. Allegrucci, et al., (2007). Gene-specific vulnerability to imprinting variability in human embryonic stem cell lines. *Genome Research, 17*(12), 1731–1742.
58. Rugg-Gunn, P.J., A.C. Ferguson-Smith, and R.A. Pedersen, (2007). *Status of genomic imprinting in human embryonic stem cells as revealed by a large cohort of independently derived and maintained lines. Human Molecular Genetics,* ***16***(2), R243–R251.
59. Stadtfeld, M., E. Apostolou, H. Akutsu, A. Fukuda, P. Follett, S. Natesan, et al., (2010). Aberrant silencing of imprinted genes on chromosome 12qF1 in mouse induced pluripotent stem cells. *Nature, 465*(7295), 175–181.
60. Inoue, K., T. Kohda, J. Lee, N. Ogonuki, K. Mochida, Y. Noguchi, et al., (2002). Faithful expression of imprinted genes in cloned mice. *Science, 295*(5553), 297.
61. Ohta, T., T.A. Gray, P.K. Rogan, K. Buiting, J.M. Gabriel, S. Saitoh, et al., (1999). Imprinting-mutation mechanisms in Prader-Willi syndrome. *American Journal of Human Genetics, 64*(2), 397–413.
62. Saitoh, S., K. Buiting, P.K. Rogan, J.L. Buxton, D.J. Driscoll, J. Arnemann, et al., (1996). Minimal definition of the imprinting center and fixation of chromosome 15q11-q13 epigenotype by imprinting mutations. *Proceedings of the National Academy of Sciences of the United States of America, 93*(15), 7811–7815.
63. Buiting, K., C. Lich, S. Cottrell, A. Barnicoat, and B. Horsthemke, (1999). A 5-kb imprinting center deletion in a family with Angelman syndrome reduces the shortest region of deletion overlap to 880 bp. *Human Genetics, 105*(6), 665–666.
64. Peery, E.G., M.D. Elmore, J.L. Resnick, C.I. Brannan, and K.A. Johnstone, (2007). A targeted deletion upstream of Snrpn does not result in an imprinting defect. *Mammalian Genome, 18*(4), 255–262.
65. Smith, E.Y., C.R. Futtner, S.J. Chamberlain, K.A. Johnstone, and J.L. Resnick, (2011). Transcription is required to establish maternal imprinting at the Prader-Willi syndrome and Angelman syndrome locus. *PLoS Genetics, 7*(12), e1002422.

66. El-Maarri, O., K. Buiting, E.G. Peery, P.M. Kroisel, B. Balaban, K. Wagner, et al., (2001). Maternal methylation imprints on human chromosome 15 are established during or after fertilization. *Nature Genetics, 27*(3), 341–344.
67. Fink, J.J., T.M. Robinson, N.D. Germain, C.L. Sirois, K.A. Bolduc, A.J. Ward, et al., (2017). Disrupted neuronal maturation in Angelman syndrome-derived induced pluripotent stem cells. *Nature Communications, 8*, 15038.
68. Kaphzan, H., S.A. Buffington, J.I. Jung, M.N. Rasband, and E. Klann, (2011). Alterations in intrinsic membrane properties and the axon initial segment in a mouse model of Angelman syndrome. *The Journal of Neuroscience, 31*(48), 17637–17648.
69. Otmakhov, N., L. Khibnik, N. Otmakhova, S. Carpenter, S. Riahi, B. Asrican, et al., (2004). Forskolin-induced LTP in the CA1 hippocampal region is NMDA receptor dependent. *Journal of Neurophysiology, 91*(5), 1955–1962.
70. Fink, J.J., J.D. Schreiner, J.E. Bloom, D.S. Baker, T.M. Robinson, R. Lieberman, et al., (2018). *Hyperexcitable phenotypes in iPSC-derived neurons from patients with 15q11-q13 duplication syndrome, a genetic form of autism.* BioRxiv. **[Preprint]**. Retrieved from https://doi.org/10.1101/286336.
71. Wang, L., K. Meece, D.J. Williams, K.A. Lo, M. Zimmer, G. Heinrich, et al., (2015). Differentiation of hypothalamic-like neurons from human pluripotent stem cells. *The Journal of Clinical Investigation, 125*(2), 796–808.
72. Rajamani, U., A.R. Gross, B.E. Hjelm, A. Sequeira, M.P. Vawter, J. Tang, et al., (2018). Super-obese patient-derived iPSC hypothalamic neurons exhibit obesogenic signatures and hormone responses. *Cell Stem Cell, 22*(5), 698–712. e9.
73. Rougeulle, C., H. Glatt, and M. Lalande, (1997). The Angelman syndrome candidate gene, UBE3A/E6-AP, is imprinted in brain. *Nature Genetics, 17*(1), 14–15.
74. Chamberlain, S.J. and C.I. Brannan, (2001). The Prader-Willi syndrome imprinting center activates the paternally expressed murine Ube3a antisense transcript but represses paternal Ube3a. *Genomics, 73*(3), 316–322.
75. Meng, L., R.E. Person, and A.L. Beaudet, (2012). Ube3a-ATS is an atypical RNA polymerase II transcript that represses the paternal expression of Ube3a. *Human Molecular Genetics, 21*(13), 3001–3012.
76. Meng, L., R.E. Person, W. Huang, P.J. Zhu, M. Costa-Mattioli, and A.L. Beaudet, (2013). Truncation of Ube3a-ATS Unsilences paternal Ube3a and ameliorates behavioral defects in the Angelman syndrome mouse model. *PLoS Genetics, 9*(12), e1004039.
77. Hsiao, J.S., N.D. Germain, A. Wilderman, C. Stoddard, L.A. Wojenski, G.J. Villafano, et al., (2019). A bipartite boundary element restricts UBE3A imprinting to mature neurons. *Proceedings of the National Academy of Sciences of the United States of America, 116*(6), 2181–2186.
78. Huang, H.S., J.A. Allen, A.M. Mabb, I.F. King, J. Miriyala, B. Taylor-Blake, et al., (2012). Topoisomerase inhibitors unsilence the dormant allele of Ube3a in neurons. *Nature, 481*(7380), 185–189.
79. Lee, H.M., E.P. Clark, M.B. Kuijer, M. Cushman, Y. Pommier, and B.D. Philpot, (2018). Characterization and structure-activity relationships of indenoisoquinoline-derived topoisomerase I inhibitors in unsilencing the dormant Ube3a gene associated with Angelman syndrome. *Molecular Autism, 9*, 45.
80. Mabb, A.M., J.M. Simon, I.F. King, H.M. Lee, L.K. An, B.D. Philpot, et al., (2016). Topoisomerase 1 regulates gene expression in neurons through cleavage complex-dependent and -independent mechanisms. *PLoS One, 11*(5), e0156439.
81. King, I.F., C.N. Yandava, A.M. Mabb, J.S. Hsiao, H.S. Huang, B.L. Pearson, et al., (2013). Topoisomerases facilitate transcription of long genes linked to autism. *Nature, 501*(7465), 58–62.
82. Mabb, A.M., P.H. Kullmann, M.A. Twomey, J. Miriyala, B.D. Philpot, and M.J. Zylka, (2014). Topoisomerase 1 inhibition reversibly impairs synaptic function. *Proceedings of the National Academy of Sciences of the United States of America, 111*(48), 17290–17295.

83. Watts, J.K. and D.R. Corey, (2012). Silencing disease genes in the laboratory and the clinic. *The Journal of Pathology, 226*(2), 365–379.
84. Geary, R.S., D. Norris, R. Yu, and C.F. Bennett, (2015). Pharmacokinetics, biodistribution and cell uptake of antisense oligonucleotides. *Advanced Drug Delivery Reviews, 87*, 46–51.
85. Beaudet, A.L. and L. Meng, (2016). Gene-targeting pharmaceuticals for single-gene disorders. *Human Molecular Genetics, 25*(R1), R18–R26.
86. Meng, L., A.J. Ward, S. Chun, C.F. Bennett, A.L. Beaudet, and F. Rigo, (2015). Towards a therapy for Angelman syndrome by targeting a long non-coding RNA. *Nature, 518*(7539), 409–412.
87. Hori, S., T. Yamamoto, and S. Obika, XRN2 (2015). XRN2 is required for the degradation of target RNAs by RNase H1-dependent antisense oligonucleotides. *Biochemical and Biophysical Research Communications, 464*(2), 506–511.
88. Lima, W.F., C.L. De Hoyos, X.H. Liang, and S.T. Crooke, (2016). RNA cleavage products generated by antisense oligonucleotides and siRNAs are processed by the RNA surveillance machinery. *Nucleic Acids Research, 44*(7), 3351–3363.
89. West, S., N. Gromak, and N.J. Proudfoot, (2004). Human 5′ --> 3′ exonuclease Xrn2 promotes transcription termination at co-transcriptional cleavage sites. *Nature, 432*(7016), 522–525.
90. Smith, S.E., Y.D. Zhou, G. Zhang, Z. Jin, D.C. Stoppel, and M.P. Anderson, (2011). Increased gene dosage of Ube3a results in autism traits and decreased glutamate synaptic transmission in mice. *Science Translational Medicine, **3***(103), 103ra97.
91. Urraca, N., J. Cleary, V. Brewer, E.K. Pivnick, K. McVicar, R.L. Thibert, et al., (2013). *The Interstitial Duplication 15q11.2-q13 Syndrome Includes Autism, Mild Facial Anomalies and a Characteristic EEG Signature. Autism Research, 6*(4), 268–279.
92. Napoli, I., V. Mercaldo, P.P. Boyl, B. Eleuteri, F. Zalfa, S. De Rubeis, et al., (2008). The fragile X syndrome protein represses activity-dependent translation through CYFIP1, a new 4E-BP. *Cell, 134*(6), 1042–1054.
93. Murphy, S.M., A.M. Preble, U.K. Patel, K.L. O'Connell, D.P. Dias, M. Moritz, et al., (2001). GCP5 and GCP6: Two new members of the human gamma-tubulin complex. *Molecular Biology of the Cell, 12*(11), 3340–3352.
94. Kuhnle, S., U. Kogel, S. Glockzin, A. Marquardt, A. Ciechanover, K. Matentzoglu, et al., (2011). Physical and functional interaction of the HECT ubiquitin-protein ligases E6AP and HERC2. *The Journal of Biological Chemistry, 286*(22), 19410–19416.
95. Martinez-Noel, G., J.T. Galligan, M.E. Sowa, V. Arndt, T.M. Overton, J.W. Harper, et al., (2012). Identification and proteomic analysis of distinct UBE3A/E6AP protein complexes. *Molecular and Cellular Biology, 32*(15), 3095–3106.
96. Svenstrup, K., R.S. Moller, J. Christensen, E. Budtz-Jorgensen, M. Gilling, and J.E. Nielsen, (2011). NIPA1 mutation in complex hereditary spastic paraplegia with epilepsy. *European Journal of Neurology, 18*(9), 1197–1199.
97. Goytain, A., R.M. Hines, A. El-Husseini, and G.A. Quamme, (2007). NIPA1(SPG6), the basis for autosomal dominant form of hereditary spastic paraplegia, encodes a functional Mg2+ transporter. *The Journal of Biological Chemistry, 282*(11), 8060–8068.
98. Cruvinel, E., T. Budinetz, N. Germain, S. Chamberlain, M. Lalande, and K. Martins-Taylor, (2014). Reactivation of maternal SNORD116 cluster via SETDB1 knockdown in Prader-Willi syndrome iPSCs. *Human Molecular Genetics, 23*(17), 4674–4685.
99. Xin, Z., M. Tachibana, M. Guggiari, E. Heard, Y. Shinkai, and J. Wagstaff, (2003). Role of histone methyltransferase G9a in CpG methylation of the Prader-Willi syndrome imprinting center. *The Journal of Biological Chemistry, 278*(17), 14996–15000.
100. Kim, Y., H.M. Lee, Y. Xiong, N. Sciaky, S.W. Hulbert, X. Cao, et al., (2017). Targeting the histone methyltransferase G9a activates imprinted genes and improves survival of a mouse model of Prader-Willi syndrome. *Nature Medicine, 23*(2), 213–222.

Using iPSC-Based Models to Understand the Signaling and Cellular Phenotypes in Idiopathic Autism and 16p11.2 Derived Neurons

Luka Turkalj, Monal Mehta, Paul Matteson, Smrithi Prem, Madeline Williams, Robert J. Connacher, Emanuel DiCicco-Bloom, and James H. Millonig

Autism Spectrum Disorder (ASD) is a neurodevelopmental disorder that is remarkably heterogeneous at the clinical, neurobiological, and genetic levels. ASD can also affect language, a uniquely human capability, and is caused by abnormalities in brain development. Traditionally obtaining biologically relevant human cells to study ASD has been extremely difficult, but new technologies including iPSC-derived neurons and high-throughput omic techniques now provide new, exciting tools to uncover the cellular and signaling basis of ASD etiology.

Luka Turkalj and Monal Mehta contributed equally with all other contributors.

L. Turkalj
Graduate Program in Neuroscience, Rutgers University, Piscataway, NJ, USA

M. Mehta
Graduate Program in Neuroscience, Rutgers University, Piscataway, NJ, USA

Center for Advanced Biotechnology and Medicine, Rutgers University, Piscataway, NJ, USA

P. Matteson
Center for Advanced Biotechnology and Medicine, Rutgers University, Piscataway, NJ, USA

S. Prem · M. Williams · R. J. Connacher
Graduate Program in Neuroscience, Rutgers University, Piscataway, NJ, USA

Department of Neuroscience and Cell Biology, Robert Wood Johnson Medical School, Rutgers University, Piscataway, NJ, USA
e-mail: premsm@rwjms.rutgers.edu

E. DiCicco-Bloom
Department of Neuroscience and Cell Biology/Pediatrics, Rutgers Robert Wood Johnson Medical School, Rutgers University, Piscataway, NJ, USA
e-mail: diciccem@rwjms.rutgers.edu

J. H. Millonig (✉)
Department of Neuroscience and Cell Biology, Center for Advanced Biotechnology and Medicine, Rutgers Robert Wood Johnson Medical School, Rutgers University, Piscataway, NJ, USA
e-mail: Millonig@cabm.rutgers.edu

E. DiCicco-Bloom, J. H. Millonig (eds.), *Neurodevelopmental Disorders*, Advances in Neurobiology 25, https://doi.org/10.1007/978-3-030-45493-7_4

1 ASD Is a Clinically Heterogeneous Disorder

Autism spectrum disorder (ASD) is a neurodevelopmental disorder characterized by deficits in social interaction and communication, and the presence of repetitive and restricted behaviors and interests (DSM V). Prevalence has risen to 1/59 in the USA and is four times more common in boys [1]. A research grade diagnosis is based upon two validated diagnostic instruments, an interview with parents or caregivers (The *Autism Diagnostic Interview*-Revised, ADI-R) plus an observation of the child's behavior (The Autism Diagnostic Observation Schedule, ADOS). The precision of currently available diagnostic tools is not absolute, which can contribute to inconsistencies in ASD diagnosis (CDC, 2019 https://www.cdc.gov/ncbddd/autism/screening.html, [2]).

ASD encompasses a broad spectrum of clinical phenotypes. Different ASD cases can exhibit deficits in different domains of behavior with various severities. For example, deficits in language ability vary considerably between individuals, ranging from within normal range abilities to individuals who are minimally verbal [3, 4]. Intellectual abilities are also widely distributed in ASD. As reported by one study, 55% percent of individuals diagnosed with ASD meet criteria for intellectual disability (IQ < 70), 17% had an IQ below average but ≥70, 25% had an average IQ (~100), and 3% were above average (>100) [5]. Furthermore, autistic behavior is usually only a part of clinical presentation and commonly co-occurs with other conditions such as intellectual disability, but also GI problems (9–70%) ADHD and anxiety (60–70%), sleep problems (50–80%), and epilepsy (25–30%) [6, 7].

2 Neurobiological Basis of ASD Is Also Heterogeneous

Not surprising given the clinical variability in ASD symptoms, when investigators examine the neurodevelopmental basis of ASD a lot of heterogeneity is also observed. To study the neurodevelopmental basis of the disorder, researchers have used mouse models combined with human postmortem studies, neuroimaging, and iPSC models. Even though great variety in the results has been observed, some themes have emerged and a number of neurobiological hypotheses for the basis of autism etiology have been formulated including: excitatory-inhibitory (E/I) imbalance in neural circuitry [8, 9], synaptic dysfunction [10], and disruption of early neurodevelopmental processes such as neurogenesis and neuron migration [11–13].

E/I imbalance is hypothesized to contribute to ASD pathology since seizures, as mentioned above, are one of the more common co-morbid ASD symptoms. Alterations in both excitatory glutamatergic and inhibitory GABAergic synaptic signaling contribute to E/I imbalance. Multiple monogenic diseases with high prevalence of ASD exhibit E/I imbalance such as GABA signaling dysfunction in both Rett Syndrome (RS) [14] and Dravet syndrome [15] or exaggerated mGluR signaling in FXS [16, 17]. On a cellular level, alterations of GABA interneuron number in postmortem autistic brains have been observed [18, 19] as well as bias in neuronal differentiation towards GABA lineage in autism patient derived-cortical organoids [20].

ASD is sometimes referred to as a synaptopathy. A number of ASD-associated genes encode for synaptic proteins, such as ion channels, neurotransmitter receptors, cell-adhesion molecules, and synaptic scaffolds [10]. For example, mutations in NLG3 and NRXN1 (cell-adhesion proteins), GRIN2B (a glutamate NMDA receptor subunit), and SHANK3 (a synaptic scaffold protein mutated in Phelan–McDermid Syndrome) have all been associated with idiopathic autism. Multiple pathogenetic mechanisms associated with ASD also converge upon the synapse including abnormalities that alter dendritic spine structure and function like dysregulation of local mRNA translation, mTOR signaling and autophagy [21–23].

Abnormalities in neurogenesis and neuronal migration are another common, pathological feature of ASD. Both macro- and microcephalic phenotypes are commonly associated with ASD and are likely the consequence, at least in part, of up- or down-regulated neurogenesis in ASD [24]. Morphological abnormalities observed in postmortem autistic brains are suggestive of defective neuronal migration including cortical dysplasia and geographically widespread heterotopias [25–27]. Also, several genes linked to ASD have a functional role in neurogenesis and/or neuronal migration, such as Ankrd11 (Ankyrin repeat domain 11), Wdfy3 (WD Repeat And FYVE Domain Containing 3), and PTEN [12]. It is hypothesized that improper neurogenesis and migration of newly generated neurons to their final destination could underlie some aspects of aberrant neuronal connectivity in ASD. It is conceivable that disruption in these critical, early-stage neurodevelopmental processes could contribute to ASD [11].

While some themes have emerged in the possible neurodevelopmental hypotheses for ASD etiology, most cases have no readily identifiable cause, which is broadly referred to as idiopathic autism. Idiopathic autism therefore includes an extremely heterogeneous clinical population with a broad spectrum of autistic behaviors and potentially very different etiologies. These different ASD etiologies could very likely also be caused by different neurodevelopmental and signaling defects. Consistent with this possibility, it is clear that ASD is an extremely heterogeneous disorder at the genetic level.

3 Autism Genetics

ASD is highly heritable. The concordance rate for ASD is higher in monozygotic (0.77–0.99) than dizygotic twins (0.22–0.65) [28], and data from Swedish population-based studies estimate ASD heritability to be around 50% [29, 30]. Finally, ASD recurrence rate in families with one already affected child is estimated to be as high as 20% for future offspring [31] and the relative risk for siblings of ASD cases is as high as tenfold [29].

The genetic architecture of ASD is also highly complex and heterogeneous, and many different classes of genetic variants play a role. These can be classified by frequency (common or rare), mode of transmission (de novo, dominant, recessive, X-linked), size (SNVs, indels, and CNVs), and penetrance (modest to high). Together these genetic variants in combination with epigenetic changes and non-genetic environmental factors, all of which can have individual and cell specific effects, contribute to ASD risk. On an individual level, common inherited variations like single-nucleotide level (SNVs) carry modest effects while rare de novo variants can have large effects. At a population level, common inherited SNVs collectively are postulated to contribute to a large fraction of ASD variance (~50%) [30, 32], but with few exceptions, remain to be identified [33]. De novo mutations include loss of function (LoF) and missense point mutations, insertion/deletion (indels), and large chromosomal duplications or deletions (CNVs). Together de novo mutations are postulated to account for an additional 3% of ASD cases [30, 34].

Large-scale whole-exome sequencing studies (WES) on cohorts of ASD families (Simons Simplex Collection, Autism Sequencing Consortium) have identified de novo LoF variants within protein-coding regions of probands. These data on de novo LoF in combination with sophisticated statistical models are used to identify high-confidence ASD genes [35–37]. Based on some estimates, there could be more than a 1000 ASD risk genes [36]. However, the data for only a minority of these genes is significantly strong enough to support their role in ASD etiology. SFARI database comprises a list of 1079 putative ASD genes classified into different confidence categories and includes a set of 91 high-confidence genes discovered to date including, for example, *CHD8, POGZ, ANK2, NRXN1* (www.sfari.org/resource/sfari-gene/). The list of high-confidence ASD genes is expected to expand in the following years as ASD cohorts grow in size.

However, individually these genes contribute to only a minority of ASD cases. For example, de novo point mutations in some of the high-confidence ASD genes like *CHD8* or *ANK2* account for only 0.21% and 0.13% of all the cases, respectively [34]. In addition to variants in protein-coding sequence, recently genome-wide discoveries of noncoding variants in ASD have been made by whole-genome sequencing (WGS) technologies. Those are estimated to account for another 4% of ASD cases [38, 39]. Despite these advances, the etiology of ASD remains unclear in the majority of cases, such that only 11% of probands had identifiable causes of ASD in one WGS study [40].

Even though genetics of ASD is extremely diverse, a degree of biological convergence is indeed observed. ASD genes are enriched in some broad functional categories including transcription, chromatin regulation, and synaptic development/plasticity [36, 37, 41]. In addition, de novo coding and noncoding variants tend to target the same genes in ASD probands, providing further line of evidence for convergence of diverse genetic causes [38, 39].

Multiple loci contribute to ASD liability in each individual. The effect of one high-impact casual mutation could be modified by other loci with additive or protective effect. For example it is postulated that 80% of ASD cases carrying a de novo CNV likely have the disorder due to the CNV, while another 20% of the cases, although they have the CNV, ASD is caused by other genetic and non-genetic factors. Likewise 57% of cases carrying a de novo LoF variant have ASD due to the mutation, while another 43% of ASD cases carry the mutation but have other genetic or non-genetic changes that contribute to ASD [30]. Thus, each individual likely has their own constellation of genetic variants that increase the risk for ASD.

Considering the complexity of interactions between different genetic factors and given that most ASD loci remain unidentified, it is clear that modeling idiopathic ASD is a difficult task. In addition the cell types and developmental ages that are affected in ASD still have to be identified, and the signaling pathways and neurodevelopmental cellular processes affected in these cell types also need to be better understood. All of this need to be characterized so we can understand the etiology of ASD and develop new therapies for the disorder.

In sum, the clinical, neurobiological, and genetic heterogeneity of ASD may suggest that the disorder is composed of different "autisms," with different signaling and cellular defects causing different phenotypes. How can we identify the signaling and cellular basis of these different "autisms" so we can better understand their etiology and develop therapies? One way is to leverage iPSC-derived autism neurons to identify the signaling and neurodevelopmental defects that may be observed, and determine whether there may be different autism subgroups that warrant targeted interventions.

4 Monogenic and CNV ASD-Associated Diseases Are Used as Model Systems for Studying Idiopathic Autism

As discussed the genetic architecture of ASD is enormously complex with the type and number of variants contributing to ASD liability differing greatly among affected individuals in the population. Epigenetic and non-genetic environmental factors also contribute to idiopathic ASD risk. Altogether this heterogeneity makes it extremely difficult to study the etiology of idiopathic autism in model organisms, which are typically used to study mutants inherited in a Mendelian fashion. Nevertheless, model organisms are extremely tractable and their use has contributed enormously to our understanding of disease mechanism for many different disorders.

Because model organisms are so tractable and Mendelian diseases are easier to study compared to complex common disorders, researchers have turned to a number rare human single-gene disorders to study the etiology of idiopathic autism. These diseases include, for example, Fragile X Syndrome (FXS), Tuberous Sclerosis Complex (TSC), and Rett Syndrome (RS). Importantly, affected individuals exhibit disease-specific phenotypes that sometimes include autism. For example individuals with FXS can have seizures (5–30%), a long and narrow face, large ears, a prominent jaw and forehead (30–79%), unusually flexible fingers, flat feet, macroorchidism (80–99%), but only 15–30% reach diagnostic criteria for autism (rarediseases.

info.nih.gov/diseases/6464/fragile-x-syndrome). The same is true for RS. Individuals with RS can exhibit microcephaly, seizures (80–99%), respiratory impairments, scoliosis (30–79%), muscle weakness, and hepatomegaly (5–29%) while only 30–79% are diagnosed with autism (rarediseases.info.nih.gov/diseases/5696/rett-syndrome). Further the genes responsible for these specific diseases (FMRP, MeCP2) do not contribute significantly to idiopathic ASD, based upon extensive genetic studies (www.sfari.org/resource/sfari-gene/).

Thus these single-gene disorders are unlike idiopathic ASD in that affected individuals exhibit additional phenotypes in addition to autism, and only sometimes are diagnosed with ASD. This specific constellation of phenotypes helps define the particular disease. However because these diseases are inherited in a Mendelian fashion, they are much more leverageable experimentally. As a result, an enormous amount of progress has been made in understanding the signaling and cellular bases of these diseases. For example, mouse and iPSC models exist for all of these disorders. In FXS we have discovered a link between hyperactive mTOR and exaggerated local cap-dependent RNA translation and protein synthesis at the synapse, while mouse models of TSC were leveraged to dissect the role of the hyperactive mTOR, an autophagy-dependent synaptic pruning deficit in TSC. And perhaps surprisingly, these disease phenotypes can be reversed in adult mouse models of FXS and TSC, implying that the genes may contribute to both developmental programs and ongoing synaptic dysfunction in the adult [17, 21–23].

Another leverageable model to study ASD pathogenesis are copy number variants (CNVs). In the last decade with the emergence of high-throughput genomic profiling, CNVs below the detection of karyotyping have been identified. This advance along with an increased clinical focus on these genomic rearrangements has identified a number of CNVs that have been linked with autism. These include: 1q21.1, 3q29, 7q11.23, 15q11–13, 15q13.3, 16p11.2, 22q11, 17q12 (Krishnan 2016). All of these CNVs delete or duplicate a certain complement of genes. One of the most tractable CNVs is 16p11.2.

Both a duplication or a deletion of the 16p11.2 region is associated with autism and estimates suggest the CNV contributes to 0.3–1% of idiopathic autism [7]. Individuals with the deletion have 16p11.2 Deletion Syndrome (16pDS) (ghr.nlm.nih.gov/condition/16p112-deletion-syndrome) and display developmental coordination disorder, phonologic processing disorder, expressive and receptive language disorders. 15–26% of 16pDS individuals are reported to have ASD [42, 43]. Very few genetic mutations have such a large effect size (OR = 40) for idiopathic autism [43], making it an attractive genetic mutation to study. In addition, the 16p11.2 CNV has several advantages over other CNVs for studying autism. It is a relatively small CNV spanning only 555 kb; only 28 genes are located within the CNV, which is less than most CNVs, and unlike many CNVs, it has consistent breakpoints. Mouse models for the deletion and duplication have been generated and The Simons Foundation has recruited over 200 individuals with the CNV. Cells from these individuals and unaffected controls are cryobanked and iPSCs have been generated, now available for distribution. This is not the case for many of the other CNVs associated with autism.

Despite the advantages of these rare Mendelian diseases and CNVs to model idiopathic autism, the research community still needs other systems for studying the signaling and neurodevelopmental basis of idiopathic autism, which is the vast majority of all cases. Fortunately, the advent of iPSC technology and the ability to generate neurons from iPSCs now for the first time make it possible to study the developmental and molecular basis of idiopathic autism, which will hopefully lead to greater understanding of ASD etiology and the identification of more potential therapeutic targets. But how can we leverage iPSC-derived neurons to answer fundamental questions about the signaling pathways and cell types affected in autism? To tackle this enormous problem, maybe the autism field can learn from other fields that have made more progress in this area.

5 Lessons Learned from Cancer Biology?

If the genetic and phenotypic heterogeneity of ASD is so complex, maybe the ASD field can learn from other diseases that are equally heterogeneous but the research is further ahead concerning the signaling pathways and cellular processes that contribute to the disease. One such disease is cancer, which is clinically and genetically diverse, but the signaling pathways and cellular processes affected in cancer have been well defined. Although cancers exhibit remarkable heterogeneity, they share some common features that apply to most, if not all cancers, indicating some level of convergence of complex genetic, molecular, and cellular events between different cancer types. Possibly these successes in the cancer field could suggest directions for ASD research in untangling the etiology of ASD.

Like ASD, cancer is not a single disease but is clinically extremely diverse. For example, a multitude of different cancer entities are recognized by the WHO brain tumor classification [44]. For decades, tumor diagnosis was established mainly based on histopathological findings. However, as gene expression profiling has allowed detailed molecular characterization of tumors, it is now widely accepted that certain tumors with similar histopathology (and which were as such considered to be a single disease) actually represent different biological entities. Some cancers, like medulloblastoma, are no longer considered a single disease but can now be further divided into distinct molecularly defined subgroups (e.g. SHH, WNT, type 3 and type 4)—each with their own distinct genetic mutations, signaling pathways involved, and clinical outcomes [45–48]. This same paradigm has been observed in many of other cancers including breast cancer, lung cancer, and lymphomas [49–51].

Consistent with this heterogeneity at the clinical and gene expression level, the genetics of cancer is also very complex, with over three million somatic mutations contributing to cancer phenotype. Of these three million somatic mutations, 125 driver genes have been identified to contribute to human cancer out of 20,000 genes in the human genome [52]. Driver genes are those genes that confer a selective growth advantage to cancer cells. Typically, 3–6 driver genes are mutated per tumor in common adult cancers.

Even though genetics of cancer is heterogeneous and many different genes are involved, protein products of these 125 driver genes interact with each other and converge to dysregulate certain signaling pathways. Indeed, the 125 driver genes are classified into one or more of only 12 different signaling pathways affected in cancer, which when disrupted confer selective growth advantage to the cancer cell. The 12 signaling pathways affect three core cellular processes: cell fate, cell survival, and genome stability [52]. In cancer, cell fate is shifted towards cell division. The 125 driver mutations also converge and promote cancer cell survival by, for example, inducing tumor vascularization [53] or evading the immune system [54]. Finally, genome maintenance is also a key cellular process affected in cancer with chromosomal instability and accumulation of mutations typically happening in cancer cells [55, 56].

Thus, even though extraordinary complexity is observed on the clinical, genetic, and molecular level in cancer, there are common disease features on the cellular level. Hundreds of different mutations funnel into only a dozen dysregulated signaling pathways. Furthermore, these 12 signaling pathways affect only 3 core cellular processes, with the same outcome of selective growth advantage. This results in cancerous behavior of cells that divide abnormally, invade adjacent tissue, and metastasize to distant sites. It is noteworthy to mention the exclusivity principle here. Mutations in driver genes that function within the same core pathway are mutually exclusive, i.e. they rarely occur together within a single tumor. This highlights the importance of pathway level analysis over mutations in single genes in understanding the biology of genetically heterogeneous diseases [52, 57–59]. Targeting these downstream dysregulated pathways that are affected in many cancers are then possibly more efficacious than the upstream mutations. For example, there has been tremendous success targeting PDL-1/PD-1 interactions, rather than upstream mutations, so that immune evasion does not occur [54].

Maybe the same principles exist in autism with many divergent mutations converging on a few signaling pathways and cell biological processes? Maybe if we identify these signaling pathways and cellular processes, this will lead to new targets and therapies for ASD? If this is the case, then how can we identify the signaling pathways and cellular processes affected in autism—a disorder that is human-specific and affects brain development? The advent of iPSC technology and the ability to drive the differentiation of human iPSCs to neurons now allow us for the first time to characterize developmental processes abnormal in autism and leverage omic approaches to understand the molecular and signaling defects in autism. Maybe we can use this iPSC technology to identify the signaling pathways and cellular processes commonly affected in autism.

6 Evidence for Convergence in Signaling and Cellular Defects in ASD

Given the heterogeneity of ASD, is there any evidence for convergence in the signaling pathways and cellular processes as in cancer? Or are the signaling and cellular defects much more personalized in ASD so convergence is not observed? Summarized below are some common defects that have been consistently reported in a subset of individuals with idiopathic autism. These include abnormalities in mTOR signaling, macrocephaly, and similar transcriptional signatures in postmortem samples.

6.1 mTOR Signaling

When examining monogenic diseases associated with ASD, such as TSC and FXS, common perturbations in mTOR signaling have been reported. In Tsc2+/− mutant mice, researchers observed autophagy-dependent synaptic pruning deficits and autism-like behavioral defects, which they were able to rescue by using the mTOR inhibitor, rapamycin [60, 61]. In FXS a link has been discovered between hyperactive mTOR signaling, synaptic protein synthesis, and GluR activation. Consistent with the hypothesis, reducing mGluR5 gene dosage by 50% in *Fmr1*$^{-/y}$ mutant mice was able to rescue multiple FXS phenotypes [16, 17]. The *Fmr1* mutation also causes exaggerated synaptic protein synthesis, which could be corrected by inhibition of mGlur5. Sharma et al. [62] was able to link elevated mTOR signaling in *Fmr1*$^{-/y}$ mutant mice to mGluR overactivation and exaggerated mGluR-LTD [63, 64].

Interestingly mutations in other genes that affect mTOR signaling like PTEN and NF1 cause rare monogenic disorders that are also sometimes associated with autism. For example PTEN is a negative regulator of the AKT/mTOR pathway and mutations in PTEN result in Cowden Syndrome. Affected individuals have a myriad of phenotypes including cancer, ataxia, macrocephaly, and ID with 5–29% being diagnosed with autism. PTEN knockout mice display similar phenotypes with brain overgrowth, neuronal hypertrophy, and ASD-like behavioral phenotypes (reduced social activity, increased anxiety, and sporadic seizures), which are rescued by rapamycin treatment [65]. Neurofibromatosis type 1 (NF1) negatively regulates Ras as part of the ERK pathway. The ERK pathway inhibits TSC1/2 and therefore affects Akt/mTOR signaling. Individuals with NF1 have benign and cancerous tumors, café au lait skin spots, macrocephaly, and about 30% of individuals with NF1 also have autism ([66]; rarediseases.info.nih.gov/diseases/7866/neurofibromatosis-type-1). NF1 knockout mice display learning and memory and social recognition deficits [67, 68]. Together these data indicate that mutations in the mTOR pathway can lead to autism phenotypes in both humans and mice.

mTOR signaling promotes cap-dependent translation [69], which may also contribute to some ASD deficits [21]. For example, *Tsc2*$^{+/-}$ mutant mice display decreased synaptic protein synthesis while *Fmr1*$^{-/y}$ mutant mice result increased mRNA translation [17]. In addition, exaggerated cap-dependent translation in the brain results in ASD-like phenotypes in mice including repetitive/perseverative behaviors, altered communication, and deficits in social interactions. Pharmacological inhibition of cap-dependent translation can reverse these ASD phenotypes [70, 71]. Therefore, abnormal mTOR signaling can lead to increased or decreased local protein synthesis, both of which can contribute to autism phenotypes.

mTOR signaling defects have also been observed in idiopathic ASD. In a 2015 study, Nicolini et al., investigated whether the Akt/mTOR pathway is altered in idiopathic autism using postmortem brain tissue samples. Protein expression results, assessed through western blotting, revealed 9 of the 11 ASD cases exhibited downregulation through the mTOR pathway, specifically via the downstream effector pathway of p70S6K/eIF4B [72]. This is in direct contrast to what has been seen in monogenic forms of ASD, FXS, TSC, neurofibromatosis type 1, and PTEN-related macrocephaly, all which show increased mTOR pathway components [72]. Taken together these results support the hypothesis that abnormal mTOR signaling, in either direction, can contribute to autistic characteristics. In addition, when examining patients ranging from mild to severe idiopathic ASD, levels of mTOR pathway components: total levels of S6, and TSC1 proteins, and phosphorylated forms of p-eIF4E and p-MNK1, appear to provide a molecular signature of severity [73]. These four proteins could be used to discriminate between controls, mild ASD, and severe ASD, with the severe cases exhibiting increased protein levels. Consistent with these observations mTOR signaling pathway regulates a number of developmental processes which are thought to play a role in pathology of ASD: neural progenitor proliferation, neuronal migration, and various stages of neural circuitry formation (axon growth, formation, and elimination of synapses) [22].

6.2 Macrocephaly

Another possible convergence point for autism is macrocephaly, defined as a head circumference measurement greater than two standard deviations above the population mean, and is present in approximately 20% of individuals with ASD [74, 75]. Brain imaging and head circumference studies have reported that brain enlargement in ASD is not present at birth, but emerged around 2–3 years of age [76, 77]. A longitudinal and cross-sectional MRI study of male and female ASD brain size from 2 to 50 years show three different periods of brain development: first, an initial abnormally accelerated overgrowth, followed by a period of abnormally slow/arrested growth between young childhood and preadolescence, followed by a premature and accelerated rate of decline from adolescence to middle age [78, 79]. It is unclear what the biological basis of the macrocephaly is during the development of the human brain but could conceivably come from a wide variety of sources including dysregulated proliferation or apoptosis, decreased post-natal spine pruning, or altered gliogenesis or white matter tract formation [76, 77].

Researchers have begun to identify what regions of the brain show enlargement. A longitudinal MRI study found generalized cortical enlargement in children with ASD, with specific disproportionate enlargement in temporal lobe white matter [80]. It has been hypothesized that proliferation of neural progenitor cells is a major early determinant in affecting cortical dimensions [77]. When macrocephaly is modeled in vitro using iPSC-derived organoids, researchers have found an accelerated cell cycle and overproduction of GABAergic inhibitory neurons, caused by an overexpression of the FOXG1 transcription factor. Though the cause of macrocephaly is yet to be determined, it is one of the most consistently observed ASD phenotypes, and has been related to poorer clinical outcomes [20].

6.3 *Postmortem Transcriptome Studies*

Finally, transcriptome-wide gene expression studies have also been conducted in postmortem brains and have uncovered ASD-associated transcriptomic signatures. This includes a down-regulation of neuronal/synaptic gene pathways and up-regulation of glial/inflammatory gene pathways that interestingly partially overlaps with schizophrenia and bipolar disorder, two other neuropsychiatric disorders [41, 81, 82]. These data suggest that some biological pathways are consistently disrupted in ASD and some might even be relevant across neurodevelopmental/neuropsychiatric disorders. The existence of shared molecular pathology in postmortem ASD brains supports the idea of biological pathway convergence, much like cancer. However, it remains unknown how many of these transcriptional changes are causative due to genetic variants or a further downstream effect of the disorder.

Given the number of genes that have now been associated with ASD and the intense efforts to map transcriptional networks in developing and adult human brain (see Allen Brain Atlas), researchers can use bioinformatic tools to determine if ASD-associated gene expression is enriched in certain neuronal cell types or brain regions. Interestingly ASD risk genes are enriched in both upper- and lower-layer cortical glutamatergic neurons during neocortical development [83, 84], and more recently in neural progenitor cells [85], suggesting a possible convergence of ASD-related biological pathways onto specific cell types and early developmental periods.

Thus some level of convergence is observed in idiopathic autism, consistent with the notion of autism subgroups. To further investigate this important question and to interrogate for the first time how neurodevelopment is altered in ASD, we and others have turned to iPSC technology. By generating iPSCs from individuals with autism and then deriving neurons we can begin to answer some fundamental questions. Given the heterogeneity of the disorder do we observe personalized individualized phenotypes? Or instead is the disorder like cancer and the heterogeneity filters down to some common dysregulated pathways? And can we rescue these autism neurodevelopmental phenotypes by targeting these downstream pathways?

7 iPSCs as a Model System: Advantages and Disadvantages

In 2006 Yamanaka and his colleagues transduced somatic cells with just four factors and reprogrammed these cells into induced pluripotent stem cells (iPSCs). Amazingly within a short time, this technology has rapidly become a standard protocol for researchers around the globe. Blood cells, fibroblasts, and other somatic human cells can now be transduced with the Yamanaka factors or treated in a variety of other ways [86–88] and within a matter of a month or two, iPSCs clones are available to study. iPSCs have introduced cell based disease modeling and regenerative medicine as reasonable approaches for many human disorders.

For autism and other neurodevelopmental diseases, iPSC-based research has become transformative. We as a research community can now for the first time identify developmental processes and signaling pathways that are abnormal in affected individuals. Prior to iPSCs, there was virtually no way to investigate how human neurons developed abnormally in disorders like autism. The research community can now generate iPSCs from individuals with autism, determine how autism neurons develop incorrectly, and apply standard biochemical and molecular approaches as well as standard and emerging omic approaches to identify the signaling pathways responsible for the abnormal development. Obviously, as described earlier for cancer, understanding the signaling pathways and cellular processes that are affected in autism is key for understanding its etiology and for developing new therapies for the disorder.

Very importantly iPSCs also provide a tractable system to model human idiopathic, polygenic disorders like autism. As discussed previously, common polygenic disorders like autism are due to numerous rare and common genetic variants that contribute to disease risk. Due to this genetic complexity, it is nearly impossible to model idiopathic polygenic disorders in mice and other model systems that rely upon Mendelian inheritance. iPSCs, on the other hand, are generated from a patient's cell, which captures all the genetic variants that contribute to disease risk in that individual. So if autism risk is due to multiple genetic loci that function in an epistatic, synergistic way, we now have a model system to investigate how these genetic variants work in concert to affect neuronal development. Thus not only do iPSCs allow us to look at development of the affected neurons, but they also allow us to capture the genetic heterogeneity for each individual that is a hallmark of idiopathic autism.

Despite these incredible positives, all systems have negatives. For iPSCs, these negatives include the expense and labor to generate iPSCs and iPSC-derived neurons. The system is also very demanding, with the cells needing care every single day and considerable technical expertise. The system also requires substantial time to learn and prior culture experience with both immortalized cell lines and primary rodent neurons is almost a pre-requisite. Different clones from each individual could potentially contain clone-specific differences in the reprograming so multiple clones from each individual need to be studied for all subsequent developmental and molecular phenotypes. The most rigorous analysis requires multiple derivations of neurons from each clone, exponentially increasing the amount of work that is both time-consuming and expensive. Nevertheless the positives of being able to interrogate, for the first time, the neurodevelopmental events that are abnormal in autism and to perform these studies in patient cells that capture all the genetic variants that contribute to disease risk remains transformative for the field.

8 Selection of iPSCs: Strategies to Reduce Heterogeneity

The selection of individuals, from whom iPSCs will be derived, gives researchers an opportunity to reduce clinical and genetic heterogeneity. This in turn might reduce the heterogeneity of the signaling and developmental phenotypes that are observed in iPSC-derived neurons. Investigators have used multiple approaches to reduce heterogeneity. These strategies include anatomical endophenotypes like macrocephaly [20, 24], stratifying on a particular genetic variants like SHANK3 [89, 90], or recruiting patients and families with a particular clinical phenotype in addition to autism.

We decided to use a clinical phenotype to reduce heterogeneity. We worked with our long-term collaborator, Linda Brzustowicz MD, who has recruited families for two language disorders: autism and Specific Language Impairment (SLI). These families have two probands: one with autism and another with SLI. SLI defects are specific to the language domain and include an inability to master spoken and written language expression and comprehension. This occurs even though SLI-affected individuals have normal nonverbal intelligence, hearing acuity, and speech motor skills. They also have no overt physical disability or recognized disorder like autism, or other mitigating medical factors known to cause language disorders in children [91]. Thus, diagnosis of SLI is largely exclusionary—requiring the absence of these specific symptoms. Additionally, a speech pathologist also assesses a child's language skills through direct observation, questionnaires, and implementing standardized tests of language performance. Phenotypically individuals with SLI and autism have overlapping language defects including delayed language onset, impairments in syntax, and deviant or absent language babble [92]. Interestingly there is also a genetic overlap between autism and SLI. For example in the families recruited by Dr. Brzustowicz, several genetic loci are shared between autism and SLI [93–95]. By taking this recruitment strategy, it is reasonable to speculate that the clinical phenotypes would be less heterogeneous.

To further reduce potential heterogeneity, we used the additional following steps. One, importantly all families were diagnosed by Dr. Brzustowicz's team over the years reducing the clinical diagnostic variability that can occur between sites. All family members received the same battery of tests so controls were diagnosed as unaffected individuals and were not simply self-reported as normal. Two, to further reduce heterogeneity, we only selected same-sex male sibpairs to study, because we wanted initially to avoid random X inactivation in females that would cause intrinsic variability. Finally, we also only selected individuals with a strict autism diagnosis (no broad phenotypes) and originally our studies only included families with a severely affected individual with autism with little or no language. By taking this approach, we hoped to reduce the genetic and clinical heterogeneity, which would in turn result in consistent signaling and developmental phenotypes. Interestingly our preliminary data does show a convergence of signaling and developmental phenotypes between these families suggesting that the above strategy could be decreasing heterogeneity.

9 16p11.2 Deletion CNV: A Gene-Centric Approach

To complement our idiopathic autism analysis where there is likely to be significant genetic heterogeneity, we decided to utilize the 16p11.2 iPSCs provided by the Simons Foundation Autism Research Initiative (SFARI). The 16p11.2 CNV deletion affects only 28 genes, reducing genetic heterogeneity considerably. Inheritance of the CNV results in a 20× higher risk of autism and the deletion accounts for approximately 1% of non-syndromic autism risk [96]. Other advantages of the system have been described previously (see Monogenic and CNV ASD-associated diseases are used as model systems for studying idiopathic autism), but one significant advantage is that a mouse mutant for the deletion has been engineered and is available from Jackson Laboratories. The mouse mutants display macro/microcephaly, hyperactivity, and repetitive motor behaviors, phenocopying defects observed in affected individuals with the CNV [97, 98]. Thus the 16p11.2 system provides a unique resource where investigators can move between the mouse mutant and human iPSC-derived neurons to investigate the developmental and molecular basis of phenotypes.

Another significant advantage of the 16p11.2 system is that the CNV only affects 28 genes. This makes studying the underlying molecular and signaling basis of neurodevelopmental phenotypes much more straightforward than idiopathic autism. For example, we have applied high throughput, multiplex assays to determine which of the 28 genes are expressed in human 16p11.2 NPCs and various mouse embryonic brain regions. Working with Affymetrix, a Luminex multiplex assay was established for the 28 genes deleted in the CNV. Using the mouse model, three brain regions (cortex, basal ganglia, and hindbrain) were isolated from wild-type and heterozygous mice at E13.5. Fourteen genes are expressed in all brain regions and all of the genes were expressed at ~50% levels in the 16p11.2 mutants (Fig. 1). We have performed a similar analysis for human NPCs, determined which genes are expressed, and are currently performing knockdown studies to investigate if these genes contribute to the 16p11.2 NPC phenotypes.

We and others have been investigating the neurodevelopmental phenotypes associated with the 16p11.2 CNV. Pucilowska et al. evaluated cortical development in 16p11.2 heterozygotes for the deletion, and discovered that the mutant mice have increased progenitor proliferation during early and mid-neurogenesis (E12.5 and E14.5) [99]. The group had previously investigated the role of ERK signaling in

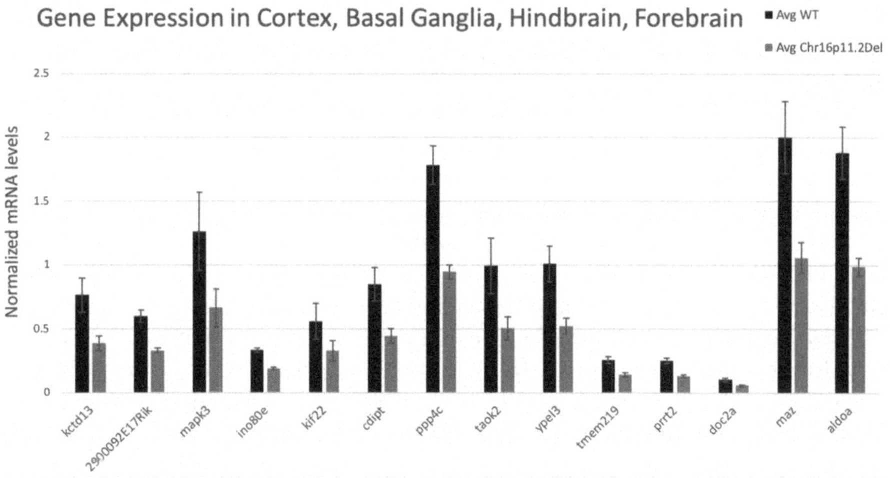

Fig. 1 Multiplex mRNA expression analysis for E13.5 WT and Chr16p11.2Del mouse brain regions. mRNA expression analysis was performed using a Luminex multiplex assay for the 27 genes deleted in the Chr16p11.2 CNV. Three different E13.5 brain regions were isolated (cortex, basal ganglia, and hindbrain). Fourteen genes were found to be expressed in all three brain regions in both WT and 16p11.2 +/deletion embryos. As expected, all 14 genes were expressed at ~50% levels in 16p11.2 +/deletion embryos compared to WT embryos

cortical neurogenesis and they hypothesized that this proliferation increase is due to the heterozygous loss of *Mapk3* in the CNV [100]. Consistent with this possibility, recent work demonstrated that a novel ERK pathway inhibitor during a critical period of brain development rescues molecular, anatomical, and behavioral deficits in the 16p11.2 mutant mice [97].

In a recent human iPSC-based study for Chr16p11.2, iPSCs were generated from deletion carriers and controls. Forebrain cortical neurons were then assessed for proliferation, cell growth, differentiation, and synaptic development. Increased soma size and dendrite length but decreased synaptic density were observed for Chr16p11.2 deletion carriers, suggesting that these defects may contribute to the macrocephaly. They found no differences in proliferation or apoptosis [101]. Our groups have obtained 16p11.2 iPSCs from SFARI. These individuals have the deletion and have been diagnosed with autism. We observe consistent neurite and migration phenotypes in numerous individuals with the deletion. Interestingly the 16p11.2 developmental phenotypes converge with our idiopathic results, with all affected individuals displaying reduced neurites and migration.

10 Rigor and Reproducibility of iPSC Studies

Given how transformative and common iPSC-based research has become, it is essential that the studies are performed in a rigorous manner. Importantly, rigorous recommendations have been established. Several issues that affect the quality of iPSC research are cell line identity, genomic instability, pluripotency, and residual reprogramming factors [102]. Best practices for each of these items are outlined below.

To test for cell line identity, short-tandem repeat (STR) or SNP genotyping, or genomic sequencing should be performed. If this analysis is done early and is repeated over time, it will allow researchers to establish a base-line identity and can be used to confirm that the iPSCs do not change considerably over time. This type of analysis also addresses cell line authenticity as required by the NIH (NOT-OD-17-068). Another issue with reprogramming is genomic instability, which is addressed by performing G-banding karyotyping or chromosomal arrays.

When deciding the cell type for generating iPSCs, most somatic cells successfully differentiate into neurons [103]. One potential issue that can limit iPSC differentiation is the epigenomic profile of the iPSC clone. One possible way to test for this epigenetic limitation is to perform stochastic embryoid body (EB) differentiation protocols and test if the iPSC clone has a propensity for neuronal differentiation.

When reprogramming somatic cells, researchers should avoid methods involving integration of reprogramming genes into the donor-cell genome. No residual reprogramming factors should remain in the iPSCs, which can be accomplished by passing iPSCs at least ten times and using rigorous PCR detection methods to test for any remaining reprogramming genes [102]. In addition, alternative reprogramming approaches, such RNA or chemical-based protocols, can be employed [86–88].

Post-reprogramming, pluripotency assessment should be performed to establish this defining stem cell property, and can be done through evaluation of pluripotency marker expression, EB formation and differentiation into the three germ layers, plus in vivo teratoma assays. Finally, because the effect of reprogramming on genetic and epigenetic changes is not well understood, all iPSC lines should be derived in the same method. Using these guidelines and best practice standards for stem cell generation will allow researchers to ensure cell line authenticity and improve iPSC research quality.

11 Selection of Controls and Cell Types to Study for Neurodevelopmental Studies

Time should also be taken to consider the controls used for iPSC studies. We have used two types of controls for our studies: unaffected same-sex siblings and unrelated individuals that are sex and ethnicity/race matched. Both controls have advantages and disadvantages. In our idiopathic families, the unaffected same-sex siblings are controls and are diagnosed as unaffected. All family members underwent a battery of tests so the same-sex siblings were diagnosed as having no language phenotypes or associated disorders (e.g. autism, hearing impairment). In contrast self-reported controls with no official diagnosis can lead to unintentional errors in clinical phenotyping that can be a confound for the study. However, one disadvantage of using the unaffected sibling as a control is that they could have a genetic load that is sub-threshold for autism and language disorders. While this genetic load does not manifest itself in a clinical phenotype, it could be sufficient to display a signaling or developmental defect in culture. To complement the above approach, we have also acquired iPSC controls from the NIMH Regenerative Medicine Program. These cells have been generated from newborn umbilical cord blood. These individuals had no gross developmental phenotypes and are matched to affected individuals based upon sex and race/ethnicity. Yet a negative of the NIMH controls is that they have not been followed clinically so it is unknown if later in life they presented with a relevant disorder. Because both controls have advantages and disadvantages, we typically employ both especially when beginning to define a developmental phenotype.

We have also applied additional rigor in the generation of neuronal cell types for our studies. For our developmental studies, we have focused on two types of cells: neural precursor cells (NPCs) and induced neurons (iNs). In generating NPCs, we needed an appropriate methodology to faithfully produce these cells to near homogeneity and in a timely fashion. For this reason we have adopted the Gibco Neural Induction Protocol (https://assets.thermofisher.com/TFS-Assets/LSG/manuals/MAN0008031.pdf). NPCs are induced in approximately 7 days when starting with iPSCs, after which they are expanded until passage 3 before being used for experiments. During this time period, NPCs begin to express neurogenic markers (Nestin, Pax6, Sox1, and Sox2), while pluripotency markers (Oct4, Dnmt3a, Sox2) show reduced expression levels. The NPCs that are generated represent a nearly homogeneous population of dividing neural cells with the potential to generate both neurons and glia. This has been determined by both immunostaining for some of the above markers and performing qRTPCR for a panel of pluripotency and neurogenic markers. We have also performed functional assays to further demonstrate that the NPCs are multi-potent and have the potential to differentiate into neurons, astrocytes, and oligodendrocytes. The NPCs generated by this procedure are similar to forebrain progenitors based upon RNA-seq analysis [104]. Numerous studies have linked forebrain neurogenesis, as mentioned previously, with autism making this population of NPCs relevant for disease modeling [13, 105, 106]. Finally, we generate NPCs three separate times per experimental assay to ensure that phenotypes are not due to unforeseen and unnoticed variations in NPC generation.

To complement NPCs, induced neurons (iNs) have been employed to investigate phenotypes in fully differentiated neurons. In this assay the forced expression of a single transcription factor, Ngn2, is sufficient to generate glutamatergic neurons to nearly 95% homogeneity [107]. Differentiated neurons with extensive dendrites and axons are present after a week and further culturing for about 3 weeks on mouse glia is needed for the formation of functional synapses. Transcriptome and marker analysis suggest that these glutamatergic iNs, which express Brn-2, Cux1, and FoxG1 at high levels, represent layer 2/3 excitatory cortical neurons making them a relevant cell type for studying autism [107]. Both NPC and iN protocols generate a biologically relevant, nearly homogenous population of cells that can be generated in a relatively short time frame and used for molecular and omic analysis to define the developmental and signaling phenotypes underlying autism.

12 Developmental Analysis of Idiopathic Autism

Given the genetic heterogeneity of autism, we first simply wanted to leverage our iPSC-derived NPCs and iNs to investigate if consistent neurodevelopmental phenotypes could be observed in our idiopathic autism dataset. We hypothesized two opposing scenarios. One possibility would be a convergence of phenotypes such that the same downstream signaling and developmental processes would be affected in all ASD cases (similar to what is observed in cancer), or alternatively there would be no convergence but instead person-specific phenotypes.

To investigate this important question, we wanted to develop assays for different steps in neurodevelopment that were both high throughput and reliable. For this reason, we adopted a series of protocols for neurogenesis and early differentiation that allowed us to quantify developmental phenotypes in a consistent but in a relatively rapid manner. These assays have been applied to rodent cultures for over a decade and these in vitro studies have been validated by comparing to in vivo models [108–113]. These protocols are described in detail in [114] and in the accompanying chapter (see Prem et al). Briefly for neurogenesis, we employed thymidine incorporation and EdU labeling, cell counts and cell death assays including activated caspase-3 immunostaining. For neurite assays, we plate neurons at low density and count the number of neurites with processes twice the length of the cell body that develop over 2 days. Finally for neuronal migration, we plate neurospheres on a matrix and calculate the distance the cell "carpet" traveled over a time period. For all these assays, results can be generated in about a week and can be easily quantified. Interestingly we have not observed person-specific effects. Instead differences in neurogenesis, neuronal migration, and neurite outgrowth have been detected in all idiopathic autism families as well as the 16p11.2 CNV NPCs, suggesting possible convergence in neurodevelopmental phenotypes.

There are of course limitations to the above approaches. Even though our in vitro assays have been extensively used in rodent neuronal cultures and validated in vivo using mouse models [108–113], the assays still do not recapitulate the three dimensionality and histological organization of the developing brain. Fortunately the development of in vitro organoids and mini-brains allows a complementary approach ([20, 115]; see chapters by Christian et al. and Lunden et al. for more information). However these systems can take months to generate and analyze. Thus a combinatorial approach might be best to thoroughly characterize developmental phenotypes. The two-dimensional higher throughput assays could be used to initially characterize phenotypes, which would then validated in the more time-consuming 3-D systems.

A significant limitation for all these studies is the workload needed to analyze the developmental phenotypes. Typically images are taken on a confocal microscope and then ImageJ or another similar program is used to quantify morphometric features such as neurite length or dendritic branching. Multiple clones from multiple affected individuals and controls need to be analyzed in a similar way. It can take weeks to months to measure one or two morphometric features. Because this analysis is so labor-intensive, measurements of other morphometric features that could be informative are not quantified. Also the number of neurons assayed in this way is limited affecting statistical power. Finally unintentional human error and bias in the analysis can also be a confound. To circumvent these issues new high content imaging, from ThermoFisher (CellInsight CX7) and Molecular Devices, is available that automates both image acquisition and data processing. In this way hundreds of neurons or cells can be assayed simultaneously for up to 40 different morphometric features in an afternoon. This type of automation, if available, could significantly decrease labor while increasing statistical power, the number of morphometric features analyzed. and productivity.

13 Pathway Analysis

The next step is to understand the underlying signaling pathways that are contributing to the autism neurodevelopmental phenotypes. To address this important question, we have combined a traditional candidate pathway approach with different unbiased omic strategies.

For our candidate approach, possible pathways that are affected in the ASD cells have been identified by adding extracellular factors (EFs) to our cultures and performing the described neurodevelopmental assays. Cells continuously respond to their environment by binding to EFs and then interpret these environmental cues by initiating signaling cascades. During brain development, these EFs can be, for example, growth factors (FGF, EGF), neurotrophins (BDNF, NGF), neurotransmitters (5HT). These EFs stimulate all aspects of development including neurogenesis, migration, and neurite outgrowth. We have asked if our autism NPCs respond differently from control NPCs to these EFs in our assays. In our extensive preliminary studies we have found that idiopathic NPCs fail to respond to numerous EFs while 16p11.2 NPCs still retain their ability to respond to EFs and the addition of EFs is sufficient to rescue the autism phenotypes.

This analysis identified candidate downstream pathways that may be affected and contribute to the autism phenotype. For instance, the idiopathic NPCs fail to respond to the EF, PACAP, which regulates neurogenesis and neuritogenesis [108, 109, 113]. PACAP signals through the AKT/mTOR and ERK pathways so we analyzed the phosphorylation status of important effectors in these pathways. In this way we found consistent defects in S6, a ribosomal protein that is downstream of mTORC1 phosphorylation. We then used available small molecules that target the AKT/mTOR pathway to rescue the autism phenotypes and to reproduce the autism phenotypes in controls, demonstrating that this pathway mechanistically contributes to the autism phenotypes.

We are also applying numerous unbiased omic approaches to investigate the molecular basis of the different ASD neurodevelopmental phenotypes. Because the phenotypes seem to have some signaling abnormalities, we decided to apply a phospho-proteomic (P-proteome) approach coupled with standard proteomics. This strategy captures in an unbiased way the phosphorylation (P) differences in the proteome as well as any protein fold differences in the proteome. P-proteomics has not been typically applied to neurodevelopmental iPSC-based studies but importantly it has the capacity to identify in an unbiased way the candidate signaling pathways that could contribute to phenotypes. It has been used successfully in SHANK3 deficient patient-derived cells to discover that CDC-like kinase 2 contributes to synaptic defects in both patient neurons and mouse models [116]. Some surprises have emerged from our studies. Few differences were observed at the proteome level but many changes were detected for the P-proteome, consistent with signaling differences underlying our autism phenotypes. We also observe a considerable overlap between the 16p11.2 and idiopathic P-proteome datasets, again supporting a possible convergence in the downstream pathways that are responsible for the neurite and migration defects observed in both datasets.

Importantly the AKT/mTOR defects that were characterized by the candidate approach were also observed in the P-proteome. Interestingly, known downstream pathways regulated by mTOR signaling such as mRNA translation and cytoskeletal regulation were enriched in the P-proteome. mRNA translation has emerged as a key player in neurodevelopment and as mentioned previously, mRNA translation has been consistently implicated in monogenic forms of ASD [17, 21, 69]. In addition manipulation of mRNA translation is sufficient to rescue autism-like phenotypes in mice [70, 71]. Possibly mRNAs that are typically translationally regulated are now dysregulated in autism. If this is the case, then new omic methodologies like Ribo-seq or polysome RNA-seq [117–119] could be used to identify these mRNAs. Finally, microtubule and actin regulation has been implicated many times in neurite development and neuronal migration [120–122] so the identification of these pathways is satisfying given our autism phenotypes. The P-proteome results are now being mined to identify individual P-protein differences that may contribute to these neurite and migration phenotypes.

We are also applying standard next-generation sequencing to identify rare nonsynonymous genetic variants that may contribute to these developmental phenotypes. Interestingly when we interrogate the next-generation sequencing data for the AKT, mTOR, and ERK pathways, some overlap with our P-proteome results is observed but most of the genetic variants seem to be situated further upstream including numerous ligands and receptors. This may explain in part the observed signaling defects, which can now be tested using standard gene editing approaches.

Other omic approaches such as single cell RNA-seq and metabolomics could also be applied to further define our autism phenotypes. Single cell RNA-seq analysis is now being used to more precisely define affected cellular subpopulations that cannot be parsed by standard approaches [123, 124]. Applying this strategy to our cells will help determine more specifically the different cellular populations that are affected in autism and if these populations are similar or different among I-ASD individuals and between I-ASD and 16p11.2 datasets. Metabolomics is currently being applied routinely to other fields like cancer biology and characterizes the effect of disease on cellular metabolites (nucleotides, lipids, sugars) [125]. Changes in metabolites are often an indication of cellular health and can reflect the ultimate outcome of the disease or disorder. Cellular metabolism has also been linked to cellular function and manipulation of metabolites can alter cellular function. For instance during iPSC reprogramming, somatic cells shift from oxidative-phosphorylation based respiration to aerobic glycolysis, and this metabolic switch is needed for efficient reprogramming [126]. In addition many of the enzymes that regulate metabolism are excellent drug targets due to their natural allosteric interactions with small molecules that regulate enzyme function. Initial studies by our group have found interesting differences in nucleotides and lipids, which is consistent with our signaling defects. Altogether these different candidate and omic approaches will help define the molecular basis of the neurodevelopmental defects.

It will be interesting as we and others use iPSC-based approaches to define autism phenotypes if convergence is noted for similar neurodevelopmental processes and signaling pathways. If consistent results are observed, then these pathways become possible targets for future therapeutic intervention.

In summary, the clinical, genetic, and developmental heterogeneity of ASD plus the developmental basis of the disorder makes it very difficult to study. Fortunately the advent of iPSC-based approaches now makes it possible to study the development of autism neurons. This in combination with new omic strategies will allow the research community to define more precisely the cell types and signaling pathways affected in autism. It will be interesting if a convergence of phenotypes is observed, and if different autism subgroups emerge from this analysis. The affected developmental processes and signaling pathways could define these subgroups, and if this is the case then the different subgroups may then warrant their own specific interventions and lead to group or personalized therapies as in cancer. Importantly the iPSC-derived NPCs or neurons can be used to test small molecule libraries or candidates for their ability to rescue phenotypes in patient derived cells. Using these combined approaches, we will hopefully over time be able to define the etiology of autism and develop new treatments for the disorder.

Acknowledgements This work was supported by the New Jersey Governor's Council for Medical Research and Treatment of Autism (CAUT13APS010; CAUT14APL031; CAUT15APL041, CAUT19APL014) and Nancy Lurie Marks Family Foundation for Dr. Millonig and Dr. DiCicco-Bloom; NJ Health Foundation (PC 63-19) for Dr. Millonig; Mindworks Charitable Lead Trust, and the Jewish Community, Foundation of Greater MetroWest, Rutgers School of Graduate Studies NJ for Dr. DiCicco-Bloom.

References

1. Baio, J., Wiggins, L., Christensen, D. L., Maenner, M. J., Daniels, J., Warren, Z., et al. (2018). Prevalence of autism spectrum disorder among children aged 8 years - autism and developmental disabilities monitoring network, 11 sites, United States, 2014. *MMWR Surveillance Summaries, 67*(6), 1–23. https://doi.org/10.15585/mmwr.ss6706a1
2. Falkmer, T., Anderson, K., Falkmer, M., & Horlin, C. (2013). Diagnostic procedures in autism spectrum disorders: A systematic literature review. *European Child and Adolescent Psychiatry, 22*(6), 329–340. https://doi.org/10.1007/s00787-013-0375-0

3. Kjelgaard, M. M., & Tager-Flusberg, H. (2001). An investigation of language impairment in autism: Implications for genetic subgroups. *Language and Cognitive Processes, 16*(2–3), 287–308.
4. Tager-Flusberg, H. (2015). Defining language impairments in a subgroup of children with autism spectrum disorder. *Science China Life Sciences, 58*(10), 1044–1052. https://doi.org/10.1007/s11427-012-4297-8
5. Charman, T., Pickles, A., Simonoff, E., Chandler, S., Loucas, T., & Baird, G. (2011). IQ in children with autism spectrum disorders: Data from the Special Needs and Autism Project (SNAP). *Psychological Medicine, 41*(3), 619–627. https://doi.org/10.1017/S0033291710000991
6. Levy, S. E., Mandell, D. S., & Schultz, R. T. (2009). Autism. *Lancet, 374*(9701), 1627–1638. https://doi.org/10.1016/S0140-6736(09)61376-3
7. Tye, C., Runicles, A. K., Whitehouse, A. J. O., & Alvares, G. A. (2019). Characterizing the interplay between autism spectrum disorder and comorbid medical conditions: An integrative review. *Frontiers in Psychiatry, 9*, 751. https://doi.org/10.3389/fpsyt.2018.00751
8. Rubenstein, J. L. (2010). Three hypotheses for developmental defects that may underlie some forms of autism spectrum disorder. *Current Opinion in Neurology, 23*, 118–123. https://doi.org/10.1097/WCO.0b013e328336eb13
9. Gao, R., & Penzes, P. (2015). Common mechanisms of excitatory and inhibitory imbalance in schizophrenia and autism spectrum disorders. *Current Molecular Medicine, 15*(2), 146–167.
10. Schaaf, C. P., & Zoghbi, H. Y. (2011). Solving the autism puzzle a few pieces at a time. *Neuron, 70*(5), 806–808. https://doi.org/10.1016/j.neuron.2011.05.025
11. Casanova, M. F., van Kooten, I. A., Switala, A. E., van Engeland, H., Heinsen, H., Steinbusch, H. W., et al. (2006). Minicolumnar abnormalities in autism. *Acta Neuropathologica, 112*(3), 287–303.
12. Packer, A. (2016). Neocortical neurogenesis and the etiology of autism spectrum disorder. *Neuroscience and Biobehavioral Reviews, 64*, 185–195. https://doi.org/10.1016/j.neubiorev.2016.03.002
13. Connacher, R. J., DiCicco-Bloom, E., & Millonig, J. H. (2018). Using human induced neural precursor cells to define early neurodevelopmental defects in syndromic and idiopathic autism. *Current Pharmacology Reports, 4*(6), 422–435. https://doi.org/10.1007/s40495-018-0155-0
14. Chao, H. T., Chen, H., Samaco, R. C., Xue, M., Chahrour, M., Yoo, J., et al. (2010). Dysfunction in GABA signalling mediates autism-like stereotypies and Rett syndrome phenotypes. *Nature, 468*(7321), 263–269. https://doi.org/10.1038/nature09582
15. Han, S., Tai, C., Westenbroek, R. E., Yu, F. H., Cheah, C. S., Potter, G. B., et al. (2012). Autistic-like behaviour in Scn1a+/− mice and rescue by enhanced GABA-mediated neurotransmission. *Nature, 489*(7416), 385–390. https://doi.org/10.1038/nature11356
16. Dölen, G., Osterweil, E., Rao, B. S., Smith, G. B., Auerbach, B. D., Chattarji, S., et al. (2007). Correction of fragile X syndrome in mice. *Neuron, 56*(6), 955–962.
17. Auerbach, B. D., Osterweil, E. K., & Bear, M. F. (2011). Mutations causing syndromic autism define an axis of synaptic pathophysiology. *Nature, 480*(7375), 63–68. https://doi.org/10.1038/nature10658
18. Hashemi, E., Ariza, J., Rogers, H., Noctor, S. C., & Martínez-Cerdeño, V. (2017). The number of parvalbumin-expressing interneurons is decreased in the prefrontal cortex in autism. *Cerebral Cortex, 27*(3), 1931–1943. https://doi.org/10.1093/cercor/bhw021
19. Blatt, G. J., & Fatemi, S. H. (2011). Alterations in GABAergic biomarkers in the autism brain: Research findings and clinical implications. *The Anatomical Record, 294*(10), 1646–1652. https://doi.org/10.1002/ar.21252
20. Mariani, J., Coppola, G., Zhang, P., Abyzov, A., Provini, L., Tomasini, L., et al. (2015). FOXG1-dependent dysregulation of GABA/glutamate neuron differentiation in autism spectrum disorders. *Cell, 162*(2), 375–390. https://doi.org/10.1016/j.cell.2015.06.034
21. Kelleher 3rd, R. J., & Bear, M. F. (2008). The autistic neuron: Troubled translation? *Cell, 135*(3), 401–406. https://doi.org/10.1016/j.cell.2008.10.017

22. Winden, K. D., Ebrahimi-Fakhari, D., & Sahin, M. (2018). Abnormal mTOR activation in autism. *Annual Review of Neuroscience, 41*, 1–23. https://doi.org/10.1146/annurev-neuro-080317-061747
23. Tang, G., Gudsnuk, K., Kuo, S. H., Cotrina, M. L., Rosoklija, G., Sosunov, A., et al. (2014). Loss of mTOR-dependent macroautophagy causes autistic-like synaptic pruning deficits. *Neuron, 83*(5), 1131–1143. https://doi.org/10.1016/j.neuron.2014.07.040
24. Marchetto, M. C., Belinson, H., Tian, Y., Freitas, B. C., Fu, C., Vadodaria, K., et al. (2017). Altered proliferation and networks in neural cells derived from idiopathic autistic individuals. *Molecular Psychiatry, 22*(6), 820–835. https://doi.org/10.1038/mp.2016.95
25. Wegiel, J., Flory, M., Kuchna, I., Nowicki, K., Ma, S. Y., Imaki, H., et al. (2014). Stereological study of the neuronal number and volume of 38 brain subdivisions of subjects diagnosed with autism reveals significant alterations restricted to the striatum, amygdala and cerebellum. *Acta Neuropathologica Communications, 2*, 141. https://doi.org/10.1186/s40478-014-0141-7
26. Bailey, A. J. (2008). Postmortem studies of autism. *Autism Research, 1*(5), 265. https://doi.org/10.1002/aur.51
27. Varghese, M., Keshav, N., Jacot-Descombes, S., Warda, T., Wicinski, B., Dickstein, D. L., et al. (2017). Hof PR autism spectrum disorder: Neuropathology and animal models. *Acta Neuropathologica, 134*(4), 537–566. https://doi.org/10.1007/s00401-017-1736-4
28. Colvert, E., Tick, B., McEwen, F., Stewart, C., Curran, S. R., Woodhouse, E., et al. (2015). Heritability of autism spectrum disorder in a UK population-based twin sample. *JAMA Psychiatry, 72*(5), 415–423. https://doi.org/10.1001/jamapsychiatry.2014.3028
29. Sandin, S., Lichtenstein, P., Kuja-Halkola, R., Hultman, C., Larsson, H., & Reichenberg, A. (2017). The heritability of autism spectrum disorder. *JAMA, 318*(12), 1182–1184. https://doi.org/10.1001/jama.2017.12141
30. Gaugler, T., Klei, L., Sanders, S. J., Bodea, C. A., Goldberg, A. P., Lee, A. B., et al. (2014). Most genetic risk for autism resides with common variation. *Nature Genetics, 46*(8), 881–885. https://doi.org/10.1038/ng.3039
31. Ozonoff, S., Young, G. S., Carter, A., Messinger, D., Yirmiya, N., Zwaigenbaum, L., et al. (2011). Recurrence risk for autism spectrum disorders: A Baby Siblings Research Consortium study. *Pediatrics, 128*(3), e488–e495. https://doi.org/10.1542/peds.2010-2825
32. Klei, L., Sanders, S. J., Murtha, M. T., Hus, V., Lowe, J. K., Willsey, A. J., et al. (2012). Common genetic variants, acting additively, are a major source of risk for autism. *Molecular Autism, 3*(1), 9. https://doi.org/10.1186/2040-2392-3-9
33. Grove, J., Ripke, S., Als, T. D., Mattheisen, M., Walters, R. K., Won, H., et al. (2019). Identification of common genetic risk variants for autism spectrum disorder. *Nature Genetics, 51*(3), 431–444. https://doi.org/10.1038/s41588-019-0344-8
34. Ramaswami, G., & Geschwind, D. H. (2018). Genetics of autism spectrum disorder. *Handbook of Clinical Neurology, 147*, 321–329. https://doi.org/10.1016/B978-0-444-63233-3.00021-X
35. Iossifov, I., Ronemus, M., Levy, D., Wang, Z., Hakker, I., Rosenbaum, J., et al. (2012). De novo gene disruptions in children on the autistic spectrum. *Neuron, 74*(2), 285–299. https://doi.org/10.1016/j.neuron.2012.04.009
36. De Rubeis, S., He, X., Goldberg, A. P., Poultney, C. S., Samocha, K., Cicek, A. E., et al. (2014). Synaptic, transcriptional and chromatin genes disrupted in autism. *Nature, 515*(7526), 209–215. https://doi.org/10.1038/nature13772
37. Sanders, S. J., He, X., Willsey, A. J., Ercan-Sencicek, A. G., Samocha, K. E., Cicek, A. E., et al. (2015). Insights into autism spectrum disorder genomic architecture and biology from 71 risk loci. *Neuron, 87*(6), 1215–1233. https://doi.org/10.1016/j.neuron.2015.09.016
38. Williams, S. M., An, J. Y., Edson, J., Watts, M., Murigneux, V., Whitehouse, A. J. O., et al. (2019). An integrative analysis of non-coding regulatory DNA variations associated with autism spectrum disorder. *Molecular Psychiatry, 24*(11), 1707–1719. https://doi.org/10.1038/s41380-018-0049-x

39. Devanna, P., Chen, X. S., Ho, J., Gajewski, D., Smith, S. D., Gialluisi, A., et al. (2018). Next-gen sequencing identifies non-coding variation disrupting miRNA-binding sites in neurological disorders. *Molecular Psychiatry, 23*(5), 1375–1384. https://doi.org/10.1038/mp.2017.30
40. Yuen, R. K., Merico, D., Cao, H., Pellecchia, G., Alipanahi, B., Thiruvahindrapuram, B., et al. (2016). Genome-wide characteristics of *de novo* mutations in autism. *NPJ Genomic Medicine, 1*, 160271–1602710.
41. Voineagu, I., & Eapen, V. (2013). Converging pathways in autism spectrum disorders: Interplay between synaptic dysfunction and immune responses. *Frontiers in Human Neuroscience, 7*, 738. https://doi.org/10.3389/fnhum.2013.00738
42. D'Angelo, D., Lebon, S., Chen, Q., Martin-Brevet, S., Snyder, L. G., Hippolyte, L., et al. (2016). Defining the effect of the 16p11.2 duplication on cognition, behavior, and medical comorbidities. *JAMA Psychiatry, 73*(1), 20–30. https://doi.org/10.1001/jamapsychiatry.2015.2123
43. Niarchou, M., SJRA, C., Doherty, J. L., Maillard, A. M., Jacquemont, S., Chung, W. K., et al. (2019). Psychiatric disorders in children with 16p11.2 deletion and duplication. *Translational Psychiatry, 9*(1), 8. https://doi.org/10.1038/s41398-018-0339-8
44. Louis, D. N., Perry, A., Reifenberger, G., von Deimling, A., Figarella-Branger, D., Cavenee, W. K., et al. (2016). The 2016 World Health Organization classification of tumors of the central nervous system: A summary. *Acta Neuropathologica, 131*(6), 803–820. https://doi.org/10.1007/s00401-016-1545-1
45. Northcott, P. A., Jones, D. T., Kool, M., Robinson, G. W., Gilbertson, R. J., Cho, Y. J., et al. (2012). Medulloblastomics: The end of the beginning. *Nature Reviews Cancer, 12*(12), 818–834. https://doi.org/10.1038/nrc3410
46. Northcott, P. A., Dubuc, A. M., Pfister, S., & Taylor, M. D. (2012). Molecular subgroups of medulloblastoma. *Expert Review of Neurotherapeutics, 12*(7), 871–884. https://doi.org/10.1586/ern.12.66
47. Northcott, P. A., Korshunov, A., Pfister, S. M., & Taylor, M. D. (2012). The clinical implications of medulloblastoma subgroups. *Nature Reviews Neurology, 8*(6), 340–351. https://doi.org/10.1038/nrneurol.2012.78
48. Kool, M., Korshunov, A., Remke, M., Jones, D. T., Schlanstein, M., Northcott, P. A., et al. (2012). Molecular subgroups of medulloblastoma: An international meta-analysis of transcriptome, genetic aberrations, and clinical data of WNT, SHH, Group 3, and Group 4 medulloblastomas. *Acta Neuropathologica, 123*(4), 473–484. https://doi.org/10.1007/s00401-012-0958-8
49. Hon, J. D., Singh, B., Sahin, A., Du, G., Wang, J., Wang, V. Y., et al. (2016). Breast cancer molecular subtypes: From TNBC to QNBC. *American Journal of Cancer Research, 6*(9), 1864–1872.
50. Rudin, C. M., Poirier, J. T., Byers, L. A., Dive, C., Dowlati, A., George, J., et al. (2019). Molecular subtypes of small cell lung cancer: A synthesis of human and mouse model data. *Nature Reviews Cancer, 19*(5), 289–297. https://doi.org/10.1038/s41568-019-0133-9
51. Chapuy, B., Stewart, C., Dunford, A. J., Kim, J., Kamburov, A., Redd, R. A., et al. (2018). Molecular subtypes of diffuse large B cell lymphoma are associated with distinct pathogenic mechanisms and outcomes. *Nature Medicine, 24*(5), 679–690. https://doi.org/10.1038/s41591-018-0016-8
52. Vogelstein, B., Papadopoulos, N., Velculescu, V. E., Zhou, S., Diaz Jr., L. A., & Kinzler, K. W. (2013). Cancer genome landscapes. *Science, 339*(6127), 1546–1558. https://doi.org/10.1126/science.1235122
53. Carmeliet, P. (2005). Angiogenesis in life, disease and medicine. *Nature, 438*(7070), 932–936.
54. Sharpe, A. H., & Pauken, K. E. (2018). The diverse functions of the PD1 inhibitory pathway. *Nature Reviews Immunology, 18*(3), 153–167. https://doi.org/10.1038/nri.2017.108
55. Taylor, A. M. R., Rothblum-Oviatt, C., Ellis, N. A., Hickson, I. D., Meyer, S., Crawford, T. O., et al. (2019). Chromosome instability syndromes. *Nature Reviews Disease Primers, 5*(1), 64. https://doi.org/10.1038/s41572-019-0113-0

56. Harada, S., & Morlote, D. (2020). Molecular pathology of colorectal cancer. *Advances in Anatomic Pathology, 27*(1), 20–26. https://doi.org/10.1097/PAP.0000000000000247
57. Yeang, C. H., McCormick, F., & Levine, A. (2008). Combinatorial patterns of somatic gene mutations in cancer. *The FASEB Journal, 22*(8), 2605–2622. https://doi.org/10.1096/fj.08-108985
58. Ciriello, G., Cerami, E., Sander, C., & Schultz, N. (2012). Mutual exclusivity analysis identifies oncogenic network modules. *Genome Research, 22*(2), 398–406. https://doi.org/10.1101/gr.125567.111
59. Vogelstein, B., & Kinzler, K. W. (2015). The path to cancer—Three strikes and you're out. *The New England Journal of Medicine, 373*(20), 1895–1898. https://doi.org/10.1056/NEJMp1508811
60. Ehninger, D., Han, S., Shilyansky, C., Zhou, Y., Li, W., Kwiatkowski, D. J., et al. (2008). Reversal of learning deficits in a Tsc2+/− mouse model of tuberous sclerosis. *Nature Medicine, 14*(8), 843–848. https://doi.org/10.1038/nm1788
61. Inoki, K., Corradetti, M. N., & Guan, K. L. (2005). Dysregulation of the TSC-mTOR pathway in human disease. *Nature Genetics, 37*(1), 19–24.
62. Sharma, A., Hoeffer, C. A., Takayasu, Y., Miyawaki, T., McBride, S. M., Klann, E., et al. (2010). Dysregulation of mTOR signaling in fragile X syndrome. *The Journal of Neuroscience, 30*(2), 694–702. https://doi.org/10.1523/JNEUROSCI.3696-09.2010
63. Bagni, C., & Zukin, R. S. (2019). A synaptic perspective of fragile X syndrome and autism spectrum disorders. *Neuron, 101*(6), 1070–1088. https://doi.org/10.1016/j.neuron.2019.02.041
64. Huber, K. M., Klann, E., Costa-Mattioli, M., & Suzanne Zukin, R. (2015). Dysregulation of mammalian target of rapamycin signaling in mouse models of autism. *The Journal of Neuroscience, 35*(41), 13836–13842. https://doi.org/10.1523/JNEUROSCI.2656-15.2015
65. Zhou, J., Blundell, J., Ogawa, S., Kwon, C. H., Zhang, W., Sinton, C., et al. (2009). Pharmacological inhibition of mTORC1 suppresses anatomical, cellular, and behavioral abnormalities in neural-specific Pten knock-out mice. *The Journal of Neuroscience, 29*(6), 1773–1783. https://doi.org/10.1523/JNEUROSCI.5685-08.2009
66. Korf, B. R., & Bebin, E. M. (2017). Neurocutaneous disorders in children. *Pediatrics in Review, 38*(3), 119–128. https://doi.org/10.1542/pir.2015-0118
67. Maloney, S. E., Chandler, K. C., Anastasaki, C., Rieger, M. A., Gutmann, D. H., & Dougherty, J. D. (2018). Characterization of early communicative behavior in mouse models of neurofibromatosis type 1. *Autism Research, 11*(1), 44–58. https://doi.org/10.1002/aur.1853
68. Molosh, A. I., Johnson, P. L., Spence, J. P., Arendt, D., Federici, L. M., Bernabe, C., et al. (2014). Social learning and amygdala disruptions in Nf1 mice are rescued by blocking p21-activated kinase. *Nature Neuroscience, 17*(11), 1583–1590. https://doi.org/10.1038/nn.3822
69. Nandagopal, N., & Roux, P. P. (2015). Regulation of global and specific mRNA translation by the mTOR signaling pathway. *Translation (Austin), 3*(1), e983402. https://doi.org/10.4161/21690731.2014.983402
70. Santini, E., Huynh, T. N., MacAskill, A. F., Carter, A. G., Pierre, P., Ruggero, D., et al. (2013). Exaggerated translation causes synaptic and behavioural aberrations associated with autism. *Nature, 493*(7432), 411–415. https://doi.org/10.1038/nature11782
71. Gkogkas, C. G., Khoutorsky, A., Ran, I., Rampakakis, E., Nevarko, T., Weatherill, D. B., et al. (2013). Autism-related deficits via dysregulated eIF4E-dependent translational control. *Nature, 493*(7432), 371–377. https://doi.org/10.1038/nature11628
72. Nicolini, C., Ahn, Y., Michalski, B., Rho, J. M., & Fahnestock, M. (2015). Decreased mTOR signaling pathway in human idiopathic autism and in rats exposed to valproic acid. *Acta Neuropathologica Communications, 3*, 3. https://doi.org/10.1186/s40478-015-0184-4
73. Rosina, E., Battan, B., Siracusano, M., Di Criscio, L., Hollis, F., Pacini, L., et al. (2019). Disruption of mTOR and MAPK pathways correlates with severity in idiopathic autism. *Translational Psychiatry, 9*(1), 50. https://doi.org/10.1038/s41398-018-0335-z

74. Sacco, R., Gabriele, S., & Persico, A. M. (2015). Head circumference and brain size in autism spectrum disorder: A systematic review and meta-analysis. *Psychiatry Research, 234*(2), 239–251. https://doi.org/10.1016/j.pscychresns.2015.08.016
75. Vaccarino, V., & Müller, S. K. (2009). Increased brain size in autism—What it will take to solve a mystery. *Biological Psychiatry, 66*(4), 313–315. https://doi.org/10.1016/j.biopsych.2009.06.013
76. Courchesne, E., Carper, R., & Akshoomoff, N. (2003). Evidence of brain overgrowth in the first year of life in autism. *Journal of the American Medical Association, 290*(3), 337–344.
77. Piven, J., Elison, J., & Zylka, M. (2017). Toward a conceptual framework for early brain and behavior development in autism. *Molecular Psychiatry, 22*, 1385–1394. https://doi.org/10.1038/mp.2017.131
78. Redcay, E., & Courchesne, E. (2005). When is the brain enlarged in autism? A meta-analysis of all brain size reports. *Biological Psychiatry, 58*(1), 1–9.
79. Carper, R. A., Moses, P., Tigue, Z. D., & Courchesne, E. (2002). Cerebral lobes in autism: Early hyperplasia and abnormal age effects. *NeuroImage, 16*(4), 1038–1051.
80. Hazlett, H. C., Poe, M. D., Gerig, G., Styner, M., Chappell, C., Smith, R. G., et al. (2011). Early brain overgrowth in autism associated with an increase in cortical surface area before age 2 years. *Archives of General Psychiatry, 68*(5), 467–476. https://doi.org/10.1001/archgenpsychiatry.2011.39
81. Parikshak, N. N., Swarup, V., Belgard, T. G., Irimia, M., Ramaswami, G., Gandal, M. J., et al. (2016). Genome-wide changes in lncRNA, splicing, and regional gene expression patterns in autism. *Nature, 540*(7633), 423–427. https://doi.org/10.1038/nature20612
82. Gandal, M. J., Zhang, P., Hadjimichael, E., Walker, R. L., Chen, C., Liu, S., et al. (2018). Transcriptome-wide isoform-level dysregulation in ASD, schizophrenia, and bipolar disorder. *Science, 362*(6420), eaat8127. https://doi.org/10.1126/science.aat8127
83. Willsey, A. J., Sanders, S. J., Li, M., Dong, S., Tebbenkamp, A. T., Muhle, R. A., et al. (2013). State MW. Coexpression networks implicate human midfetal deep cortical projection neurons in the pathogenesis of autism. *Cell, 155*(5), 997–1007. https://doi.org/10.1016/j.cell.2013.10.020
84. Parikshak, N. N., Luo, R., Zhang, A., Won, H., Lowe, J. K., Chandran, V., et al. (2013). Integrative functional genomic analyses implicate specific molecular pathways and circuits in autism. *Cell, 155*(5), 1008–1021. https://doi.org/10.1016/j.cell.2013.10.031
85. Schafer, S. T., Paquola, A. C. M., Stern, S., Gosselin, D., Ku, M., Pena, M., et al. (2019). Pathological priming causes developmental gene network heterochronicity in autistic subject-derived neurons. *Nature Neuroscience, 22*(2), 243–255. https://doi.org/10.1038/s41593-018-0295-x
86. Yoshioka, N., Gros, E., Li, H. R., Kumar, S., Deacon, D. C., Maron, C., et al. (2013). Efficient generation of human iPSCs by a synthetic self-replicative RNA. *Cell Stem Cell, 13*(2), 246–254. https://doi.org/10.1016/j.stem.2013.06.001
87. Kogut, I., McCarthy, S. M., Pavlova, M., et al. (2018). High-efficiency RNA-based reprogramming of human primary fibroblasts. *Nature Communications, 9*, 745. https://doi.org/10.1038/s41467-018-03190-3
88. Chen, G., Gulbranson, D. R., Hou, Z., Bolin, J. M., Ruotti, V., Probasco, M. D., et al. (2011). Chemically defined conditions for human iPSC derivation and culture. *Nature Methods, 8*(5), 424–429. https://doi.org/10.1038/nmeth.1593
89. Gouder, L., Vitrac, A., Goubran-Botros, H., Danckaert, A., Tinevez, J. Y., André-Leroux, G., et al. (2019). Altered spinogenesis in iPSC-derived cortical neurons from patients with autism carrying de novo SHANK3 mutations. *Scientific Reports, 9*(1), 94. https://doi.org/10.1038/s41598-018-36993-x
90. Kathuria, A., Nowosiad, P., Jagasia, R., Aigner, S., Taylor, R. D., Andreae, L. C., et al. (2018). Stem cell-derived neurons from autistic individuals with SHANK3 mutation show morphogenetic abnormalities during early development. *Molecular Psychiatry, 23*(3), 735–746. https://doi.org/10.1038/mp.2017.185

91. Evans, J. L., Saffran, J. R., & Robe-Torres, K. (2009). Statistical learning in children with specific language impairment. *Journal of Speech, Language, and Hearing Research, 52*(2), 321–335. https://doi.org/10.1044/1092-4388(2009/07-0189)
92. Williams, D., Botting, N., & Boucher, J. (2008). Language in autism and specific language impairment: Where are the links? *Psychological Bulletin, 134*(6), 944–963. https://doi.org/10.1037/a0013743
93. Bartlett, C. W., Flax, J. F., Logue, M. W., Vieland, V. J., Bassett, A. S., Tallal, P., et al. (2002). A major susceptibility locus for specific language impairment is located on 13q21. *American Journal of Human Genetics, 71*(1), 45–55.
94. Bartlett, C. W., Flax, J. F., Logue, M. W., Smith, B. J., Vieland, V. J., Tallal, P., et al. (2004). Examination of potential overlap in autism and language loci on chromosomes 2, 7, and 13 in two independent samples ascertained for specific language impairment. *Human Heredity, 57*(1), 10–20.
95. Bartlett, C. W., Flax, J. F., Fermano, Z., Hare, A., Hou, L., Petrill, S. A., et al. (2012). Gene × gene interaction in shared etiology of autism and specific language impairment. *Biological Psychiatry, 72*(8), 692–699. https://doi.org/10.1016/j.biopsych.2012.05.019
96. Walsh, K. M., & Bracken, M. B. (2011). Copy number variation in the dosage-sensitive 16p11.2 interval accounts for only a small proportion of autism incidence: A systematic review and meta-analysis. *Genetics in Medicine, 13*(5), 377–384. https://doi.org/10.1097/GIM.0b013e3182076c0c
97. Pucilowska, J., Vithayathil, J., Pagani, M., Kelly, C., Karlo, J. C., Robol, C., et al. (2018). Pharmacological inhibition of ERK signaling rescues pathophysiology and behavioral phenotype associated with 16p11.2 chromosomal deletion in mice. *The Journal of Neuroscience, 38*(30), 6640–6652. https://doi.org/10.1523/JNEUROSCI.0515-17.2018
98. Horev, G., Ellegood, J., Lerch, J. P., Son, Y. E., Muthuswamy, L., Vogel, H., et al. (2011). Dosage-dependent phenotypes in models of 16p11.2 lesions found in autism. *Proceedings of the National Academy of Sciences of the United States of America, 108*(41), 17076–17081. https://doi.org/10.1073/pnas.1114042108
99. Pucilowska, J., Vithayathil, J., Tavares, E. J., Kelly, C., Karlo, J. C., & Landreth, G. E. (2015). The 16p11.2 deletion mouse model of autism exhibits altered cortical progenitor proliferation and brain cytoarchitecture linked to the ERK MAPK pathway. *The Journal of Neuroscience, 35*(7), 3190–3200. https://doi.org/10.1523/JNEUROSCI.4864-13.2015
100. Pucilowska, J., Puzerey, P. A., Karlo, J. C., Galán, R. F., & Landreth, G. E. (2012). Disrupted ERK signaling during cortical development leads to abnormal progenitor proliferation, neuronal and network excitability and behavior, modeling human neuro-cardio-facial-cutaneous and related syndromes. *The Journal of Neuroscience, 32*(25), 8663–8677. https://doi.org/10.1523/JNEUROSCI.1107-12.2012
101. Deshpande, A., Yadav, S., Dao, D. Q., Wu, Z. Y., Hokanson, K. C., Cahill, M. K., et al. (2017). Cellular phenotypes in human iPSC-derived neurons from a genetic model of autism spectrum disorder. *Cell Reports, 21*(10), 2678–2687. https://doi.org/10.1016/j.celrep.2017.11.037
102. Yaffe, M. P., Noggle, S. A., & Solomon, S. L. (2016). Raising the standards of stem cell line quality. *Nature Cell Biology, 18*(3), 236–237. https://doi.org/10.1038/ncb3313
103. Bock, C., Kiskinis, E., Verstappen, G., Gu, H., Boulting, G., Smith, Z. D., et al. (2011). Reference maps of human ES and iPS cell variation enable high-throughput characterization of pluripotent cell lines. *Cell, 144*(3), 439–452. https://doi.org/10.1016/j.cell.2010.12.032
104. Shang, Z., Chen, D., Wang, Q., Wang, S., Deng, Q., Wu, L., et al. (2018). Single-cell RNA-seq reveals dynamic transcriptome profiling in human early neural differentiation. *GigaScience, 7*(11), giy117. https://doi.org/10.1093/gigascience/giy117
105. Kaushik, G., & Zarbalis, K. S. (2016). Prenatal neurogenesis in autism spectrum disorders. *Frontiers in Chemistry, 4*, 12. https://doi.org/10.3389/fchem.2016.00012
106. Gilbert, J., & Man, H. Y. (2017). Fundamental elements in autism: From neurogenesis and neurite growth to synaptic plasticity. *Frontiers in Cellular Neuroscience, 11*, 359. https://doi.org/10.3389/fncel.2017.00359

107. Zhang, Y., Pak, C., Han, Y., Ahlenius, H., Zhang, Z., Chanda, S., et al. (2013). Rapid single-step induction of functional neurons from human pluripotent stem cells. *Neuron, 78*(5), 785–798. https://doi.org/10.1016/j.neuron.2013.05.029
108. Yan, Y., Zhou, X., Pan, Z., Ma, J., Waschek, J. A., & DiCicco-Bloom, E. (2013). Pro- and anti-mitogenic actions of pituitary adenylate cyclase-activating polypeptide in developing cerebral cortex: Potential mediation by developmental switch of PAC1 receptor mRNA isoforms. *The Journal of Neuroscience, 33*(9), 3865–3878. https://doi.org/10.1523/JNEUROSCI.1062-12.2013
109. Lu, N., & DiCicco-Bloom, E. (1997). Pituitary adenylate cyclase-activating polypeptide is an autocrine inhibitor of mitosis in cultured cortical precursor cells. *Proceedings of the National Academy of Sciences of the United States of America, 94*(7), 3357–3362.
110. Mairet-Coello, G., Tury, A., Van Buskirk, E., Robinson, K., Genestine, M., & DiCicco-Bloom, E. (2012). p57 (KIP2) regulates radial glia and intermediate precursor cell cycle dynamics and lower layer neurogenesis in developing cerebral cortex. *Development, 139*(3), 475–487. https://doi.org/10.1242/dev.067314
111. Tury, A., Mairet-Coello, G., & DiCicco-Bloom, E. (2011). The cyclin-dependent kinase inhibitor p57Kip2 regulates cell cycle exit, differentiation, and migration of embryonic cerebral cortical precursors. *Cerebral Cortex, 21*(8), 1840–1856. https://doi.org/10.1093/cercor/bhq254
112. Mairet-Coello, G., Tury, A., & DiCicco-Bloom, E. (2009). Insulin-like growth factor-1 promotes G(1)/S cell cycle progression through bidirectional regulation of cyclins and cyclin-dependent kinase inhibitors via the phosphatidylinositol 3-kinase/Akt pathway in developing rat cerebral cortex. *The Journal of Neuroscience, 29*(3), 775–788. https://doi.org/10.1523/JNEUROSCI.1700-08.2009
113. Rossman, I. T., Lin, L., Morgan, K. M., Digiovine, M., Van Buskirk, E. K., Kamdar, S., et al. (2014). Engrailed2 modulates cerebellar granule neuron precursor proliferation, differentiation and insulin-like growth factor 1 signaling during postnatal development. *Molecular Autism, 5*(1), 9. https://doi.org/10.1186/2040-2392-5-9
114. Williams, M., Prem, S., Zhou, X., Matteson, P., Yeung, P. L., Lu, C. W., et al. (2018). Rapid detection of neurodevelopmental phenotypes in human neural precursor cells (NPCs). *Journal of Visualized Experiments, 133*, e56628.
115. Paşca, A. M., Sloan, S., Clarke, L. E., Tian, Y., Makinson, C., Huber, N., et al. (2015). Functional cortical neurons and astrocytes from human pluripotent stem cells in 3D cultures. *Nature Methods, 12*, 671–678.
116. Bidinosti, M., Botta, P., Krüttner, S., Proenca, C. C., Stoehr, N., Bernhard, M., et al. (2016). CLK2 inhibition ameliorates autistic features associated with SHANK3 deficiency. *Science, 351*(6278), 1199–1203. https://doi.org/10.1126/science.aad5487
117. Calviello, L., & Ohler, U. (2017). Beyond read-counts: Ribo-seq data analysis to understand the functions of the transcriptome. *Trends in Genetics, 33*(10), 728–744. https://doi.org/10.1016/j.tig.2017.08.003
118. Kraushar, M. L., Thompson, K., Wijeratne, H. R., Viljetic, B., Sakers, K., Marson, J. W., et al. (2014). Temporally defined neocortical translation and polysome assembly are determined by the RNA-binding protein Hu antigen R. *Proceedings of the National Academy of Sciences of the United States of America, 111*, E3815–E3824.
119. Kraushar, M. L., Viljetic, B., Wijeratne, H. R., Thompson, K., Jiao, X., Pike, J. W., et al. (2015). Thalamic WNT3 secretion spatiotemporally regulates the neocortical ribosome signature and mRNA translation to specify neocortical cell subtypes. *The Journal of Neuroscience, 35*, 10911–10926.
120. Da Silva, J. S., & Dotti, C. G. (2002). Breaking the neuronal sphere: Regulation of the actin cytoskeleton in neuritogenesis. *Nature Reviews Neuroscience, 3*(9), 694.
121. Kawauchi, T., Chihama, K., Nabeshima, Y. I., & Hoshino, M. (2006). Cdk5 phosphorylates and stabilizes p27 kip1 contributing to actin organization and cortical neuronal migration. *Nature Cell Biology, 8*(1), 17.

122. Teng, J., Takei, Y., Harada, A., Nakata, T., Chen, J., & Hirokawa, N. (2001). Synergistic effects of MAP2 and MAP1B knockout in neuronal migration, dendritic outgrowth, and microtubule organization. *The Journal of Cell Biology, 155*, 65–76.
123. Bardy, C., Van Den Hurk, M., Kakaradov, B., Erwin, J. A., Jaeger, B. N., Hernandez, R. V., et al. (2016). Predicting the functional states of human iPSC-derived neurons with single-cell RNA-seq and electrophysiology. *Molecular Psychiatry, 21*(11), 1573.
124. Kim, D. H., Marinov, G. K., Pepke, S., Singer, Z. S., He, P., Williams, B., et al. (2015). Single-cell transcriptome analysis reveals dynamic changes in lncRNA expression during reprogramming. *Cell Stem Cells, 16*(1), 88–101.
125. Serkova, N. J., & Glunde, K. (2009). Metabolomics of cancer. *Methods in Molecular Biology, 520*, 273–295. https://doi.org/10.1007/978-1-60327-811-9_20
126. Shyh-Chang, N., Daley, G. Q., & Cantley, L. C. (2013). Stem cell metabolism in tissue development and aging. *Development, 140*(12), 2535–2547. https://doi.org/10.1242/dev.091777

Dysregulation of Neurite Outgrowth and Cell Migration in Autism and Other Neurodevelopmental Disorders

Smrithi Prem, James H. Millonig, and Emanuel DiCicco-Bloom

1 Overview

Despite decades of study, elucidation of the underlying etiology of complex developmental disorders such as autism spectrum disorder (ASD), schizophrenia (SCZ), intellectual disability (ID), and bipolar disorder (BPD) has been hampered by the inability to study human neurons, the heterogeneity of these disorders, and the relevance of animal model systems. Moreover, a majority of these developmental disorders have multifactorial or idiopathic (unknown) causes making them difficult to model using traditional methods of genetic alteration. Examination of the brains of individuals with ASD and other developmental disorders in both post-mortem and MRI studies shows defects that are suggestive of dysregulation of embryonic and early postnatal development. For ASD, more recent genetic studies have also suggested that risk genes largely converge upon the developing human cerebral cortex between weeks 8 and 24 in utero. Yet, an overwhelming majority of studies in autism rodent models have focused on postnatal development or adult synaptic transmission

S. Prem
Graduate Program in Neuroscience, Rutgers University, Piscataway, NJ, USA

Department of Neuroscience and Cell Biology, Robert Wood Johnson Medical School, Rutgers University, Piscataway, NJ, USA
e-mail: premsm@rwjms.rutgers.edu

J. H. Millonig
Department of Neuroscience and Cell Biology, Center for Advanced Biotechnology and Medicine, Rutgers Robert Wood Johnson Medical School, Rutgers University, Piscataway, NJ, USA
e-mail: Millonig@cabm.rutgers.edu

E. DiCicco-Bloom (✉)
Department of Neuroscience and Cell Biology/Pediatrics, Rutgers Robert Wood Johnson Medical School, Rutgers University, Piscataway, NJ, USA
e-mail: diciccem@rwjms.rutgers.edu

E. DiCicco-Bloom, J. H. Millonig (eds.), *Neurodevelopmental Disorders*, Advances in Neurobiology 25, https://doi.org/10.1007/978-3-030-45493-7_5

defects in autism related circuits. Thus, studies looking at early developmental processes such as proliferation, cell migration, and early differentiation, which are essential to build the brain, are largely lacking. Yet, interestingly, a few studies that did assess early neurodevelopment found that alterations in brain structure and function associated with neurodevelopmental disorders (NDDs) begin as early as the initial formation and patterning of the neural tube. By the early to mid-2000s, the derivation of human embryonic stem cells (hESCs) and later induced pluripotent stem cells (iPSCs) allowed us to study living human neural cells in culture for the first time. Specifically, iPSCs gave us the unprecedented ability to study cells derived from individuals with idiopathic disorders. Studies indicate that iPSC-derived neural cells, whether precursors or "matured" neurons, largely resemble cortical cells of embryonic humans from weeks 8 to 24. Thus, these cells are an excellent model to study early human neurodevelopment, particularly in the context of genetically complex diseases. Indeed, since 2011, numerous studies have assessed developmental phenotypes in neurons derived from individuals with both genetic and idiopathic forms of ASD and other NDDs. However, while iPSC-derived neurons are fetal in nature, they are post-mitotic and thus cannot be used to study developmental processes that occur before terminal differentiation. Moreover, it is important to note that during the 8–24-week window of human neurodevelopment, neural precursor cells are actively undergoing proliferation, migration, and early differentiation to form the basic cytoarchitecture of the brain. Thus, by studying NPCs specifically, we could gain insight into how early neurodevelopmental processes contribute to the pathogenesis of NDDs. Indeed, a few studies have explored NPC phenotypes in NDDs and have uncovered dysregulations in cell proliferation. Yet, few studies have explored migration and early differentiation phenotypes of NPCs in NDDs. In this chapter, we will discuss cell migration and neurite outgrowth and the role of these processes in neurodevelopment and NDDs. We will begin by reviewing the processes that are important in early neurodevelopment and early cortical development. We will then delve into the roles of neurite outgrowth and cell migration in the formation of the brain and how errors in these processes affect brain development. We also provide review of a few key molecules that are involved in the regulation of neurite outgrowth and migration while discussing how dysregulations in these molecules can lead to abnormalities in brain structure and function thereby highlighting their contribution to pathogenesis of NDDs. Then we will discuss whether neurite outgrowth, migration, and the molecules that regulate these processes are associated with ASD. Lastly, we will review the utility of iPSCs in modeling NDDs and discuss future goals for the study of NDDs using this technology.

2 Early Neurodevelopment

A developing embryo starts out as a fertilized egg which, following multiple rounds of division that produces a "ball of cells" (morula), cavitates to form a blastocyst. The blastocyst is composed of an inner cell mass of pluripotent cells which ultimately becomes the embryo/fetus and a layer of outer trophoblast cells which later constitutes the placenta. The inner cell mass of pluripotent cells continues to divide and ultimately differentiates to form 3 "germ" layers known as the endoderm, mesoderm, and ectoderm [1, 2]. The ectoderm, which is the outer most layer of the embryo, gives rise to the entire nervous system of a vertebrate organism. During the process of neurulation, the ectoderm thickens to form the neural plate which, following further thickening and elevation of its edges, fuses in the midline to form the neural tube. The rostral neural tube becomes the brain and the caudal neural tube becomes the spinal cord. The rostral portion of the neural tube then rapidly expands to form three primary brain vesicles. These vesicles ultimately enlarge and segment to form the forebrain, midbrain, and hindbrain, and all the structures within these brain regions. The formation of the neural tube occurs approximately mid-gestation in rodents (E10.5-11 in rats and E9-9.5 mice) and during early gestation in humans (3–4 weeks) [3–5].

3 Cortical Neurodevelopment

The adult neocortex is characterized by a complex 6-layered architecture containing projection neurons and interneurons which can be distinguished by specific molecular markers and distinct axonal projections [6]. The deep layers of the cortex (V, VI) have axons that project subcortically while the superficial layers (I-IV) are composed of neurons that project within the same hemisphere or to the contralateral cortical hemisphere. As the neural tube is closing during development, the layer of neuroepithelial cells (NECs) that is closest to the hollow part of the tube (which ultimately becomes the ventricles) forms the proliferative ventricular zone (Fig. 1). These NECs divide symmetrically to form more NECs which establish the

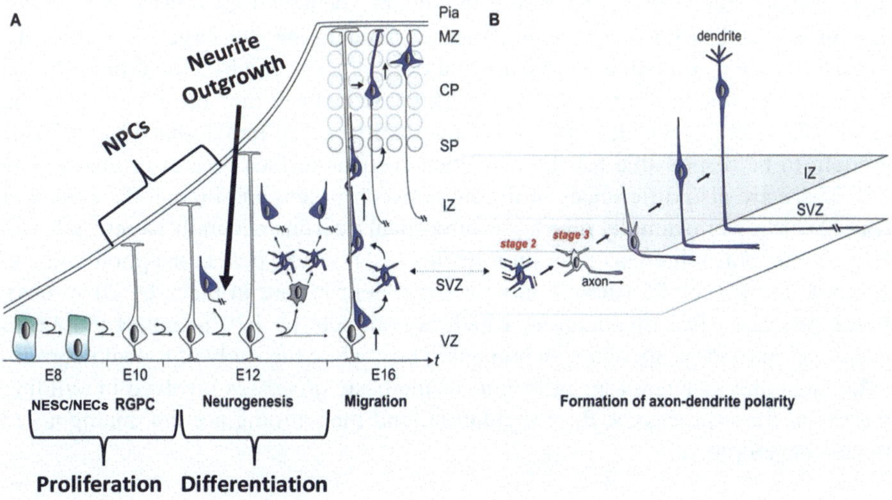

Fig. 1 Stages of cortical neurodevelopment [11]

progenitor pool needed to form the billions of neurons and glia that comprise the brain [7, 8]. In fact, limiting division of these NECs greatly reduces the size and thickness of the cortex [9]. Around E10.5-E11.5 in the mouse, NECs differentiate into another progenitor cell type known as apical radial glia [10]. Radial glial cells are distinguished by the expression of the transcription factor PAX-6 and unlike NECs, they divide asymmetrically to produce a proliferating radial glia and an immature neuron or a radial glia and an intermediate progenitor cell (IPC). The newly produced IPCs move into the subventricular zone and continue to divide while the earliest neurons migrate further to form the 6 cortical layers in an inside-out manner. That is, deeper layers of the cortex form first and then newly formed neurons migrate through already established layers to form the superficial layers of the cortex. While migrating, immature neurons extend and retract processes known as neurites which ultimately become axons and dendrites. These axons and dendrites allow for neurons to connect and communicate cortically and subcortically. Once the neurons have reached their destination, they continue to mature, form synapses, and become electrically active to ultimately produce the pyramidal excitatory neurons of the neocortex. While steps such as the early formation of neurons and neural migration largely occur in utero, many steps of cortical development such as the production of glia, synapse formation, and later synaptic pruning continue to occur postnatally. Interneuron production occurs in different subcortical regions such as the medial, lateral, and caudal ganglion eminences; however, these cells also migrate to reach the cortex.

3.1 Unique Feature of Human Neocortical Development

The literature reviewed above pertains to the rodent neocortex. While rodent and human neurodevelopment are similar, human neurodevelopment is lengthier and more complicated [12, 13]. Moreover, in humans, the neocortex is 90% of the cortical surface and is the brain region that has undergone the largest evolutionary expansion when compared to rodents and even other apes [14]. In addition to the VZ/SVZ, humans have an additional neurogenic layer in the cortex known as the outer SVZ (OSVZ) which contains another distinct set of radial glia. The OSVZ is thought to be responsible for the expanded cortical surface area in primates [15–17]. There are also differences in timing between rodent and human development (Fig. 2) [18]. For example, migration of cortical neurons occurs between E19 and E22 in rats, but in humans migration begins at 18 weeks post-conception and can continue until week 36 (though most cortical neurons are in place by 24 weeks). Some processes like myelination, which is complete by P20 in rats, continue to occur well past 20 years of age in humans. Thus, while the study of rodent neurodevelopment and its alterations inform us of the basic processes involved in building the brain, these processes, their regulation, and their timing are not analogous to human development.

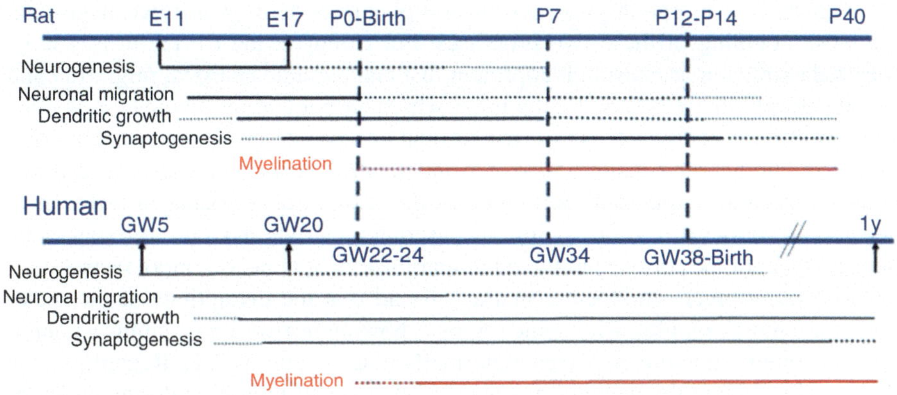

Fig. 2 Time windows of important developmental processes in human and rat [18]

4 Migration and Neurite Outgrowth in the Developing Cortex

The proper layered structure of the cortex is dependent upon migration of immature neurons from the ventricular and subventricular zones to the appropriate layers. Moreover, the elaboration and retraction of early neurites occurs while neurons are migrating. Studies have found that there are multiple modes of migration and multiple morphologies of neurons in the developing cortex [19–22]. One mode of migration, known as somal translocation occurs during early corticogenesis to form the deeper layers of the cortex. During somal translocation, neurons extend a leading process (neurite) into a region of the developing cortex known as the marginal zone above the VZ/SVZ. Shortening and lengthening of the leading processes allows the soma of new neurons to move into the deep layers of the cortex. Later, as the neocortex becomes thicker, neurons migrate by locomotion. In this process, as newborn neurons start to move out of the VZ, they begin to extend and retract multiple processes dynamically while their soma "wanders" around. These multipolar cells then become bipolar and use fibers of radial glia to migrate past the deep layers to form the superficial layers. While migration is occurring, cells also determine their neurite polarization—that is specification of the axon and dendrite. Interestingly, we see that alterations in neurite outgrowth are occurring while cells are migrating showing the intimate coupling of migration to differentiation/neurite outgrowth in the developing brain.

4.1 Regulation of Neurite Outgrowth and Migration by Extracellular Factors

Both migration and neurite outgrowth are regulated by numerous extracellular factors (EF), by adhesion molecules like cadherins, and by cytoskeleton regulating molecules [23]. Moreover, mutations in regulatory molecules that alter migration often change the morphology or neurite complexity of neurons, further illustrating the close coupling of these two processes. For example, the EF reelin is a large secreted extracellular matrix glycoprotein that can regulate neuronal migration and neurite morphology [24–26]. Reeler mice, which have mutations in the Reelin gene, show multiple defects such as inverted cortical laminar structure, abnormal orientation of cell bodies, and changes in axonal and dendrite projection. Classically, it was believed that the "inside-out" formation of the cortex did not occur in reeler mice because newborn neurons fail to migrate past older neurons and therefore stay in the deeper layers of the cortex leading to laminar inversion. However, more recent studies with more precise molecular markers found that the disruptions of the reeler cortex are more complex with some regions showing mirror-image laminar phenotype and others showing dispersed abnormally placed neurons [27]. Regardless, the deficiencies seen in the reeler cortex are due to errors of migration during development. In addition to changes in laminar structure, alterations in reelin change the cellular polarization of migrating neurons, which ultimately leads to dendritic abnormalities in the brain [26, 28]. Interestingly, defects seen in the reeler mice can also be caused by mutations in APOE2 and Dab1 both of which are downstream mediators of reelin signaling to the nucleus. This suggests that disruptions in molecules that convey EF signals to the nucleus can also lead to errors of migration and neurite outgrowth [29–31].

Another EF, IGF-1, is most known for its role in regulation of NEC and NPC proliferation [32]. Mice that are knockouts for the IGF-1R often die at birth and are found to have significantly reduced brain sizes with "gross morphological abnormalities" [33]. The presence of these morphological abnormalities is indicative of potential migration or neurite defects in these animals. A recent study found that IGF-1 is involved in the regulation of neuronal polarity and migration in the cerebral cortex [34]. When E15 embryos were electroporated with shRNA against IGFR-1, their neurons failed to migrate to the upper cortical layers and accumulated at the VZ/SVZ. Moreover, neurons lacking the IGF-1R were stuck in the multipolar morphology leading to the formation of highly disorganized tissue. With further differentiation, the cells lacking IGF-1R were also unable to form axons indicating that the fidelity of early neurite outgrowth is important for neuronal polarization. Thus, we see that molecules such as IGF-1, that are traditionally known to regulate proliferation, also regulate neurite outgrowth and migration. Later in development, IGF-1 also regulates the formation of synapses. This suggests that certain molecules and their mediating receptors have multiple functions throughout development. As seen with reelin, pathways downstream of IGF-1 such as mTOR and ERK are important regulators of development and an abundance of literature demonstrates that disruption of these pathways has also been known to alter neurite outgrowth and migration in the developing brain.

4.2 Regulation of Neurite Outgrowth and Migration by Adhesion Molecules

Changes in cell surface adhesive properties are extremely important throughout neurodevelopment and regulate everything from the initial segregation and shaping of neural tissue in the ectoderm, to the migration of neuroblasts, to the elongation of axons, to modulating interactions between differentiated neurons to form synapses [35]. There are three broad categories of adhesion molecules including calcium dependent adhesive molecules mediated by cadherins, calcium independent systems modulated by the CAMs, and extracellular matrix proteins like laminin and fibronectin. Of these, the cadherins have been more extensively studied in neurodevelopment (particularly in mammalian systems). N-cadherin is the classical cadherin subtype found in the nervous system [36]. N-cadherin expression is switched on during neurulation and is believed to be important for the segregation of the neuroectoderm from the epithelium. Cadherins play a central role in the balance of NPC maintenance and differentiation as many processes depend on the appropriate assembly and disassembly of cadherin-mediated adhesions. While N-cadherin knockouts die in the early embryonic stage, conditional (Nestin-Cre) knockout of N-cadherin in NPCs results in grossly disorganized brains with no proper cortical lamination and disorganized scattered radial glia and intermediate progenitor cells across the cortex [37]. The loss of N-cadherin results in a failure of radial glial cells to extend processes from the apical surface to the basal lamina of the cortical layer. This radial glial process is used by newly formed neuronal cells to migrate across the cortex [37]. Moreover, even when radial glial processes are present, the loss of N-cadherin also leads to a failure of migrating neuroblasts to attach to the radial glia thereby disrupting proper migration [38]. Thus, disruption in cadherins can lead to improper cell population layering within the cortex. The few studies of early neurite outgrowth have been conducted in vitro or in zebrafish and chicks and have found that N-cadherin promotes neurite extension in the brain and in retinal ganglion cells. There are, however, multiple studies showing the central roles of N-cadherin in specification of neuronal polarity and axon patterning in post-mitotic neurons [39–44]. Cadherins are also linked to β-catenin which connects the cadherin molecules to the cell's actin cytoskeleton [45]. Moreover β-catenin is a central signaling molecule in the WNT pathway, which is commonly implicated in neurodevelopmental disorders. Alterations of β-catenins or the association of cadherins with catenins leads to increased neuronal production, morphological defects in brain regions such as failure of anterior commissure to cross the midline, and alterations in cortical cytoarchitecture, hippocampal neuronal morphology and organization, and Purkinje cell migration [46]. Unsurprisingly, both cadherins and catenins are involved in later neurodevelopment including synapse formation and dendrite outgrowth [47]. Ultimately, we find that adhesion is incredibly important to the normal regulation of neurodevelopment from early neurulation all the way to synaptic plasticity.

4.3 Regulation of Neurite Outgrowth and Migration by the Cytoskeleton

In addition to EFs, signaling molecules, and adhesion molecules, the cytoskeleton and cytoskeletal regulators are major players in the orchestration of neurodevelopment. In general, neurons are dependent on the proper function of the microtubule (MT), actin, and intermediate filament based cytoskeleton for many different processes during embryonic neurodevelopment including proliferation, migration, axon guidance, and synapse formation [48]. As such, there are numerous molecules that regulate the cytoskeleton and disruptions in these molecules or the cytoskeleton itself lead to a variety of different neurodevelopmental disorders. Another important thing to note is that neurons are unique in that while migrating they also maintain a dynamic leading process. Thus, a neuron needs to coordinate the extension and branching of their leading processes, the movement of the cell with axon specification, and the attachment and recycling of adhesion proteins. All of this require the coordination between different cytoskeletal elements. We will focus on the microtubule (MT) and actin cytoskeleton and their regulators as these molecules have been shown to have diverse and well-established roles in neurodevelopment.

4.4 The Role of Microtubules and Their Regulators in Neurodevelopment

MTs are dynamic, polarized intracellular filaments formed by the dimerization of globular alpha and beta tubulin proteins. The MTs play a key role in many cellular processes such as mitosis, centrosome dynamics, and cargo transport within the cells. They also participate in many of the large-scale morphogenetic events that shape the embryo including the formation of the neural tube [49]. In early neurogenesis, MTs are essential for orchestrating normal proliferation of radial glia and NECs by forming mitotic spindles, aiding in the separation of sister chromatids, and by playing an active role in cytokinesis by regulating interkinetic nuclear migration—a unique feature of radial glia and NECs [50]. Upon exit from the cell cycle, MTs are essential for the regulation of neural migration by two major mechanisms: (1) surrounding the nuclei of neuroblasts to generate the energy needed to propel the nucleus forward and (2) moving cytoplasm forward with dynamic depolarization and repolarization. Likewise, MTs and their motor proteins (like kinesin and dynein) are some of the major players which influence the formation of neurites by creating bundles that provide mechanical force to push the cytoplasm in multiple directions. Neurites with more stable MTs end up becoming axons. Indeed, the use of the MT stabilizing drug taxol leads to the formation of multiple axons in developing neurons.

Interestingly, the tubulin subunits of MTs on their own also have important roles in the regulation of neurodevelopment [51]. One α-tubulin isoform, Tuba1a, has been associated with numerous disorders of brain development and has been linked with defects in migration and neurite outgrowth. A study by Belvindrah et al. found that mice that had missense mutations in Tuba1a had neurons accumulate in the SVZ and glial cell dispersion along the rostral migratory stream [52]. Live imaging of the TubA1A mutant neurons revealed slowed migration and increased neuronal branching which correlated with perturbed nucleus-centrosome coupling. Further studies have found that alterations in TubA1A also lead to impairment of dynein motor protein binding to MTs which further alters neuronal function [53]. More recently, a human iPSC study assessed neurospheres derived from a patient with a missense TubA1A mutation and found that radial extension of glial fibers and neurites were impaired and that neurites were shorter when compared to control cells [54]. Ultimately, we see that the tubulin molecules and the MT cytoskeleton itself are fundamental to regulate basic developmental processes.

The normal function of the MT cytoskeleton relies on multiple microtubule associated proteins including doublecortin (DXC), Lis1, and microtubule associated proteins (MAPs), which are essential for ensuring the normal progression of migration and neurite outgrowth [48, 55, 56]. DCX is well known for its importance in regulating cortical neuroblast migration [57, 58]. DCX is thought to bind to microtubules in migrating cells to promote their movement. The lack of DCX leads to improper formation of cortical layers and the accumulation of multipolar cells in the cortex of rats. On the other hand, overexpression of DCX leads to an overproduction of cells with bipolar morphology along with altered lamination of the cortex [59, 60]. More recently, studies in NPCs derived from human embryonic stem cells show that DCX overexpression leads to a significant decrease in proliferation, an increase in migration, and an increase in fiber growth showing the important role of DCX in proliferation, migration, and neurite outgrowth in both humans and rodents [61]. Likewise, LIS1 is also well known for its role in regulating migration. RNAi experiments targeting LIS1 prevent or slow down the migration of neurons from the VZ and SVZ leading to accumulation of neurons in these compartments [62]. Knockdown of LIS1 also led to the accumulation of multipolar neurons. Another important regulator of the MT cytoskeleton are the MAP proteins which can be divided into three distinct groups including MAP1, MAP2, and Tau. The MAP proteins generally have functional redundancy and work synergistically. For example, Takei et al. found that mice that are knockouts for MAP1B have hypoplastic commissural axon tracts and disorganized neuronal layering suggesting a role for MAP1B in neurodevelopment. Mice that were double mutants for MAP1B and Tau showed more severe phenotypes and cultured primary neurons from double mutants showed inhibited axonal elongation and delayed migration indicating the synergistic function of these two MAP proteins [63]. Likewise, MAP2 and MAP1B also act synergistically to regulate neuronal migration, neurite outgrowth, and microtubule organization [64]. Surprisingly, MAP2 deficient mice were found not to have any gross abnormalities, indicating functional redundancy. However, mice negative for both MAP1B and MAP2 died in the perinatal period and had fiber tract malformation and disrupted cortical patterning caused by impaired migration [64]. Analysis of primary hippocampal cultures from these double mutants showed inhibited microtubule bundling and impaired neurite elongation which ultimately impacted dendrites more than axons. Thus, microtubule regulators are essential for the regulation of tubulin dynamics and their absence leads to cortical defects reminiscent of several NDDs.

4.5 Actin and Actin Regulators in Neurodevelopment

The actin cytoskeleton and its regulators also play just as significant a role in neurodevelopment as MTs. The polymerization of actin has multiple functions ranging from assisting with subcellular localization of proteins important for polarization, to generation of motive forces, interaction with cell adhesion molecules, alteration of membrane shape, and intracellular protein trafficking. Actin is best known for producing protrusive and contractive forces through coordinated polymerization and thus, in neurodevelopment, its best-defined roles are in neural migration and neurite generation. During migration, translocation of the nucleus helps move the cell forward. It is believed that nuclear translocation requires the traction created by the microtubule cage tethered to the centrosome and the contractile force created by actin-myosin complexes. The role of actin in neuronal migration was first demonstrated by Rivas and Hatten who used time-lapse images of cerebellar granule neurons to show that inhibition of actin depolymerization caused a loss of motility in the leading process and halted the movement of these neurons. Moreover, dysregulations in the actin cytoskeleton have been shown to lead to heterotopias or cortical disorganization that is associated with failures in migration. Thus, we see how a failure in extension of the leading process (neurite) leads to the disruption of migration. Neurites are largely composed of microtubules. However, extension and navigation of neurites is normally driven by actin-rich growth cones which are found at the tips of neurites [63, 65]. Yet, neurons treated with actin depolymerizing drugs are still capable of elongation. These neurons, however, cannot respond to guidance cues and can become misrouted or form axonal loops [66, 67]. Thus, actin dynamics seem to be particularly important for growth cone exploration of the local environment.

One of the key regulators of actin dynamics are Rho GTPases such as Rac1, RhoA, Rnd2, and cdc42 [68]. Of the Rho GTPases, Rac1, Rnd2, and Rnd3 are the ones with specific function in the control of migration and neurite extension in the developing cortex. Rac1 regulates leading process formation, Rnd2 mediates the multi to bipolar transition required for migration, and Rnd3 is important for the nuclear centrosome coupling during locomotion [69–72]. The deletion of Rac1 using either FoxG1-Cre or Emx1-Cre delays radial migration in mice [69, 70, 73]. On the other hand, using Rac1 shRNA in E13 mice blocks radial migration and disrupts the formation of the leading process in cortical neurons [74]. Neurons derived from FoxG1-Cre and Emx-1 CRE Rac1 mice do not seem to have a defect in neuritogenesis as cultured cells can extend processes [73, 75]. However, the brains of both these mutant mice have an absent anterior commissure and have axons in the corpus callosum and hippocampal commissure that fail to cross midline [73, 75]. This suggests that Rac1 likely controls axon guidance rather than neurite outgrowth. The loss of another Rho GTPase, Rnd2 leads to a failure of neurons to leave the IZ. Moreover, these neurons also seem to be stuck in the multipolar morphological phase [71, 72]. Rnd2 is thought to regulate neuronal migration by inhibiting RhoA. Indeed, studies suggest that inhibition of RhoA is imperative to promote radial migration and neurite extension/polarization in cortical pyramidal neurons [76, 77]. Yet, while Emx-1 Cre RhoA mice have migration defects, the KO neurons migrate and extend processes normally when they are transplanted into a wild-type environment [78]. Further analyses showed that in the KO background, abnormalities in radial glia contributed to the defects of migration in the KO mice. Moreover, the normal migration of the KO neurons in a wild-type environment is thought to be because the related GTPases RhoB and RhoC could substitute for the lack of RhoA in KO mice thereby normalizing migration [79]. RhoA has also been shown to play an important role in regulating dendritic length and inhibition of RhoA in adult mice NSCs and PC12 cells lead to an increase in neurite outgrowth [80–83]. Lastly, the loss of RhoGTPase Cdc42 in migrating cortical neurons leads to a mild impairment of radial migration (fewer neurons stuck in IZ than in Rac1 mutants) [70]. However, Cdc42 seems to be required for efficient axonal growth and for the establishment of axonal polarity [84]. Indeed, Nestin-Cre Cdc42 KO mice showed a global reduction of axonal tracts and decreases in dendrite branching and complexity in layer II/III pyramidal neurons [85, 86].

In the end, we see that changes in EFs, cellular adhesions systems, and cytoskeleton can all impact the specification and structure of early neurites that becomes dendrites and axons in mature neurons. It is also evident that migration and neurite outgrowth are often regulated in conjunction and alterations in either process can alter cortical organization and cytoarchitecture.

5 NDDs that Result from Disrupted Migration and Neurite Outgrowth

Thus far, we have seen that disruptions in molecules that regulate neurodevelopment such as EFs, adhesion molecules, and cytoskeletal regulators lead to abnormal brain architecture in rodents due to disruption of basic developmental processes. Thus, it would be expected that mutations in these genes or aberrancies in these regulatory proteins in humans would also lead to neurodevelopmental disorders or fetal demise. Indeed, there are some very severe NDDs that are caused by mutations of genes that regulate neurites and migration during early development [87, 88]. One such disorder, lissencephaly, leads to a brain that is "smooth" and with reduced or absent sulci and gyri. Lissencephalic brains also lack the traditional 6-layered cortical structure and often have a thickened four-layer cortex. Lissencephaly is thought to be caused by the failure of neurons to reach proper positions leading to a disorganized cortex and it is linked to mutations in several genes associated with microtubule regulation, the most classic being Lis1 [89, 90]. Mutations in microtubule (MT) regulators DCX and TubA1A also lead to lissencephaly [91]. DCX is located on the X-chromosome and as such, its mutation in males leads to lissencephaly but a milder phenotype is found in females with a single mutant DCX. While Lis1 mutations lead to a more severe phenotype in the posterior brain regions, DCX mutations show a more severe phenotype in the frontal brain regions. TubA1A mutations are more similar to Lis1 mutations with posterior predominance. However, the TubA1A subtype of lissencephaly has the unique finding of dysgenesis of the anterior limb of the internal capsule suggesting altered neurite outgrowth and/or pathfinding. Mutations in Lis1 and DCX are thought to account for about 75% of lissencephaly cases while TubA1A is thought to account for 4% of cases. Mutations in TubA1A are also associated with polymicrogyria which is characterized by excess cortical folding and malformations of cortical layering. Thus, we find that the alterations in neurodevelopmental processes caused by tubulin associated protein dysregulation lead to devastating neurological disorders in humans.

Mutations in other regulatory molecules such as, reelin, IGF-1, and actin binding proteins also alter the brain and lead to developmental disorders [92]. Interestingly, the EF reelin is also associated with lissencephaly but is also accompanied by cerebellar hypoplasia and hippocampal abnormalities [93, 94]. Similarly, IGF-1 deficiencies have been associated with cerebellar hypoplasia and microcephaly [95]. Mutations in actin binding proteins such as Filamin A have been associated with the cortical malformation, periventricular heterotopia, which is characterized by reductions in brain size, ectopic clusters of neurons along the lateral ventricle, abnormal connectivity of neurons, and symptoms such as epilepsy and dyslexia [96, 97]. Mutations in actin genes (ActB and ActG1) lead to Baraitser–Winter syndrome whose main brain abnormality is pachygyria (reduced cortical gyrations) and abnormal cortical layering [98, 99]. Again, we see that disruption of the very molecules that are involved in regulating basic developmental processes can lead to brain abnormalities in humans.

6 Alterations in Migration and Neurite Outgrowth in ASD

In Sect. 5 above, we see that dysregulation of neurite outgrowth and migration can lead to some rare but severe neurodevelopmental disorders like lissencephaly. Yet, more common NDDs like ASD, ID, SCZ often exhibit less severe defects and are not characterized by gross abnormalities in brain structure. Yet, for all these NDDs, genetic and animal studies suggest that altered neurodevelopment plays a role. Thus, it is possible that neurite outgrowth, migration, and the molecules that regulate these processes may be abnormal in ASD. In the following section, we will explore evidence of disruptions of these developmental processes in human studies of ASD.

Post-mortem studies are incredibly important for taking an in-depth microscopic and macroscopic look at the brain [100]. However, unlike disorders such as SCZ and Alzheimer's, the availability of post-mortem tissue has been severely limited for ASD. In fact, from 1980 to 2010 only about 100 post-mortem ASD brains were analyzed [101]. Over the years, many studies have observed alterations in the brains of individuals with ASD. The most consistent differences have been observed in the cerebellum, where almost 60% of the brains studied to date have shown some sort of cerebellar pathology such as reduced cell numbers, altered cell size, or altered dendrites [101–107]. Changes in regions of the brain such as the amygdala, hippocampus, striatum, and brainstem have also been observed in ASD. However, alterations in these regions are not always replicated or consistent across groups [108]. Some groups have also found altered cell numbers, presence of disorganized patches of neuronal cells, and changes in cell packing density in the cerebral cortex of individuals with ASD. Yet, about half of ASD brains analyzed to date do not have cortical alterations [104, 108, 109]. Regardless, all these studies indicate that the defects seen in the ASD brain are likely caused by alterations in basic developmental processes. This means that the pathogenetic mechanisms that contribute to ASD act way before birth, during the embryonic and fetal periods, and cannot occur only due to postnatal events. For example, the changes in brain size and cell numbers observed in ASD can be caused by changes/abnormalities/alterations in cell proliferation, cell death, or synapse formation. A study by Wegiel et al. found subcortical, periventricular, hippocampal, and cerebellar heterotopias in 4 out of 13 brains studied, suggesting abnormal neuronal migration. Likewise, cerebral dysplasias were found in the neocortex of 4 brains while cerebellar dysplasias were observed in 12 out of 13 of the ASD brains studied which reflect multi-regional dysregulation of neurogenesis, neuronal migration, and maturation in ASD [110]. Likewise, post-mortem studies in SCZ patients (even samples derived from unmedicated patients) have shown evidence of abnormally positioned cell clusters in the entorhinal cortex and other regions, decreased size and arborization of hippocampal and neocortical pyramidal neurons, and decreases in number of cells in regions such as the thalamus, all of which again suggest dysregulation in neurogenesis, migration, and neuronal maturation [111].

Unlike post-mortem studies, imaging techniques give us insight into the live brain, allow us to collect both structural and functional data, and provide the opportunity to monitor and compare developmental progression of disorders. Structural data on ASD brains has largely been derived from MRI. Again, much like the post-mortem studies, there seems to be no consistent defect that has been uncovered in all cases of ASD. Rather, some common patterns are observed in ASD brains, though none are pathognomonic. One commonly reported finding is increased brain volume from ages 2 to 4 that seems to occur in both grey and white matter [112–115]. Indeed, current data suggests that about 20% of individuals with ASD are macrocephalic, meaning their head sizes are greater than two standard deviations above the mean [116]. Yet, about 10–15% of individuals with ASD have the opposite phenotype, that is, microcephaly [117]. Another commonly reported finding is decreased volume of the corpus callosum suggesting there are reduced hemisphere to hemisphere connections in ASD [118–121]. Indeed, diffusion tensor imaging studies that give insight into myelinated axon bundles in the brain have also reported changes in the corpus callosum [119, 121–123]. However, it is unclear how common corpus callosum defects are in ASD patients. While some studies have observed changes such as decreased volumes in cerebellar vermis or alterations in volume of the amygdala, these findings were often not confirmed by other groups. Moreover, reports of altered cortical thickness, volume, and surface area have also been inconsistent [118].

In summary, a look at the brain of individuals with ASD through both postmortem and imaging studies has shown us that, individuals with ASD have alterations in brain structure and function. However, there is no single defining pathological or functional change seen in ASD brain. Yet, the changes observed in the ASD brain suggest abnormalities of neurodevelopment.

7 Neurodevelopmental Genes: Associations with ASD and Abnormal Behavior

While neuropathological and imaging studies reveal brain abnormalities in NDDs, it is important to note that NDDs like ASD, ADHD, ID, and SCZ are not characterized or diagnosed by neuropathology but rather are classified and diagnosed using clinical symptoms. Thus, while rare NDDs like lissencephaly may be associated with severe neuropathology secondary to dysregulated neurodevelopment, to truly understand the pathogenesis of more common and often less severe NDDs, we need to explore whether abnormalities in developmental regulatory genes actually lead to more common NDDs in humans and if disruption of these genes are associated with both neurobiological and behavioral abnormalities in rodents. There are multiple rodent models of dysregulated neurodevelopment that show abnormalities in behavior reminiscent of behavioral phenotypes found in individuals with ASD, ADHD, and SCZ. Furthermore, disruptions in developmental regulators have been associated with some rare cases of disorders like ID, ASD, and SCZ [23]. For example, TBR1 is a transcription factor that is expressed in early born neurons of the deep layers of the cortex during development and disruptions in TBR1 have been found in case studies of individuals and families with ASD and ID [23, 124–128]. Moreover, mouse models with either deletions or alterations in TBR1 often show behavioral symptoms reminiscent of ASD and ID. Even though TBR1 is thought to be expressed after the developmental process of migration has occurred, mice that are KOs for TBR1 displayed a vast change in cortical laminar structure [129, 130]. Moreover, the alterations were varied across different parts of the cortex (instead of uniform alterations like reeler mice). A study by Hevner et al., used BrdU pulses at E10.5, 11.5, and 13.5 to track the survival and localization of neurons across multiple embryonic and postnatal time points in wildtype (WT) and TBR1 knockout mice (KO). In mutant KO mice, early born neurons largely occupied superficial (rather than deep) positions when dams were injected with BrdU label at E10.5 and analyzed at E18.5 [30]. There were also clusters of ectopic cells born at different times accumulating in the caudal cortex. Thus, by tracing neurons across time, Hevner et al. were able to show abnormal migration in this mouse model with abnormal behavioral phenotypes. Likewise, TBR1 KO mice had alterations in embryonic axon growth and targeting. Mouse models with mutation in Ankrd11, a chromatin regulatory molecule, have mislocalization of cortical neurons with excess cells accumulating in the VZ and SVZ during embryonic development and alterations in the positioning of upper and lower layer neurons in the postnatal brain of mice [131]. Interestingly, these mice displayed reduced social interaction and increased repetitive behaviors, again suggesting ASD like behavioral abnormalities. Humans with either a deletion or mutation of one allele of the ANKRD11 gene have been found to have cognitive dysfunction and ASD. Some individuals with this gene mutation also show abnormalities in the corpus callosum and periventricular heterotopias. Defects in cortical lamination that are suggestive of aberrant migration have also been seen in other mouse models including those with altered PTEN, CNTNAP2, Bcl11a, and Baf170 [130, 132–136]. These rodent models also have abnormal behaviors including reduced social interactions, increased repetitive behavior anxiety, and memory and cognitive issues. Likewise, humans with defects in PTEN and CNTNAP2, for example, have shown altered cortical lamination and symptoms such as seizures, autism, and hyperactivity, which are paralleled in the mice [137–140]. Mouse mutants of many genes that show migration defects also exhibit alterations in early neurite outgrowth, specifically the early transition of neurons from a multipolar to a bipolar morphology. This includes genes such as Bcl11a, Auts2, Foxg1, Foxp1, LDS1, and FMR1, as well as members of the WNT pathway [130, 132–134, 141–145]. In particular, AUTS2 is associated with ASD and numerous other disorders such as ID, ADHD, SCZ, epilepsy, and developmental delay [146]. Thus, AUTS2 has important roles in development and its alteration can lead to changes in brain and behavior. A study by Hori et al. found that in cortices that were electroporated with Aut2s shRNA and GFP at E14.5, cells with GFP expression failed to migrate to the appropriate layers and ectopic cells were found across the cortex [147]. During early migration to the cortex, the shRNA-expressing neurons also displayed twisted, irregular leading processes with abnormal branching. Upon reaching adulthood, the Aut2s animals showed reduced exploratory activity in novel circumstances, impaired memory, reduced social interactions, repetitive behaviors, and increased addictive behaviors. In aggregate, these observations indicate that early neurite outgrowth and migration defects often are commonly dysregulated in animal models that have abnormal behaviors that are characteristic of NDDs like ASD, SCZ, ID, and ADHD. Moreover, we find that genes that regulate neurodevelopment and are associated with abnormal behaviors in mouse models are associated with human NDDs. This suggests that early neurodevelopmental processes such as migration and early maturation may play key roles in the pathogenesis of NDDs and that dysregulation of neuro development can lead to abnormal behavior.

8 Large-Scale Genetic Studies: What Do they Tell us About ASD and Other NDDs?

In Sects. 4, 5, 6, and 7 we provided an overview of regulatory molecules that are involved in the orchestration of neurodevelopment, we discussed the role of altered neurodevelopment in both rare and common NDDs and we reviewed both the neurobiological and behavioral consequences that result from dysregulated neurodevelopment. In Sect. 7 above, we saw that mutations in genes such as PTEN, AUTS2, and TBR1 were associated with disorders such as ASD, ID, and SCZ. In the case of ASD, studies suggest that genetics play a large role in disease susceptibility with monozygotic concordance rates between 77 and 95% and dizygotic concordance rates between 10 and 30%. Yet, despite this, almost 70% of cases of ASD are genetically undefined and studies have shown that only about 1–5% of cases of ASD are due to mutations in single genes such AUTS2, CNTNAP2, TBR1, and MECP2 [148–150]. In some instances, ASD cases associated with highly penetrant single genes like AUTS2 and MECP2 are not associated with idiopathic forms of ASD. Thus, as rodent studies of ASD and other NDDs have largely focused on models with alterations in single highly penetrant rare variant genes, our insight into the pathophysiology of the majority of ASD cases has been limited. Do the over 1000 different ASD risk genes, over half of which are low-risk conferring common variants, play a role in neurodevelopment? Do these genes regulate common functions? Newer genetic studies have used pathway analysis techniques to uncover whether ASD risk genes could be involved in regulating common process and if these genes play a role in neurodevelopment. Reviews that have analyzed numerous genetic studies to find points of convergence have uncovered 4–8 categories that ASD risk genes can fall into [151–159]. Alterations have been observed in genes and proteins that are in the following categories (1) proteins that can alter neural activity or have activity dependent expression such as ion channels (2) regulators of postsynaptic translation such as FMR1 (3) proteins involved in cell adhesion such as CNTNAP2, neurexins, neuroligins, and cadherins (4) genes that specify or determine the ratio of excitatory to inhibitory neurons like neuroligin-2 (5) cytoskeletal proteins such as tubulins, RhoA, (6) members of signaling pathways such as MAPK and P13K-mTOR, (7) chromatin regulators such as CHD8, and (8) immune-associated molecules. Convergent pathways can help us narrow down what processes may be altered in ASD. However, convergence alone does not necessarily indicate causality. These common pathways that are implicated in ASD could be confounds or a non-causal consequence of ASD. For example, excess microglia (or immune activation) is commonly observed in ASD [160]. The excess microglia could be causal, that is higher levels of microglia could lead to autism or it could be a confound—a change in chromatin modifiers could lead to both ASD and excess microglia. Alternatively, the excess microglial phenotype could be seen as a consequence of ASD, where having the disorder leads to the activation of microglia. Thus, while convergence studies help narrow down potential etiologies of ASD, it is important to remain cautious about the results of these studies. Furthermore, even though pathway analyses have helped focus attention to processes which are altered in ASD, many of these genes have a vast range of functions and do not always provide insight into what developmental processes are altered in the brain to cause ASD. Moreover, the "categorization" of many of these genes is artificial or limiting. For example, an ASD risk gene categorized as a cell adhesion molecule may also play roles in synapse formation. Additionally, if we look at an earlier time point in development, the cell adhesion molecule that regulates synapse formation in the postnatal brain may be important for promoting cell migration in embryonic development. Many early convergence studies utilized protein-protein interaction (PPI) networks to uncover convergence. However, PPI networks are often built upon general protein-protein interactions without considering tissue or cell type and thus may not fully capture the brain-specific functional relationships among cells. In other convergence studies, brain-specific network-based approaches were utilized but these studies did not always consider the dynamic patterns of gene relationships that occur during brain development, thereby limiting the potential for uncovering disease pathogenesis. Interestingly, a recent paper has suggested that 70% of genes found on the SFARI list of syndromic developmental disorders converged onto the regulation of two basic developmental cellular processes: proliferation and differentiation of neural stem cells [161]. Indeed, many other reviews of genetic literature are suggesting that psychiatric illnesses including ASD begin with abnormal specification, growth, expansion, and differentiation of embryonic neural precursor cells [162, 163]. Thus, while pathway analyses are showing that ASD genes can be categorized by their putative functions in adults, it is important to look at where in development these genes are expressed and what roles in development these genes play.

Newer genetic studies have begun to focus on the expression of ASD risk genes in development. A study by Willsey et al. explored convergence of 9 ASD associated mutations in relation to their expression by brain region, cell type, and time points in human development [164]. The study found that the risk genes converged onto glutamatergic projection neurons in layer 5 and 6 of the human mid-fetal (12–26 weeks post-conception) prefrontal and primary motor-somatosensory cortex. This study and others have also found a temporal convergence to early development, including in the mid-fetal striatum as well as perinatal thalamus and cerebellum, which are regions that are implicated in ASD. Since the small number of genes analyzed in the Willsey study limited the points of convergence uncovered, larger studies will be necessary to fully understand the spatiotemporal pattern of ASD risk genes. Another study by Parikshak et al., which used a much larger set of ASD risk genes, found that ASD genes often converged onto pathways related to translational and transcriptional regulation in glutamatergic projection neurons in the cortex [165]. Expression of these genes ranged from fetal to early neonatal developmental period. More recently, a study by [166] used a meta-analysis approach to test for functional convergence among ASD candidate genes (1002 genes) across multiple different convergence approaches. This study found that while ASD risk genes have multiple functional roles including many of the 8 categories discussed above, the most striking finding was that most ASD disease genes were involved in the regulation of neurodevelopment. In particular, top candidate genes (ex: DSCAM, DYRK1A, CHD8) were involved in neuronal migration, neuronal maturation, and synapse formation. Moreover, the importance of very early cortical development was recenty confirmed by the large scale (over 35,000 samples) whole genome sequencing study by [167] that shows that 80% of 102 high risk ASD genes are expressed in the forebrain by 23 weeks gestation and regulate maturing or mature neurons of both excitatory and inhibitory lineages. Thus, while ASD risk genes may have different roles in the adult brain, a majority of these risk genes seem to play a key role in the regulation of neurodevelopment. Interestingly, genetic studies have also found significant overlap among genes implicated in the pathogenesis of SCZ, BPD, and other NDDs. In fact, a recent study by Gandal et al. used transcriptome analyses to study post-mortem samples derived from individuals with SCZ, BPD, ASD, alcohol addiction, and major depressive disorder and found that there was a high degree of overlap in the transcriptome profiles of SCZ, ASD, and BPD [168]. This suggests that many developmental disorders may have common molecular pathology and that dysregulated neurodevelopment plays a key role in disease pathogenesis.

In conclusion, genetic studies illustrate that autism is highly heritable and that the genes associated with ASD ultimately seem to converge onto the regulation of neurodevelopmental processes during embryonic/fetal and early postnatal development. While genetic studies can show us that ASD risk genes are expressed in certain areas of the brain during development, these studies alone cannot show us what functions these genes may play in development. Moreover, genetic models alone cannot elucidate how gene alterations can lead to the changes in brain structure and behavior that cause ASD. Lastly, genetic studies detect dynamic changes in the brain such as alterations in EFs or signaling molecules that may be the downstream consequence of a genetic alteration. For example, recent studies have found dysregulations in the mTOR pathway in multiple subtypes of ASD where genetic mutations are not coding for mTOR pathway molecules. For example, studies have found that lymphocytes derived from individuals with Fragile-X syndrome had alterations in the mTOR pathway [169]. Likewise, abnormalities in mTOR have also been noted in Rett syndrome and in models even with cases of idiopathic ASD [170–173]. What this suggests is that mutations in genes that are not functionally related can lead to downstream changes in cells that alter specific signaling hubs. Ultimately, uncovering these points of signaling convergence would allow us to subtype ASD and other developmental disorders by molecular pathology rather than by clinical phenotypes. However, without studying different gene mutations and assessing signaling and neurobiological outcomes, we could not know whether mutations or alterations of two disparate molecules could lead to a common outcome. Consequently, to learn more about autism, we need a way to model and study autism, particularly in a human system. Luckily, iPSC technology provides us with the opportunity to study human neuronal cells and their development.

9 Why Use iPSCs to Model Neurodevelopment and NDDs?

As discussed, a developing embryo first starts out as a fertilized egg undergoing rapid proliferation. Ultimately, this simple fertilized egg develops into a blastocyst which consists of a layer of trophoblast cells that become the placenta and an inner cell mass of pluripotent stem cells which ultimately becomes the embryo [33]. These pluripotent stem cells are characterized by a unique ability to ultimately form almost any cell in the body with the exception of placental tissues. Moreover, these early blastocyst cells also divide and self-renew. In 1998, Thomson et al. derived the first human ES cells from surplus fertilized eggs donated by couples who were undergoing fertility treatment [174]. The derivation of these ES cells provided researchers with the opportunity to study the live embryonic development of human tissue. Until the derivation of these ES cells, human developmental studies were restricted to analyzing tissue samples derived from donated abortion materials. Of course, both ethical concerns and donor sparsity very much limited these studies. Thus, the study of live cellular development in humans was restricted to cancer cell lines like SH-SY5Y. While cancer lines did have some shared similarities to their tissue of origin, they also had unstable karyotypes and other genetic alterations that made them less than ideal models. For neuroscientists, ES technology provided an avenue to study live human neurons in culture—a nearly impossible feat before the generation of ES lines.

In 2006, Takahashi and Yamanaka revolutionized the pluripotent stem cell field by showing that ESC-like cells could be derived from adult mice fibroblasts by infecting them with retroviral vectors that contained just 4 pluripotency inducing factors: Oct3/4, Sox-2, c-MYC, and Klf4 [175–177]. These ESC-like cells, known as induced pluripotent stem cells (iPSCs), could form all 3 germ layers of a developing embryo and could contribute to different tissues in mice when injected into the blastocyst of a developing embryo. Moreover, they expressed similar markers and had similar morphology to ESCs. This showed that mature, somatic adult cells could be reprogrammed into cells that were "embryonic" and pluripotent. By 2007, the Yamanaka lab and other groups showed that similar techniques could be used to reprogram human fibroblasts into ESC-like cells [178, 179]. Further advances in iPSC technology now allow for the derivation of these cells from numerous tissue types including white blood cells, keratinocytes, melanocytes, liver, and neural cells using a variety of reprogramming methods. Moreover, iPSCs have now been derived from numerous species including humans, mice, rats, and other primates. One of the main advantages of iPSCs is that they retain the genetic signature of the individual from whom they are derived. Thus, iPSCs have given us the first opportunity to study idiopathic disorders, which encompasses the majority of cases of ASD, SCZ, and BPD.

For developmental disorders like ASD, iPSC-derived neural cells can provide us the opportunity to understand what processes may be going awry early in development to lead to disease. As mentioned, in mouse models, there are few studies of early neurodevelopmental processes and thus, iPSCs could be utilized to fill the knowledge gap. Since 2010, numerous studies have been published on developmental pathology in everything from monogenic diseases to idiopathic disorders like schizophrenia, ASD, and BPD. Excitingly, these studies have found fascinating differences in disease neural cells that were not always seen in mouse models. Moreover, some of these studies were even able to test potential therapeutics in iPSC models. Yet, many human iPSC studies of NDDs have focused on in-depth analysis of post-mitotic neurons. Indeed, while these studies have found aberrations in these neurons, NPCs and their functions have been largely overlooked. Overall, unlike mouse models, iPSCs have provided us with the ability to study idiopathic disorders and human neurodevelopment in a simplified culture model system.

10 iPSC Studies of ASD and Other NDDs

10.1 Monogenic NDDs/ASDs

Some of the earliest studies of human iPSC models of neuropsychiatric diseases were conducted on the highly penetrant monogenic developmental disorders such as Rett Syndrome (RTT), Fragile-X syndrome, and Timothy Syndrome [180–182]. These studies largely derived NPCs and neurons using the embryoid body (EB) method. In 2010 Marchetto et al. derived iPSCs from the fibroblasts of 4 female patients with Rett syndrome and 5 control individuals [180]. Using the NPCs that were generated in this study, the researchers examined the percentage of cells in G1, S, and G2/M stages of the cell cycle at a single passage by performing FACs analysis. No differences were found in the percentage of cells in each phase between WT and RTT NPCs. However, no other cellular assays or assessment of NPC morphology was conducted. Significantly, the neurons derived from Rett patients had (1) reduced glutamatergic synapses, (2) smaller soma sizes, (3) lower number of dendritic spines, and (4) changes in the frequency and amplitude of mIPSCs and mESPCs. Further, the RTT neuronal phenotype could be reversed with IGF-1 treatment and with increasing MECP2 expression. Thus, even though Rett Syndrome does not manifest until 6–18 months of age, fetal-like iPSC-derived neurons from Rett patients showed abnormalities. Shortly thereafter, a study on Timothy Syndrome (TS) was published by Pasca et al. in 2011 [181]. TS is a syndromic ASD caused by a mutation in an L-type calcium (Ca2+) channel. Surprisingly, in addition to the expected defective Ca2+ signaling, iPSC-derived neurons from TS patients displayed changes in catecholamine biosynthetic enzyme, tyrosine hydroxylase (TH), which subsequently led to increased norepinephrine (NE) and dopamine (DA). Excitingly, excess NE and DA production was reversed by treatment with an atypical L-type Ca2+ channel blocker, roscovitine. However, when the TS mutant channels were expressed in transgenic mice, the Ca2+ signaling defects were observed but the TH dependent changes were not found, illustrating the value of using a human model to study neuropsychiatric disorders. Further studies in

2013 on TS iPSCs from the same group uncovered dendritic spine abnormalities and abnormalities in cytoskeletal protein RhoA, which as we saw earlier is an important regulator of neurodevelopment [183]. In the original 2011 study, NPCs were derived from TS patients. The transcriptome and gene expression of these NPCs were compared to those of iPSCs and neurons, and this showed that NPC gene expression was more similar to neuronal rather than iPSC cells. Interestingly, NPCs from a single TS patient were also assessed for proliferation and migration defects and none were found. However, the 2014 transcriptome analysis of TS NPCs and neurons suggested that genes involved in proliferation were upregulated in TS NPCs while differentiation and morphogenesis genes were downregulated, suggesting that more careful analysis of NPCs could uncover earlier developmental defects [184]. Neurobiological analysis of neurons derived from Fragile-X syndrome (FXS) is fairly sparse. Transcriptome analysis of NPCs and neurons derived from FXS patients suggests defects in NPCs differentiation and neuronal maturation [185–187]. Specifically, the one study that assessed the neurobiology of FXS did find that there were neurite outgrowth differences in post-mitotic FXS neurons [188]. More recently, a study of iPSC-derived neural cells from patients with tuberous sclerosis utilized NPCs to assess proliferation and found increased proliferation along with other alterations in glial and neuronal development [188]. Thus, in summary, iPSC studies of monogenic disorders were able to uncover interesting changes in post-mitotic, differentiated neurons derived from individuals with disease. This indicated that as early as fetal development individuals with these diseases have abnormal neurons. Moreover, in some cases, studies in a human model were able to uncover defects that rodent models were not able to show, indicating the value of iPSCs.

10.2 Studies of NDDs Associated with CNVs and Common Variant Gene Alterations

In addition to monogenic disorders, other genetically defined forms of autism including disorders caused by copy number variations (CNVs) such as the 16p11.2 (16p) deletion (del) & duplication (dup) syndrome and Phelan McDermid Syndrome (PMS) have been examined. There are also defined forms of autism that arise in families due to the presence of a rare genetic variant in genes such as neuroligins. In 2013, Shcheglovitov et al. studied iPSC-derived neurons from 2 patients with PMS [189]. PMS is usually caused by heterozygous deletions in the 22q13.3 region which contains the SHANK3 gene that encodes a scaffolding protein in the postsynaptic density. Individuals with this deletion have a higher risk for ASD. The iPSCs were generated from fibroblasts and neural induction was done through monolayer methods. Again, this study focused on post-mitotic neurons and found reduced amplitude and frequency of spontaneous synaptic events, decreased number of synapses, and reduced expression of glutamate receptors.

Restoration of SHANK-3 levels and treatment with IGF-1 both rescued the excitatory synaptic transmission defect in these neurons. In 2016, the Sudhof lab also studied neurons derived from iPSCs of individuals with PMS and found reductions in neurite length, number of primary processes, and neurite branching, and alterations in dendritic arborization, intrinsic electrical properties, and synaptic transmission [190]. More recently, the same group that published the 2013 studies on PMS chose to study neurodevelopmental phenotypes in immature neurons rather than in mature-synapse forming and electrophysiologically active neurons to better understand the role of SHANK3 in early development [191]. However, the immature 22q13.3 neurons had smaller cell bodies, more extensively branched neurites, and reduced motility (migration) when compared to controls. Thus, Shank-3 has a critical role in early neurodevelopment including regulation of early neurite outgrowth, cell size, and migration. In 2017 Deshpande et al. published the first study of CNV 16p11.2 del and dup patient derived neurons [192]. NPCs were derived in this study and proliferation was studied by Edu incorporation assay. Surprisingly, despite macro- and microcephalic phenotypes observed in the deletion and duplication patients, respectively, there were no changes in cell proliferation in either set of NPCs. Neurons from the 16p-del cells showed increased soma size, increased dendrite length, increased mEPSC amplitude, and reduced synaptic density. On the other hand, while 16p-dup neurons showed reduced soma size and dendrite length, they also exhibited increased mEPSC amplitude and reduced synaptic density. Thus, the increased and decreased dosage of the genes in the 16p11.2 locus lead to common functional and synaptic defect outcomes, while upstream morphological features were bidirectional in nature, consisting of either too little or too much, a concept proposed for both Rett and FXS syndromes. Lastly, there are few iPSC studies on rare variant genes associated with ASD; however, there are groups currently working on these models—though data has not yet been published. One study looked at one individual with heterozygous deletions in CNTNAP2 [193]. Compared to NPCs from unaffected parents along with 3 unrelated controls, NPCs derived from the individuals with CNTNAP2 deletions showed significantly reduced radial migration. Thus, by using NPCs, this group detected an aberration that could not be found by using mature neurons. In summary, much like the monogenic diseases, CNV disorders associated with ASD and SCZ can also be modeled using iPSCs. These studies have revealed abnormalities in post-mitotic electrophysiologically active neurons similar to the monogenic disorders suggesting perhaps that excitatory synaptic transmission changes may be a common phenotype in NDDs. Moreover, by studying deletions and duplications together, studies have uncovered the effects of altered gene dose on human neural development and suggest that bidirectional abnormalities may lead to common functional outcomes.

10.3 iPSC Studies of Idiopathic ASD and Other NDDs

Of course, one of the most exciting things about iPSCs is that they retain the genetic signature of the individual from whom they are derived. Thus, iPSCs allow us to study idiopathic and polygenic disorders for the first time. However, initially, the heterogeneity of these diseases largely limited study of idiopathic disorders using iPSCs. In 2011, one of the first idiopathic iPSC papers was published by Brennand et al. on 4 patients with SCZ and no known genetic mutations [194]. Fibroblasts were acquired from these patients and reprogrammed into iPSCs via lentivirus. iPSCs were subsequently induced into NPCs or neurons using EB methods. This study found that SCZ neurons had reduced connectivity, which could be rescued by treatment with the antipsychotic loxapine. Moreover, SCZ neurons had a decrease in the number of neurites, and slightly decreased PSD95 synaptic density. Interestingly, gene-expression analysis of NPCs and neurons showed perturbations in glutamate, cAMP, and WNT signaling pathways. Thus, Brennand et al's study is one of the first studies of iPSC-derived neural cells that implicated signaling abnormalities in regulation of neurodevelopment. Further, despite having no known genetic similarities, idiopathic SCZ patients in this study had similar defects, some of which (but not all) were observed in previous post-mortem studies and relevant animal models. After this publication, Brennand et al. published another paper, this time focusing on the NPCs from the same set of patients in their 2012 study [195]. The NPCs derived from control and SCZ patients had no difference in doubling time, nor in the percentage of cells in G1, S, or G2/M phase, suggesting there were no differences in proliferation. Network analysis of gene expression of control and SCZ NPCs showed perturbations in genes associated with neuronal maturation and cellular adhesion. Moreover, proteomic analysis revealed alterations in oxidative stress proteins, cell adhesion proteins, and cytoskeletal remodeling proteins (like cofilins). Interestingly, the NPC gene signature overlapped significantly with the gene signature of iPSC-derived 6-week-old neurons from these patients. This suggests that the molecular events contributing to SCZ were established and are well known to participate in processes that occur before differentiation of precursors into post-mitotic neurons, providing further evidence of the value of studying NPCs. Using 3 independent assays, including neurosphere migration, microfluidic migration, and laminin spot-chaining, aberrant (reduced) migration was also observed in SCZ NPCs. The aberrant migration could not be rescued by co-culturing with control NPCs or murine tissues, indicating there was a cell autonomous defect. Characterization of the migrating cells revealed a majority of them were Tuj1 positive immature neurons though Nestin positive cells were also capable of migration. Thus, in 4 patients with idiopathic SCZ, both NPCs and neurons showed abnormalities indicative of altered neurodevelopment. This showed that idiopathic disorders, despite their heterogeneity, may have some phenotypes in common. Moreover, studies by Brennand et al. showed the value of studying NPCs neurobiology in addition to assessing post-mitotic, neuronal abnormalities.

In 2015, the first iPSC study of non-syndromic ASD was published by Griesi-Oliveira et al. [196]. In this study iPSCs were derived from a single individual with a de novo mutation that leads to disruption of the cation channel TRPC6. Before this paper, TRPC6 had not been implicated in the pathogenesis of ASD. Neurons and NPCs were derived by the EB method and NPCs were studied for proliferation differences using FACs cell sorting. Again, there were no differences in the percentage of cells in G1, S, or G2/M in TRPC6 mutant cells vs control cells. Calcium signaling revealed reduced Ca2+ influx into TRPC6 mutant NPCs. Neurons derived from this individual had shorter neurites that were less arborized and had less dendritic spines. These defects were rescued by treatment of the NPCs with a TRPC6 channel agonist or by genetically increasing TRPC6 levels via viral induction. In 2015, one of the first organoid-based studies of idiopathic ASD was published by Mariani et al. [197]. Mariani et al. focused on individuals with macrocephaly and ASD in order to pick an endophenotype that could help reduce heterogeneity within the cohort. iPSCs were derived from fibroblasts of members of 4 families that each included an ASD proband with macrocephaly, and 1–3 unaffected first-degree family members. Organoids were generated from 2 to 3 iPSC lines per person using a modified version of the free-floating SFEBq method. The organoids generated had autonomously organized layers of radial glia cells, intermediate progenitors, and neurons. After 11 days of terminal differentiation (TD11) the organoids were composed of polarized proliferating progenitors expressing radial glial markers NESTIN, SOX2, BRN2, and PAX6. The radial glia were mitotic on the luminal side and produced immature neurons expressing TUJ1 and DCX. At TD31, mature NeuN+ neurons accumulated on the basal side of the radial glia (similar to the layering found in the developing cerebral cortex). Comparison of organoid transcriptomes to the Allen BrainSpan human developmental data showed that TD11 organoids closely resembled the human brain during early fetal development (9 weeks post-conception) while TD31 organoids had significant similarities to early 2nd trimester human fetal brain samples (13–16 weeks post-conception). The cells in these organoids were more similar to human dorsal telencephalon including the cerebral cortex and hippocampus. The ASD organoids had a significant upregulation in transcription factors associated with the acquisition of neural cell fate and precursor cell proliferation. There was also upregulation of members of neural cell adhesion family and genes involved in cytoskeletal regulation of many cellular functions including neurite outgrowth, axon guidance, cell proliferation, and migration. At TD11, ASD-derived organoids showed a transient increase in size, indicating potentially increased proliferative rates in early development. This size difference normalized by TD31. Moreover, BrdU incorporation experiments showed a significant decrease in cell-cycle length in ASD-derived NPCs, suggesting more rapid proliferation. Unlike many studies that found reduced synapse formation in ASD, Mariani et al. found increases in synapsin puncta in ASD-derived neurons. Moreover, unlike other studies, glutamate synapses were unchanged yet, there was an increase in GABAergic synapses. Thus, it seems that ASD cells that have NPCs with proliferative defects seem to show different neuronal phenotypes than neurons derived from individuals whose NPCs have no proliferation defect. This could suggest that there are different subtypes of ASDs and NDDs that cluster based on NPC proliferation. In addition to these two studies, two other studies were published on idiopathic ASDs in 2016 and 2017 [198, 199]. One study by Marchetto et al. used a cohort of 9 individuals with macrocephaly and idiopathic ASD and found alterations in NPCs proliferation and WNT signaling. Neurons from these patients also displayed reduced synaptogenesis and functional defects in neuronal network [198]. Interestingly, neuronal network defects were rescued by IGF-1. In 2017 Liu et al. studied 4 male patients with idiopathic ASD excluding patients with severe intellectual disabilities, seizure disorders, or known clinical syndromes or brain malformations. iPSCs were derived from fibroblasts and differentiated into NPCs using EB methods [199]. The NPCs were differentiated into neurons and studies were conducted on neurons 80 days post-differentiation. Compared to neurons from unaffected siblings, neurons from ASD patients had slightly increased synapsin, PSD-95, and VLGUT puncta. Moreover, ASD neurons also displayed altered excitability, aberrant Na+ and K+ channel currents, and alterations in genes associated with synaptic transmission. Liu at al did not study NPCs from these patients. Thus far, the studies conducted in patients with idiopathic ASD have shown relatively similar phenotypes for patients within each study. Moreover, comparing patients from all the studies, we see that in general, ASD and other NDDs seem to be characterized by alterations in neurites, dendritic spines, synapse formation, and neuronal excitability. However, there are differences between patients in different studies on the direction of these alterations.

11 Summary of iPSCs and NPC Studies

iPSCs are truly a revolutionary technology. Not only do iPSCs allow us to study neurons and their development, these studies can be conducted in the context of the complex genetics that are often associated with neuropsychiatric disorders. In ASD and other NDDs, use of iPSC technology has revealed numerous defects in post-mitotic neurons derived from patients with monogenic, CNV, and idiopathic subtypes of ASD. Overall, these studies have uncovered some common alterations in neurites, dendritic spines, synapse formation, and electrical activity in ASD neurons. Some of these alterations have been described previously in mouse or post-mortem studies; however, a few discoveries have been unique to the live human neurons that we have been able to study for the first time. Yet, it is important to remember that all model systems have their flaws [200]. As iPSCs are a relatively new technology, there are still technical challenges in working with these cells. While there are few articles that document these challenges in quantitative fashion, in our experience, discussion among stem cell researchers suggests variability in application of iPSC methods is not uncommon and is more striking than the variability observed in primary cultures [201]. Researchers indicate that different iPSCs that are derived from the same original cells, or even among clones in the same reprogramming experiment can exhibit different cellular behavior such as cell proliferation rates or tendency to differentiate into different morphologies, suggesting distinct lineages [202, 203]. Similar variability can be observed during induction of iPSCs into cells such as NPCs. Indeed, a recent study by Carcamo-Orive et al. assessed the transcriptional variability among a large library of 317 iPSCs to look for molecular drivers of the clone-to-clone and individual-to-individual variability seen in iPSCs [204]. In this study, about 50% of genome-wide expression of variability were explained by human genetic variation across individuals. Yet, the remainder of variability seen, even among iPSCs derived from the same individuals, was largely attributed to abnormalities in chromatin regulators and genes that contribute to developmental pathways including those that contribute to central nervous system development. The authors suggest that the causes for this non-genomic variability could be attributed to the different methods of reprogramming utilized, the different initial cell source utilized, technical artifacts from the reprogramming process, and other variables of culture that have not been identified. In particular, for those studying human neurodevelopment, the variability seen in pathways associated with neurodevelopment is of concern when utilizing iPSCs. Thus, given this variability, to ensure rigorous and reproducible results, we recommend that multiple iPSC derivations or clones must be studied and multiple neural derivations from these clones be used in experiments to ensure that phenotypes are an expression of patient biology rather than a consequence of technical or culture factors. This makes these types of studies both expensive and time consuming [205].

One issue deserving attention is that iPSCs are prone to acquiring mutations in culture, and thus, this also needs to be screened for carefully. Furthermore, there are concerns that iPSCs retain "epigenetic imprints" of the cells from which they were originally derived [206–209]. A few studies have found that the cell type or origin influences the molecular and functional properties of mouse iPSCs [210]. Thus, it is possible that the characteristics of the original tissue may appear in iPSC-derived cells like neurons though no studies have explicitly found evidence beyond the retained epigenetic imprints. Thirdly, many iPSC studies have used post-mitotic neurons to study disease phenotypes. These studies sometimes have drawn conclusions that the neuronal aberrations observed in the dish may be present in the individual from who they were derived. However, iPSC-derived neurons are more similar to fetal brain cells than to adult or postnatal brain cells. Thus, these studies are largely a reflection of what may occur in the developing fetal brain. This concern is particularly relevant in the study of degenerative disorders where aged neurons would be a more appropriate model. There are also questions on selection of best patient and control groups. Early studies on iPSCs did not always consider possible confounds such as sex, age, and genetic background differences between controls and patients. Some researchers suggest a mix of both unrelated controls along with unaffected family members represent the best controls for studies. Lastly, due to the effort required to make iPSCs, and the number of clones and neural inductions needed to get consistent data, most iPSC studies are very underpowered and thus techniques to be become faster and perhaps automated are needed for more effective use of this model system [211]. There are some cases where iPSC models are not as effective as ESCs for modeling certain diseases. For example, in the cases of genetic aberrations that lead to early lethality, iPSCs derived from individuals that survive to term would not be representative as they are reflective of an exception. Likewise, in the case of disorders like Fragile-X syndrome, ESCs with this mutation often expressed FMR1 mRNA until they are differentiated into a neural fate. This is suggestive of a developmental fate based transcriptional silencing of FMR1. However, iPSCs derived from individuals with Fragile-X do not express FMR1 protein. Thus, iPSCs may not be the best model to study developmental transitions from pluripotent to differentiated cells [212]. Lastly, iPSCs derived neural models are still in vitro systems, thus the result derived from iPSC cannot fully reflect the complicated events that occur in an in vivo system. The iPSC field as a whole however has only continued to improve since its inception. Newer technology, optimization of protocols, automation, and careful experimental design are largely helping overcome the technical caveats associated with iPSCs. Moreover 3-D modeling systems are allowing iPSCs to become more similar to the in vivo tissue they model. Thus, despite some challenges, iPSCs are still the best model system we have to study the human neurodevelopmental phenotypes of complex heterogeneous diseases.

12 Future Directions

Since the inception of iPSC technology there has been an explosion of studies aiming to better define the pathogenesis and pathophysiology of NDDs. As synapses are the basic functional unit of the brain, many iPSC studies and the decades of studies in other model systems have focused on synaptic alteration in NDDs. Indeed, for a while autism was considered a "synaptopathy," that is, a disorder of impaired synapses. The synapse theory of autism was established in the early 2000s largely due to two main reasons (1) syndromic forms of autism such as Rett syndrome mouse models had dysfunctional synapses (2) genetic studies had uncovered the mutation of two neuroligin proteins in a family with ASD thereby linking a well characterized synaptic protein to ASD. In the years following these initial reports, many rare variant ASD genes were uncovered and mouse models made by mutating these rare variant genes often had defects in synapse formation which further established autism as a disorder of synapses. While there is no doubt that synapses are altered in ASD and other NDDs, our review of the literature provided numerous examples in which alterations of developmental processes that precede synapse formation fundamentally alter the brain and lead to the aberrations that are associated with NDDs. Moreover, it is important to point out that early developmental processes such as proliferation, migration, and differentiation are essential for the proper wiring of the brain and for ensuring normal synapse formation and function. Thus, what looks like a "synapse formation error" may indeed be caused by the failure of neurites to grow or the migration of cells into an incorrect position, thereby preventing the elaboration of typical neural networks and the establishment of normal synapse structure and function. More recently, with the wider use of "omic" studies and more sophisticated modeling tools, pathway analysis studies of idiopathic and syndrome-associated ASD genes have uncovered their convergence onto the cerebral cortex of the developing mid-fetal brain (8–24 weeks old) [164, 165, 167]. Interestingly, in human brain development, it is believed that the first synapses begin to form between 20 and 24 weeks of gestation and the peak of synapse formation occurs postnatally. Thus, we find that many of the aberrations that occur in ASD are taking place even before the first synapses form. Indeed, in the 8–24-week period of brain development, we find that NPCs are proliferating, migrating, and differentiating to form the brain. Considering these studies, it is important for those studying NDDs to more closely explore early neurodevelopment to better understand the pathogenesis of NDDs.

Much of our understanding about the pathogenesis of ASD comes from mouse models—many of which are made by mutating a single rare variant gene. Of course, as discussed, rare variant and even CNVs only account for a small proportion of ASD and over 80% of cases of autism are idiopathic. Yet, there have been few studies that have assessed whether syndromic/genetic ASDs have commonalities with idiopathic autism (I-ASD). In fact, the few genetic studies that compare syndromic and I-ASDs have found that the genes involved in syndromic ASDs do not contribute to I-ASD. Thus, much of our mechanistic knowledge regarding ASD is derived from syndromic ASDs which may not be generalizable to I-ASDs. Thankfully, with the advent of iPSC technology, we now can study cells derived from any individual including those with idiopathic disorders. Thus, for the first time we have the opportunity to understand the pathophysiology underlying I-ASDs. Moreover, we can attempt to understand if there are similarities among those with I-ASD and whether there may also be similarities between idiopathic and syndromic ASD. Yet, it is important to remember that ASD is a highly heterogeneous disorder both phenotypically and genetically. Thus, studying a few cases of I-ASD or syndromic ASD will not be enough to help us understand if there is a unifying pathogenic mechanism to ASD. Instead, one approach that could be considered is to study different subpopulations of individuals with ASD to look for commonalities in neurobiology or molecular pathways. This would allow us to subtype and categorize ASD based on molecular signatures—an approach that would be similar to tumor genetic profiling that is used to treat and classify cancers. Yet, unlike cancer, we cannot take samples of human brains to assess their molecular signatures and as such, iPSCs are a good model system to allow us to learn more about NDDs like ASD.

Our lab has been studying NPCs derived from 3 patients with I-ASD and their sex-matched siblings for the past few years. None of these patients have known genetic aberrations and our early genetic sequencing results do not suggest common genetic mutations among our cohort. Moreover, unlike several of the ASD iPSC studies that have been conducted, we did not select our patients based on a macrocephaly endophenotype. Instead, all of our patients have severe autism and come from families where there is a first degree relative with specific language impairment. To study early neurodevelopment, we generated NPCs from multiple iPSC clones (created using a non-integrating Sendai virus vector containing 4 Yamanaka factors) derived from both Sibs and ASD patients using the commercially available GIBCO neural induction kit. We then established methods to study proliferation, migration, and differentiation (neurite outgrowth) in these NPCs based on methods used for decades in our lab to study rodent neurodevelopment. The methods we developed were recently published in JOVE Williams et al. [213]. In brief, neurite outgrowth was studied by plating NPCs at low density into a 35 mm dish for 48 h. Then, dishes were fixed and the percentage of cells with neurites was determined by counting cells in three 1 cm rows. To study migration, neurospheres were formed by plating NPCs into an uncoated dish for 24–72 h and then collecting and plating the spheres into Matrigel coated wells for 48 h. In this time, cells migrate out from the sphere leaving a dense inner cell mass. Migration was calculated by subtracting inner cell mass area from the total sphere area. Due to ASD heterogeneity, we were not expecting extensive similarities between individuals with I-ASD. Yet, to our surprise, in each family, we found that I-ASD NPCs had a ~50% lower percentage of cells with neurites when compared to Sib NPCs. Moreover, NPCs from each I-ASD patient had a lower percentage of neurites than NPCs from each Sib in our cohort. Thus, while studies in induced neurons and post-mitotic neurons do show defects in axons, dendrites, and neurites, we find that neurite differences are already occurring earlier in development. Similarly, we found that I-ASD NPCs migrated less than Sib NPCs. On average, I-ASD NPCs migrated 50% less than Sib NPCs. Thus, by studying NPCs from patients with idiopathic autism we find that defects in migration and neurite outgrowth are present before terminal differentiation in human neural cells. Moreover, these common neurobiological aberrations occur in 3 individuals with no common genetic mutations. Interestingly, more recently we acquired iPSCs derived from 3 individuals with a genetic form of autism, the 16p11.2 deletion CNV. Surprisingly, our studies have uncovered that these 3 individuals also have defects in neurite outgrowth and migration. This indicates that dysregulation in early neurodevelopmental processes may be a common feature in ASD. Moreover, it shows that syndromic and idiopathic ASDs may indeed have common phenotypic characteristics.

In addition to studying migration and neurites in control media, we also stimulated our NPCs with extracellular factors (EFs) in order to reveal deficits that may not be apparent in control conditions and to identify dysfunctional signaling. For our studies we selected 3 EFs: Pituitary adenylate cyclase activating polypeptide (PACAP), serotonin (5-HT), and nerve growth factor (NGF). All three of these EFs have been shown to regulate processes such as neurite outgrowth or migration and studies have shown that 5-HT is dysregulated in the serum of up to one-third of ASD patients. Moreover, prior studies in our lab on the mouse model of a common gene variant associated with NDDs, EN-2, demonstrated a differential response to PACAP in En-2 knockouts, compared to wild-type mice, suggesting potential dysregulation of PACAP signaling pathways in NDDs [214]. In terms of signaling, 5-HT and PACAP function by binding to G-protein coupled receptors and through activation or inhibition of molecules like cAMP and cGMP. On the other hand, NGF binds to receptor tyrosine kinase and signals through molecules such as ERK and mTOR. By studying these EFs we not only get insight into their role in human development, but we may also get clues into potential signaling dysregulations in our ASD patients. Interestingly, NPCs derived from all 3 I-ASD patients failed to extend neurites under the stimulation of 5-HT, PACAP, and NGF whereas all Sib NPCs had increases in neurites under the stimulation of these EFs. Likewise, 5-HT and PACAP failed to elicit migration in I-ASD NPCs. We hypothesized that the lack of response to EFs could be due to impaired signaling. Thus, we decided to assess the levels of P-CREB in our I-ASD and Sib NPCs using western blot because P-CREB is downstream of PACAP, 5-HT, and NGF signaling and some studies have shown its dysregulation in NDDs. Surprisingly, in Family-1, we found that P-CREB levels were not different between Sibs and I-ASD in control conditions. However, while the Family-1 Sib exhibited a major fivefold increase in P-CREB levels in response to PACAP exposure, the I-ASD NPC response was far lower. That is, while PACAP increased levels of P-CREB in both Family-1 Sib and ASD NPCs, the level of increase was on average 60% lower in the ASD NPCs, suggesting there was diminished PACAP response and potential aberrations in signaling in autism NPCs. Likewise, preliminary studies in 2 clones each of Sib and ASD from Family-2 and Family-3 showed diminished P-CREB induction under PACAP stimulation in both I-ASD patients. This shows that impaired CREB signaling could be a common feature among our I-ASD patients. Identification of alterations in CREB provided us with a molecular target that could potentially be used to reverse the neurobiological defects in our NPCs. Thus, we tested the cAMP analogue, db-cAMP, which increases P-CREB levels, on our I-ASD patients. Interestingly, db-cAMP increased neurite outgrowth and migration in both tested I-ASD patients and in a preliminary study also increased neurite outgrowth in our third I-ASD patient. Thus, the use of EFs showed us that I-ASD cells have impaired responses to important developmental regulators and it also helped us identify dysregulations in P-CREB which could then be targeted to reverse neurobiological abnormalities.

While neurite outgrowth, migration, and EF response defects were common among all three I-ASD patients, there were some patient specific differences in behavioral, cellular, and molecular phenotypes among our patients. Of the three patients in our cohort, 2 individuals (I-ASD-1 and I-ASD-2) had severe intellectual and social impairment (per WISC and SRS scale, respectively) whereas I-ASD-3 had an above average IQ (FSIQ=118) and moderate social impairment. In terms of the NPCs, despite commonalities in two of the early neurodevelopmental processes, we found that there were patient specific differences in NPC proliferation among our cohort. Specifically, NPCs derived from two patients had reduced proliferation (when compared to their Sib) and the other I-ASD patient's NPCs were hyperproliferative compared to his Sib's NPCs. In the section above we discussed finding common impairments in P-CREB signaling in our patients. In addition to assessing signaling, however, we also explored differences in the cytoskeletal regulator, P-cofilin, in our cells. As we discussed earlier in this chapter, the cytoskeleton and its regulators are essential for the normal progression of neural development. Moreover, in multiple pathway analysis studies, defects in the cytoskeleton were a common point of convergence in ASD. Thus, with the common impairment in neurite outgrowth and migration, which are cytoskeleton dependent processes, we expected differences in the levels of P-cofilin between Sib and I-ASD NPCs. Our initial experiments were conducted in Family-1 where both immunocytochemistry and western blot studies showed increased levels of P-cofilin in I-ASD-1 NPCs compared to Sib-1 NPCs. Yet, to our surprise, when we extended our studies to I-ASD-2 and I-ASD-3 we found no differences in P-cofilin levels between these patients and their Sibs. Ultimately, we are describing the differences we uncovered between our I-ASD NPCs to show that ASD is indeed a heterogeneous disorder and each patient does have unique features both behaviorally and cellularly. Yet, remarkably, despite the considerable heterogeneity, our lab and other labs have been able to find common phenotypes among randomly selected individuals with idiopathic NDDs. What this could indicate is that with further studies, we may be able to find a few different neurobiological processes or molecular phenotypes that are common among multiple individuals with ASD.

12.1 Considering Rigor and Reproducibility

While iPSC technology has paved the way for us to more effectively study human neurodevelopment, in Sect. 11 we discussed some of the potential pitfalls of utilizing iPSC technology. As discussed, one of the biggest issues plaguing the iPSC field is the level of heterogeneity that is observed in iPSCs and iPSC-derived cells. This variability is greater than that seen with primary cell cultures or mouse model studies. Thus, when working with iPSCs it is important to study multiple clones from the same individual and multiple neural inductions from the same iPSC clone to ensure fidelity of results. For our studies, our concerns about the heterogeneity of iPSC technology and our desire to be rigorous and reproducible led us to study a minimum of 2 clones with 2 NPC inductions each from each of the families we have studied. When we started working with iPSCs, however, we focused our studies onto one Sibling pair. At the time, we wished to characterize the level of heterogeneity we observed among NPCs derived from different iPSC clones or different induction procedures to better understand the extent of replicates needed in our future experiments. Thus, in our first Sibling pair, we studied 5 different clones (C) from both the Sibling and the I-ASD patient and from each clone we conducted anywhere from 1 to 4 separate neural inductions (N). Then, multiple experiments (for both neurites and migration) were conducted on the derived NPCs as seen in Fig. 3. The first thing of note is the marked variability in percentage of NPCs with neurites among different Sib-1 clones. Sib C2, for example, has around 7% neurites, whereas Sib C3 has closer to 17% neurites—which is almost a 150% difference between the two clones. Interestingly, less variability is noted in the NPCs derived from different ASD clones. What is striking is that if we chose only to study Sib C2 and ASD C1 we would conclude that there was no difference in neurites between Sib and ASD. On the other hand, if we had just selected Sib C3 and ASD C3 the difference between ASD and Sib would be exaggerated. Similar variability was also noted when comparing NPCs derived by different inductions from the same iPSC clone.

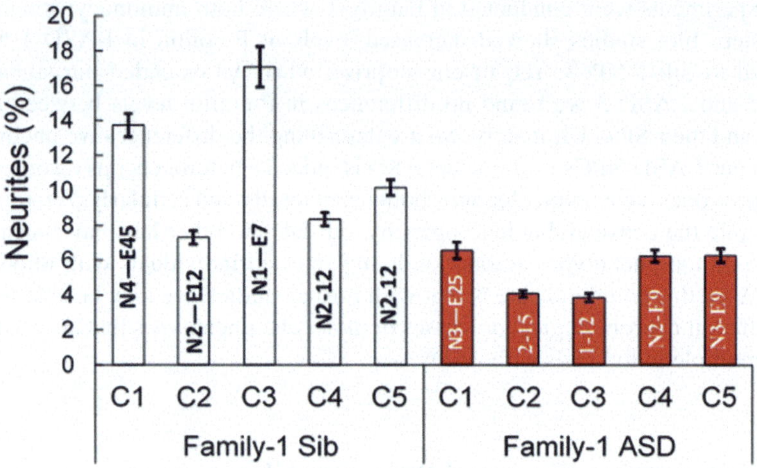

Fig. 3 Percentage of neurites broken down by NPCs derived from 5 clones (C1–C5) of Sib and ASD in Family-1. In each bar, the "N" signifies the number of distinct neural inductions conducted on the clone. The "E" indicates the number of experiments conducted for each clone. There is variation in percentage of neurites from clone-to-clone; however, on average Sib NPCs have higher neurites than ASD NPCs

Thus, we see that if multiple clones from the same patient are not evaluated, we risk making either false negative or false positive conclusions on our data.

12.2 Conclusion

Decades of studies in mouse models, analysis of human post-mortem samples and imaging studies, and results of genetic analyses have indicated that autism is a disorder of neurodevelopment. Importantly, the pathogenesis of autism seems to start in utero as early as the first trimester when neurons are just being formed and precursor cells are undergoing proliferation and migration. For many years, study of human neurodevelopment was limited as we had no model system to directly observe the development of human neural cells. Moreover, we did not have imaging means to look at the developing fetal brain in utero. Thus, much of our knowledge about neurodevelopment and NDDs came from mouse models which were made by mutating a single often rare variant gene. Yet, as we discussed, most cases of autism are idiopathic and genetic studies have shown us that there are over 1000 different autism associated genes—many of which are common variants with low penetrance. The genetic diversity of ASD is also compounded by the vast clinical heterogeneity of the disorders. Thus, uncovering the etiology of ASDs and other NDDs has been challenging. In the last decade, the advent of iPSC technology has now given us the ability to study human neurodevelopment and NDDs. iPSCs also gave us the unprecedented ability to study idiopathic diseases as the iPSC-derived cells retain the genetic signature of the individual from whom they are derived. With this powerful technology, we are in a position to move beyond studies that assess a single gene, single disorder, or a single subtype of a disorder. Instead, it would be valuable for us to identify common molecular drivers for aberrant phenotypes in NDDs. This would help us subtype NDDs by molecular pathology rather than behavioral phenotypes, it would allow us to select subgroups for clinical trials, and it would identify possible targets for drug therapies. To achieve this goal, we first need to expand our studies beyond the synapse and look at early neurodevelopment which may help us find molecules or signals that may not be readily apparent in a post-mitotic neuron. Furthermore, we need to begin studying and comparing multiple subtypes of autism and different NDDs to look for commonalities among these individuals. Indeed, a similar approach has been used in cancer biology to target molecular drivers of disease leading to improved survival for many patients. Genetic studies of autism have already started to take this approach by comparing different NDDs to look for commonly dysregulated genes. However, genetic studies do not always uncover downstream deficits that may be occurring in a system. For example, in our I-ASD patients we saw that dysregulations in P-CREB signaling and EF responses were commonly impaired in all our patients. Yet, our preliminary genetic analyses of these patients did not suggest any CREB defects nor did it show impairment in the receptors for the EFs we tested. This shows that functional molecular drivers of disease may not be readily seen in genetic studies and neurodevelopment must be modeled for further insight. Without this insight, we cannot progress in developing targeted therapeutics for NDDs. Thus, the iPSCs provide an avenue for us to better understand NDDs and human neurodevelopment which can ultimately facilitate personalized targeted medicine.

Like all model systems, there are still some caveats to utilizing iPSCs to study human neurodevelopment and NDDs. While iPSCs cells are similar to human embryonic cells, there are still concerns that the reprogramming process utilized to generate iPSCs may introduce variables that alter the behavior of iPSC and iPSC-derived cells. Indeed, there are studies that show that hESCs and iPSCs do have significant differences though we do not know the impact of these differences on differentiated cells. Secondly, as we showed with our experiments and as noted in other studies, in its current state, iPSC technology has considerable variability and thus requires more biological and technical replicates to ensure fidelity of data than with primary cultures or murine models. Thus, until the issue of variability can be better controlled, it is important that we take the effort to ensure that the data we produce are reproducible and a reflection of patient biology. This would include reproducing results among different clones and inductions in the same lab and having the data reproduced in another lab to ensure robustness. Lastly studies have shown that both neurons and NPCs derived from iPSCs are fetal in nature. Thus, what we are modeling in the dish is a neurodevelopmental window that occurred in utero and may not pertain to the patient whose brain is largely developed. Moreover, it is important to note that most iPSC systems (with the exception of organoids) do not represent the 3-D complexity of the developing brain. Thus, it is important for us to understand whether the developmental defects we see in the 2-D dish correlate with the patient's phenotype. More importantly, if our ultimate goal is to identify potential targets for therapeutic intervention, we need to know if the molecules that are dysregulated in early neurodevelopment (as modeled by iPSCs) play a role in the postnatal brain. For example, we found that P-CREB was dysregulated in the NPCs derived from our I-ASD cohort. The question would be is this P-CREB dysregulation present in the postnatal neuron and if so, will correcting this dysregulation lead to a noticeable change in behavior? Thus, we need to know if regulatory molecules have an effect on the core behavioral symptoms of NDDs. To understand this, we would need clinical trials that target molecules identified in iPSC studies to determine if behavioral issues can be alleviated or if functionality can be increased in patients with NDDs. To make such clinical studies a possibility, we need to begin taking a molecular medicine approach, learn more about iPSCs and their correlation to human brain and behavior, and we need to ensure that our studies are carefully designed and rigorous.

Acknowledgments This work was supported by the New Jersey Governor's Council for Medical Research and Treatment of Autism (CAUT13APS010; CAUT14APL031; CAUT15APL041, CAUT19APL014) and Nancy Lurie Marks Family Foundation for Dr. DiCicco-Bloom and Dr. Millonig; NJ Health Foundation (PC 63-19) for Dr. Millonig; Mindworks Charitable Lead Trust, and the Jewish Community Foundation of Greater MetroWest for Dr. DiCicco-Bloom; and the Rutgers Graduate School of Biomedical Sciences for Dr. Prem and Dr. DiCicco-Bloom.

References

1. Kandel, E. R., Schwartz, J. H., Jessell, T. M., Siegelbaum, S. A., & Hudspeth, A. J. (2000). *Principles of neural science*. New York: McGraw-Hill.
2. DiCicco-Bloom, E., & Obiorah, M. (2017). Neural development and neurogenesis. In B. J. Saddock, V. Saddock, & P. Ruiz (Eds.), *Kaplan & Sadock's comprehensive textbook of psychiatry. 1* (10th ed., pp. 39–60). Philadelphia: Wolters Kluwer.
3. Clancy, B., Finlay, B. L., Darlington, R. B., & Anand, K. J. (2007). Extrapolating brain development from experimental species to humans. *Neurotoxicology, 28*(5), 931–937.
4. Clancy, B., Darlington, R. B., & Finlay, B. L. (2001). Translating developmental time across mammalian species. *Neuroscience, 105*(1), 7–17.
5. Molnar, Z., Metin, C., Stoykova, A., Tarabykin, V., Price, D. J., Francis, F., et al. (2006). Comparative aspects of cerebral cortical development. *The European Journal of Neuroscience, 23*(4), 921–934.
6. Shipp, S. (2007). Structure and function of the cerebral cortex. *Current Biology, 17*(12), R443–R4R9.
7. Martynoga, B., Drechsel, D., & Guillemot, F. (2012). Molecular control of neurogenesis: A view from the mammalian cerebral cortex. *Cold Spring Harbor Perspectives in Biology, 4*(10), a008359.
8. Urban, N., & Guillemot, F. (2014). Neurogenesis in the embryonic and adult brain: Same regulators, different roles. *Frontiers in Cellular Neuroscience, 8*, 396.
9. Sansom, S. N., Griffiths, D. S., Faedo, A., Kleinjan, D. J., Ruan, Y., Smith, J., et al. (2009). The level of the transcription factor Pax6 is essential for controlling the balance between neural stem cell self-renewal and neurogenesis. *PLoS Genetics, 5*(6), e1000511.

10. McConnell, S. K. (1995). Constructing the cerebral cortex: Neurogenesis and fate determination. *Neuron, 15*(4), 761–768.
11. Hansen, A. H., Duellberg, C., Mieck, C., Loose, M., & Hippenmeyer, S. (2017). Cell polarity in cerebral cortex development-cellular architecture shaped by biochemical networks. *Frontiers in Cellular Neuroscience, 11*, 176.
12. Pressler, R., & Auvin, S. (2013). Comparison of brain maturation among species: An example in translational research suggesting the possible use of bumetanide in newborn. *Frontiers in Neurology, 4*, 36.
13. Stiles, J., & Jernigan, T. L. (2010). The basics of brain development. *Neuropsychology Review, 20*(4), 327–348.
14. Rubenstein, J. L. R. (2011). Development of the cerebral cortex: Implications for neurodevelopmental disorders. *The Journal of Child Psychology and Psychiatry and Allied Disciplines, 52*(4), 339–355.
15. Nicholas, C. R., Chen, J., Tang, Y., Southwell, D. G., Chalmers, N., Vogt, D., et al. (2013). Functional maturation of hPSC-derived forebrain interneurons requires an extended timeline and mimics human neural development. *Cell Stem Cell, 12*(5), 573–586.
16. Hansen, D. V., Lui, J. H., Parker, P. R., & Kriegstein, A. R. (2010). Neurogenic radial glia in the outer subventricular zone of human neocortex. *Nature, 464*(7288), 554–561.
17. LaMonica, B. E., Lui, J. H., Wang, X., & Kriegstein, A. R. (2012). OSVZ progenitors in the human cortex: An updated perspective on neurodevelopmental disease. *Current Opinion in Neurobiology, 22*(5), 747–753.
18. Vitalis, T., & Verney, C. (2017). Sculpting cerebral cortex with serotonin in rodent and primate. In K. F. Shad (Ed.), *Serotonin - A chemical messenger between all types of living cells.* Rijeka: InTech. p. Ch. 05.
19. Nadarajah, B., Alifragis, P., Wong, R. O., & Parnavelas, J. G. (2003). Neuronal migration in the developing cerebral cortex: Observations based on real-time imaging. *Cerebral Cortex, 13*(6), 607–611.
20. Stanco, A., & Anton, E. S. (2013). Chapter 17 - Radial migration of neurons in the cerebral cortex. In J. L. R. Rubenstein & P. Rakic (Eds.), *Cellular migration and formation of neuronal connections* (pp. 317–330). Oxford: Academic Press.
21. Sekine K, Tabata H, Nakajima K. Chapter 12 - Cell polarity and initiation of migration-Rubenstein, John L.R. In: Rakic P, editor. Cellular migration and formation of neuronal connections. Oxford, Academic Press; 2013. p. 231–244.
22. Noctor, S. C., Cunningham, C. L., & Kriegstein, A. R. (2013). Chapter 16 - Radial migration in the developing cerebral cortex. In J. L. R. Rubenstein & P. Rakic (Eds.), *Cellular migration and formation of neuronal connections* (pp. 299–316). Oxford: Academic Press.
23. Reiner, O., Karzbrun, E., Kshirsagar, A., & Kaibuchi, K. (2016). Regulation of neuronal migration, an emerging topic in autism spectrum disorders. *Journal of Neurochemistry, 136*(3), 440–456.
24. Tissir, F., & Goffinet, A. M. (2003). Reelin and brain development. *Nature Reviews. Neuroscience, 4*(6), 496–505.
25. Jossin, Y., Bar, I., Ignatova, N., Tissir, F., De Rouvroit, C. L., & Goffinet, A. M. (2003). The reelin signaling pathway: Some recent developments. *Cerebral Cortex, 13*(6), 627–633.
26. D'Arcangelo, G. (2014). Reelin in the years: Controlling neuronal migration and maturation in the mammalian brain. *Advances in Neuroscience, 2014*, 19.
27. Boyle, M. P., Bernard, A., Thompson, C. L., Ng, L., Boe, A., Mortrud, M., et al. (2011). Cell-type-specific consequences of reelin deficiency in the mouse neocortex, hippocampus, and amygdala. *The Journal of Comparative Neurology, 519*(11), 2061–2089.
28. Kawauchi, T., & Hoshino, M. (2008). Molecular pathways regulating cytoskeletal organization and morphological changes in migrating neurons. *Developmental Neuroscience, 30*(1–3), 36–46.

29. Bar, I., Tissir, F., Lambert de Rouvroit, C., De Backer, O., & Goffinet, A. M. (2003). The gene encoding disabled-1 (DAB1), the intracellular adaptor of the reelin pathway, reveals unusual complexity in human and mouse. *The Journal of Biological Chemistry, 278*(8), 5802–5812.
30. Hevner, R. F., Shi, L., Justice, N., Hsueh, Y., Sheng, M., Smiga, S., et al. (2001). Tbr1 regulates differentiation of the preplate and layer 6. *Neuron, 29*(2), 353–366.
31. Gilmore, E. C., & Herrup, K. (2000). Cortical development: Receiving reelin. *Current Biology, 10*(4), R162–R166.
32. O'Kusky, J., & Ye, P. (2012). Neurodevelopmental effects of insulin-like growth factor signaling. *Frontiers in Neuroendocrinology, 33*(3), 230–251.
33. Liu, J. P., Baker, J., Perkins, A. S., Robertson, E. J., & Efstratiadis, A. (1993). Mice carrying null mutations of the genes encoding insulin-like growth factor I (Igf-1) and type 1 IGF receptor (Igf1r). *Cell, 75*(1), 59–72.
34. Nieto Guil, A. F., Oksdath, M., Weiss, L. A., Grassi, D. J., Sosa, L. J., Nieto, M., et al. (2017). IGF-1 receptor regulates dynamic changes in neuronal polarity during cerebral cortical migration. *Scientific Reports, 7*(1), 7703.
35. Gennarini, G., & Furley, A. (2017). Cell adhesion molecules in neural development and disease. *Molecular and Cellular Neurosciences, 81*, 1–3.
36. Miyamoto, Y., Sakane, F., & Hashimoto, K. (2015). N-cadherin-based adherens junction regulates the maintenance, proliferation, and differentiation of neural progenitor cells during development. *Cell Adhesion & Migration, 9*(3), 183–192.
37. Kadowaki, M., Nakamura, S., Machon, O., Krauss, S., Radice, G. L., & Takeichi, M. (2007). N-cadherin mediates cortical organization in the mouse brain. *Developmental Biology, 304*(1), 22–33.
38. Shikanai, M., Nakajima, K., & Kawauchi, T. (2011). N-cadherin regulates radial glial fiber-dependent migration of cortical locomoting neurons. *Communicative & Integrative Biology, 4*(3), 326–330.
39. Takeichi, M., Inuzuka, H., Shimamura, K., Fujimori, T., & Nagafuchi, A. (1990). Cadherin subclasses: Differential expression and their roles in neural morphogenesis. *Cold Spring Harbor Symposia on Quantitative Biology, 55*, 319–325.
40. Suzuki, S. C., & Takeichi, M. (2008). Cadherins in neuronal morphogenesis and function. *Development, Growth & Differentiation, 50*(Suppl 1), S119–S130.
41. Bixby, J. L., Grunwald, G. B., & Bookman, R. J. (1994). Ca2+ influx and neurite growth in response to purified N-cadherin and laminin. *The Journal of Cell Biology, 127*(5), 1461–1475.
42. Gartner, A., Fornasiero, E. F., Munck, S., Vennekens, K., Seuntjens, E., Huttner, W. B., et al. (2012). N-cadherin specifies first asymmetry in developing neurons. *The EMBO Journal, 31*(8), 1893–1903.
43. Gartner, A., Fornasiero, E. F., & Dotti, C. G. (2012). N-cadherin: A new player in neuronal polarity. *Cell Cycle, 11*(12), 2223–2224.
44. Gartner, A., Fornasiero, E. F., & Dotti, C. G. (2015). Cadherins as regulators of neuronal polarity. *Cell Adhesion & Migration, 9*(3), 175–182.
45. Nelson, W. J., & Nusse, R. (2004). Convergence of Wnt, beta-catenin, and cadherin pathways. *Science, 303*(5663), 1483–1487.
46. Chenn, A., & Walsh, C. A. (2003). Increased neuronal production, enlarged forebrains and cytoarchitectural distortions in beta-catenin overexpressing transgenic mice. *Cerebral Cortex, 13*(6), 599–606.
47. Arikkath, J., & Reichardt, L. F. (2008). Cadherins and catenins at synapses: Roles in synaptogenesis and synaptic plasticity. *Trends in Neurosciences, 31*(9), 487–494.
48. Compagnucci, C., Piemonte, F., Sferra, A., Piermarini, E., & Bertini, E. (2016). The cytoskeletal arrangements necessary to neurogenesis. *Oncotarget, 7*(15), 19414–19429.
49. Cearns, M. D., Escuin, S., Alexandre, P., Greene, N. D., & Copp, A. J. (2016). Microtubules, polarity and vertebrate neural tube morphogenesis. *Journal of Anatomy, 229*(1), 63–74.
50. Messier, P. E. (1978). Microtubules, interkinetic nuclear migration and neurulation. *Experientia, 34*(3), 289–296.

51. Breuss, M. W., Leca, I., Gstrein, T., Hansen, A. H., & Keays, D. A. (2017). Tubulins and brain development - the origins of functional specification. *Molecular and Cellular Neurosciences, 84*, 58–67.
52. Belvindrah, R., Natarajan, K., Shabajee, P., Bruel-Jungerman, E., Bernard, J., Goutierre, M., et al. (2017). Mutation of the alpha-tubulin Tuba1a leads to straighter microtubules and perturbs neuronal migration. *The Journal of Cell Biology, 216*(8), 2443–2461.
53. Aiken, J., Moore, J. K., & Bates, E. A. (2019). TUBA1A mutations identified in lissencephaly patients dominantly disrupt neuronal migration and impair dynein activity. *Human Molecular Genetics, 28*, 1227.
54. Bamba, Y., Shofuda, T., Kato, M., Pooh, R. K., Tateishi, Y., Takanashi, J., et al. (2016). In vitro characterization of neurite extension using induced pluripotent stem cells derived from lissencephaly patients with TUBA1A missense mutations. *Molecular Brain, 9*(1), 70.
55. Reiner, O. (2013). LIS1 and DCX: Implications for brain development and human disease in relation to microtubules. *Scientifica, 2013*, 393975.
56. Reiner, O., & Sapir, T. (2013). LIS1 functions in normal development and disease. *Current Opinion in Neurobiology, 23*(6), 951–956.
57. Ayanlaja, A. A., Xiong, Y., Gao, Y., Ji, G., Tang, C., Abdikani Abdullah, Z., et al. (2017). Distinct features of doublecortin as a marker of neuronal migration and its implications in cancer cell mobility. *Frontiers in Molecular Neuroscience, 10*, 199.
58. Gleeson, J. G., Lin, P. T., Flanagan, L. A., & Walsh, C. A. (1999). Doublecortin is a microtubule-associated protein and is expressed widely by migrating neurons. *Neuron, 23*(2), 257–271.
59. Bai, J., Ramos, R. L., Ackman, J. B., Thomas, A. M., Lee, R. V., & LoTurco, J. J. (2003). RNAi reveals doublecortin is required for radial migration in rat neocortex. *Nature Neuroscience, 6*(12), 1277–1283.
60. Allen, K. M., & Walsh, C. A. (1999). Genes that regulate neuronal migration in the cerebral cortex. *Epilepsy Research, 36*(2–3), 143–154.
61. Filipovic, R., Santhosh Kumar, S., Fiondella, C., & Loturco, J. (2012). Increasing doublecortin expression promotes migration of human embryonic stem cell-derived neurons. *Stem Cells, 30*(9), 1852–1862.
62. Tsai, J. W., Chen, Y., Kriegstein, A. R., & Vallee, R. B. (2005). LIS1 RNA interference blocks neural stem cell division, morphogenesis, and motility at multiple stages. *The Journal of Cell Biology, 170*(6), 935–945.
63. Takei, Y., Teng, J., Harada, A., & Hirokawa, N. (2000). Defects in axonal elongation and neuronal migration in mice with disrupted tau and map1b genes. *The Journal of Cell Biology, 150*(5), 989–1000.
64. Teng, J., Takei, Y., Harada, A., Nakata, T., Chen, J., & Hirokawa, N. (2001). Synergistic effects of MAP2 and MAP1B knockout in neuronal migration, dendritic outgrowth, and microtubule organization. *The Journal of Cell Biology, 155*(1), 65–76.
65. Gallo, G. (2013). Mechanisms underlying the initiation and dynamics of neuronal filopodia: From neurite formation to synaptogenesis. *International Review of Cell and Molecular Biology, 301*, 95–156.
66. Lafont, F., Rouget, M., Rousselet, A., Valenza, C., & Prochiantz, A. (1993). Specific responses of axons and dendrites to cytoskeleton perturbations: An in vitro study. *Journal of Cell Science, 104*(Pt 2), 433–443.
67. Bentley, D., & Toroian-Raymond, A. (1986). Disoriented pathfinding by pioneer neurone growth cones deprived of filopodia by cytochalasin treatment. *Nature, 323*(6090), 712–715.
68. Azzarelli, R., Kerloch, T., & Pacary, E. (2014). Regulation of cerebral cortex development by Rho GTPases: Insights from in vivo studies. *Frontiers in Cellular Neuroscience, 8*, 445.
69. Kawauchi, T., Chihama, K., Nabeshima, Y., & Hoshino, M. (2003). The in vivo roles of STEF/Tiam1, Rac1 and JNK in cortical neuronal migration. *The EMBO Journal, 22*(16), 4190–4201.

70. Konno, D., Yoshimura, S., Hori, K., Maruoka, H., & Sobue, K. (2005). Involvement of the phosphatidylinositol 3-kinase/rac1 and cdc42 pathways in radial migration of cortical neurons. *The Journal of Biological Chemistry, 280*(6), 5082–5088.
71. Heng, J. I., Nguyen, L., Castro, D. S., Zimmer, C., Wildner, H., Armant, O., et al. (2008). Neurogenin 2 controls cortical neuron migration through regulation of Rnd2. *Nature, 455*(7209), 114–118.
72. Pacary, E., Heng, J., Azzarelli, R., Riou, P., Castro, D., Lebel-Potter, M., et al. (2011). Proneural transcription factors regulate different steps of cortical neuron migration through Rnd-mediated inhibition of RhoA signaling. *Neuron, 69*(6), 1069–1084.
73. Chen, L., Liao, G., Waclaw, R. R., Burns, K. A., Linquist, D., Campbell, K., et al. (2007). Rac1 controls the formation of midline commissures and the competency of tangential migration in ventral telencephalic neurons. *The Journal of Neuroscience, 27*(14), 3884–3893.
74. Yang, T., Sun, Y., Zhang, F., Zhu, Y., Shi, L., Li, H., et al. (2012). POSH localizes activated Rac1 to control the formation of cytoplasmic dilation of the leading process and neuronal migration. *Cell Reports, 2*(3), 640–651.
75. Kassai, H., Terashima, T., Fukaya, M., Nakao, K., Sakahara, M., Watanabe, M., et al. (2008). Rac1 in cortical projection neurons is selectively required for midline crossing of commissural axonal formation. *The European Journal of Neuroscience, 28*(2), 257–267.
76. Nguyen, L., Besson, A., Heng, J. I., Schuurmans, C., Teboul, L., Parras, C., et al. (2006). p27kip1 independently promotes neuronal differentiation and migration in the cerebral cortex. *Genes & Development, 20*(11), 1511–1524.
77. Tang, J., Ip, J. P., Ye, T., Ng, Y. P., Yung, W. H., Wu, Z., et al. (2014). Cdk5-dependent Mst3 phosphorylation and activity regulate neuronal migration through RhoA inhibition. *The Journal of Neuroscience, 34*(22), 7425–7436.
78. Cappello, S., Bohringer, C. R., Bergami, M., Conzelmann, K. K., Ghanem, A., Tomassy, G. S., et al. (2012). A radial glia-specific role of RhoA in double cortex formation. *Neuron, 73*(5), 911–924.
79. Ho, T. T., Merajver, S. D., Lapiere, C. M., Nusgens, B. V., & Deroanne, C. F. (2008). RhoA-GDP regulates RhoB protein stability. Potential involvement of RhoGDIalpha. *The Journal of Biological Chemistry, 283*(31), 21588–21598.
80. Newey, S. E., Velamoor, V., Govek, E. E., & Van Aelst, L. (2005). Rho GTPases, dendritic structure, and mental retardation. *Journal of Neurobiology, 64*(1), 58–74.
81. Govek, E. E., Newey, S. E., & Van Aelst, L. (2005). The role of the Rho GTPases in neuronal development. *Genes & Development, 19*(1), 1–49.
82. Gu, H., Yu, S. P., Gutekunst, C. A., Gross, R. E., & Wei, L. (2013). Inhibition of the Rho signaling pathway improves neurite outgrowth and neuronal differentiation of mouse neural stem cells. *International Journal of Physiology, Pathophysiology and Pharmacology, 5*(1), 11–20.
83. Jeon, C. Y., Moon, M. Y., Kim, J. H., Kim, H. J., Kim, J. G., Li, Y., et al. (2012). Control of neurite outgrowth by RhoA inactivation. *Journal of Neurochemistry, 120*(5), 684–698.
84. Garvalov, B. K., Flynn, K. C., Neukirchen, D., Meyn, L., Teusch, N., Wu, X., et al. (2007). Cdc42 regulates cofilin during the establishment of neuronal polarity. *The Journal of Neuroscience, 27*(48), 13117–13129.
85. Rosario, M., Schuster, S., Juttner, R., Parthasarathy, S., Tarabykin, V., & Birchmeier, W. (2012). Neocortical dendritic complexity is controlled during development by NOMA-GAP-dependent inhibition of Cdc42 and activation of cofilin. *Genes & Development, 26*(15), 1743–1757.
86. Yokota, Y., Eom, T. Y., Stanco, A., Kim, W. Y., Rao, S., Snider, W. D., et al. (2010). Cdc42 and Gsk3 modulate the dynamics of radial glial growth, inter-radial glial interactions and polarity in the developing cerebral cortex. *Development, 137*(23), 4101–4110.
87. Gleeson, J. G., & Walsh, C. A. (2000). Neuronal migration disorders: From genetic diseases to developmental mechanisms. *Trends in Neurosciences, 23*(8), 352–359.

88. Desikan, R. S., & Barkovich, A. J. (2016). Malformations of cortical development. *Annals of Neurology, 80*(6), 797–810.
89. Shu, T., Ayala, R., Nguyen, M. D., Xie, Z., Gleeson, J. G., & Tsai, L. H. (2004). Ndel1 operates in a common pathway with LIS1 and cytoplasmic dynein to regulate cortical neuronal positioning. *Neuron, 44*(2), 263–277.
90. Jiang, X., & Nardelli, J. (2016). Cellular and molecular introduction to brain development. *Neurobiology of Disease, 92*(Pt A), 3–17.
91. Lasser, M., Tiber, J., & Lowery, L. A. (2018). The role of the microtubule cytoskeleton in neurodevelopmental disorders. *Frontiers in Cellular Neuroscience, 12*, 165.
92. Parrini, E., Conti, V., Dobyns, W. B., & Guerrini, R. (2016). Genetic basis of brain malformations. *Molecular Syndromology, 7*(4), 220–233.
93. Chang, B. S., Duzcan, F., Kim, S., Cinbis, M., Aggarwal, A., Apse, K. A., et al. (2007). The role of RELN in lissencephaly and neuropsychiatric disease. *American Journal of Medical Genetics. Part B, Neuropsychiatric Genetics, 144B*(1), 58–63.
94. Crino, P. (2001). New Reln mutation associated with lissencephaly and epilepsy. *Epilepsy Currents, 1*(2), 72.
95. Riikonen, R. (2017). Insulin-like growth factors in the pathogenesis of neurological diseases in children. *International Journal of Molecular Sciences, 18*(10), 2056.
96. Sheen, V. L. (2012). Periventricular heterotopia: Shuttling of proteins through vesicles and actin in cortical development and disease. *Scientifica, 2012*, 480129.
97. Fox, J. W., Lamperti, E. D., Eksioglu, Y. Z., Hong, S. E., Feng, Y., Graham, D. A., et al. (1998). Mutations in filamin 1 prevent migration of cerebral cortical neurons in human periventricular heterotopia. *Neuron, 21*(6), 1315–1325.
98. Riviere, J. B., van Bon, B. W., Hoischen, A., Kholmanskikh, S. S., O'Roak, B. J., Gilissen, C., et al. (2012). De novo mutations in the actin genes ACTB and ACTG1 cause Baraitser-Winter syndrome. *Nature Genetics, 44*(4), 440–444, S1-2.
99. Di Donato, N., Rump, A., Koenig, R., Der Kaloustian, V. M., Halal, F., Sonntag, K., et al. (2014). Severe forms of Baraitser-Winter syndrome are caused by ACTB mutations rather than ACTG1 mutations. *European Journal of Human Genetics, 22*(2), 179–183.
100. Uppal, N., & Hof, P. R. (2013). Chapter 3.6 - Discrete cortical neuropathology in autism spectrum disorders. In *The neuroscience of autism spectrum disorders* (pp. 313–325). San Diego: Academic Press.
101. Schumann, C. M., Noctor, S. C., & Amaral, D. G. (2011). Autism spectrum disorders. In D. G. Amaral, D. Geschwind, & D. Dawson (Eds.), *Neuropathology of autism spectrum disorders: Postmortem studies*. Oxford: Oxford University Press.
102. Amaral, D. G., Schumann, C. M., & Nordahl, C. W. (2008). Neuroanatomy of autism. *Trends in Neurosciences, 31*(3), 137–145.
103. Blatt, G. J. (2012). The neuropathology of autism. *Scientifica, 2012*, 703675.
104. Bauman, M. L., & Kemper, T. L. (2005). Neuroanatomic observations of the brain in autism: A review and future directions. *International Journal of Developmental Neuroscience, 23*(2–3), 183–187.
105. Varghese, M., Keshav, N., Jacot-Descombes, S., Warda, T., Wicinski, B., Dickstein, D. L., et al. (2017). Autism spectrum disorder: Neuropathology and animal models. *Acta Neuropathologica, 134*(4), 537–566.
106. Schumann, C. M., & Nordahl, C. W. (2011). Bridging the gap between MRI and postmortem research in autism. *Brain Research, 1380*, 175–186.
107. Hampson, D. R., & Blatt, G. J. (2015). Autism spectrum disorders and neuropathology of the cerebellum. *Frontiers in Neuroscience, 9*, 420.
108. Gadad, B. S., Hewitson, L., Young, K. A., & German, D. C. (2013). Neuropathology and animal models of autism: Genetic and environmental factors. *Autism Research and Treatment, 2013*, 731935.

109. Santos, M., Uppal, N., Butti, C., Wicinski, B., Schmeidler, J., Giannakopoulos, P., et al. (2011). Von Economo neurons in autism: A stereologic study of the frontoinsular cortex in children. *Brain Research, 1380*, 206–217.
110. Wegiel, J., Kuchna, I., Nowicki, K., Imaki, H., Wegiel, J., Marchi, E., et al. (2010). The neuropathology of autism: Defects of neurogenesis and neuronal migration, and dysplastic changes. *Acta Neuropathologica, 119*(6), 755–770.
111. Fatemi, S. H., & Folsom, T. D. (2009). The neurodevelopmental hypothesis of schizophrenia, revisited. *Schizophrenia Bulletin, 35*(3), 528–548.
112. Chen, R., Jiao, Y., & Herskovits, E. H. (2011). Structural MRI in autism spectrum disorder. *Pediatric Research, 69*(5 Pt 2), 63R–68R.
113. Kucharsky Hiess, R., Alter, R., Sojoudi, S., Ardekani, B. A., Kuzniecky, R., & Pardoe, H. R. (2015). Corpus callosum area and brain volume in autism spectrum disorder: Quantitative analysis of structural MRI from the ABIDE database. *Journal of Autism and Developmental Disorders, 45*(10), 3107–3114.
114. Schumann, C. M., Bloss, C. S., Barnes, C. C., Wideman, G. M., Carper, R. A., Akshoomoff, N., et al. (2010). Longitudinal magnetic resonance imaging study of cortical development through early childhood in autism. *The Journal of Neuroscience, 30*(12), 4419–4427.
115. Zielinski, B. A., Prigge, M. B., Nielsen, J. A., Froehlich, A. L., Abildskov, T. J., Anderson, J. S., et al. (2014). Longitudinal changes in cortical thickness in autism and typical development. *Brain, 137*(Pt 6), 1799–1812.
116. Dementieva, Y. A., Vance, D. D., Donnelly, S. L., Elston, L. A., Wolpert, C. M., Ravan, S. A., et al. (2005). Accelerated head growth in early development of individuals with autism. *Pediatric Neurology, 32*(2), 102–108.
117. Fombonne, E., Roge, B., Claverie, J., Courty, S., & Fremolle, J. (1999). Microcephaly and macrocephaly in autism. *Journal of Autism and Developmental Disorders, 29*(2), 113–119.
118. Anagnostou, E., & Taylor, M. J. (2011). Review of neuroimaging in autism spectrum disorders: What have we learned and where we go from here. *Molecular Autism, 2*(1), 4.
119. Hardan, A. Y., Pabalan, M., Gupta, N., Bansal, R., Melhem, N. M., Fedorov, S., et al. (2009). Corpus callosum volume in children with autism. *Psychiatry Research, 174*(1), 57–61.
120. Frazier, T. W., & Hardan, A. Y. (2009). A meta-analysis of the corpus callosum in autism. *Biological Psychiatry, 66*(10), 935–941.
121. Stanfield, A. C., McIntosh, A. M., Spencer, M. D., Philip, R., Gaur, S., & Lawrie, S. M. (2008). Towards a neuroanatomy of autism: A systematic review and meta-analysis of structural magnetic resonance imaging studies. *European Psychiatry, 23*(4), 289–299.
122. Ameis, S. H., Fan, J., Rockel, C., Voineskos, A. N., Lobaugh, N. J., Soorya, L., et al. (2011). Impaired structural connectivity of socio-emotional circuits in autism spectrum disorders: A diffusion tensor imaging study. *PLoS One, 6*(11), e28044.
123. Hardan, A. Y., Libove, R. A., Keshavan, M. S., Melhem, N. M., & Minshew, N. J. (2009). A preliminary longitudinal magnetic resonance imaging study of brain volume and cortical thickness in autism. *Biological Psychiatry, 66*(4), 320–326.
124. Neale, B. M., Kou, Y., Liu, L., Ma'ayan, A., Samocha, K. E., Sabo, A., et al. (2012). Patterns and rates of exonic de novo mutations in autism spectrum disorders. *Nature, 485*(7397), 242–245.
125. Deriziotis, P., O'Roak, B. J., Graham, S. A., Estruch, S. B., Dimitropoulou, D., Bernier, R. A., et al. (2014). De novo TBR1 mutations in sporadic autism disrupt protein functions. *Nature Communications, 5*, 4954.
126. Huang, T. N., & Hsueh, Y. P. (2015). Brain-specific transcriptional regulator T-brain-1 controls brain wiring and neuronal activity in autism spectrum disorders. *Frontiers in Neuroscience, 9*, 406.
127. Traylor, R. N., Dobyns, W. B., Rosenfeld, J. A., Wheeler, P., Spence, J. E., Bandholz, A. M., et al. (2012). Investigation of TBR1 hemizygosity: Four individuals with 2q24 microdeletions. *Molecular Syndromology, 3*(3), 102–112.

128. Hamdan, F. F., Srour, M., Capo-Chichi, J. M., Daoud, H., Nassif, C., Patry, L., et al. (2014). De novo mutations in moderate or severe intellectual disability. *PLoS Genetics, 10*(10), e1004772.
129. Bedogni, F., Hodge, R. D., Elsen, G. E., Nelson, B. R., Daza, R. A., Beyer, R. P., et al. (2010). Tbr1 regulates regional and laminar identity of postmitotic neurons in developing neocortex. *Proceedings of the National Academy of Sciences of the United States of America, 107*(29), 13129–13134.
130. Packer, A. (2016). Neocortical neurogenesis and the etiology of autism spectrum disorder. *Neuroscience and Biobehavioral Reviews, 64*, 185–195.
131. Gallagher, D., Voronova, A., Zander, M. A., Cancino, G. I., Bramall, A., Krause, M. P., et al. (2015). Ankrd11 is a chromatin regulator involved in autism that is essential for neural development. *Developmental Cell, 32*(1), 31–42.
132. De Rubeis, S., He, X., Goldberg, A. P., Poultney, C. S., Samocha, K., Cicek, A. E., et al. (2014). Synaptic, transcriptional and chromatin genes disrupted in autism. *Nature, 515*(7526), 209–215.
133. Sanders, S. J., He, X., Willsey, A. J., Ercan-Sencicek, A. G., Samocha, K. E., Cicek, A. E., et al. (2015). Insights into autism spectrum disorder genomic architecture and biology from 71 risk loci. *Neuron, 87*(6), 1215–1233.
134. Tuoc, T. C., Narayanan, R., & Stoykova, A. (2013). BAF chromatin remodeling complex: Cortical size regulation and beyond. *Cell Cycle, 12*(18), 2953–2959.
135. Tuoc, T. C., Boretius, S., Sansom, S. N., Pitulescu, M. E., Frahm, J., Livesey, F. J., et al. (2013). Chromatin regulation by BAF170 controls cerebral cortical size and thickness. *Developmental Cell, 25*(3), 256–269.
136. Chen, Y., Huang, W. C., Sejourne, J., Clipperton-Allen, A. E., & Page, D. T. (2015). Pten mutations Alter brain growth trajectory and allocation of cell types through elevated beta-catenin signaling. *The Journal of Neuroscience, 35*(28), 10252–10267.
137. Strauss, K. A., Puffenberger, E. G., Huentelman, M. J., Gottlieb, S., Dobrin, S. E., Parod, J. M., et al. (2006). Recessive symptomatic focal epilepsy and mutant contactin-associated protein-like 2. *The New England Journal of Medicine, 354*(13), 1370–1377.
138. Bakkaloglu, B., O'Roak, B. J., Louvi, A., Gupta, A. R., Abelson, J. F., Morgan, T. M., et al. (2008). Molecular cytogenetic analysis and resequencing of contactin associated protein-like 2 in autism spectrum disorders. *American Journal of Human Genetics, 82*(1), 165–173.
139. Alarcon, M., Abrahams, B. S., Stone, J. L., Duvall, J. A., Perederiy, J. V., Bomar, J. M., et al. (2008). Linkage, association, and gene-expression analyses identify CNTNAP2 as an autism-susceptibility gene. *American Journal of Human Genetics, 82*(1), 150–159.
140. Conti, S., Condo, M., Posar, A., Mari, F., Resta, N., Renieri, A., et al. (2012). Phosphatase and tensin homolog (PTEN) gene mutations and autism: Literature review and a case report of a patient with Cowden syndrome, autistic disorder, and epilepsy. *Journal of Child Neurology, 27*(3), 392–397.
141. Wiegreffe, C., Simon, R., Peschkes, K., Kling, C., Strehle, M., Cheng, J., et al. (2015). Bcl11a (Ctip1) controls migration of cortical projection neurons through regulation of Sema3c. *Neuron, 87*(2), 311–325.
142. Li, X., Xiao, J., Frohlich, H., Tu, X., Li, L., Xu, Y., et al. (2015). Foxp1 regulates cortical radial migration and neuronal morphogenesis in developing cerebral cortex. *PLoS One, 10*(5), e0127671e.
143. Miyoshi, G., & Fishell, G. (2012). Dynamic FoxG1 expression coordinates the integration of multipolar pyramidal neuron precursors into the cortical plate. *Neuron, 74*(6), 1045–1058.
144. La Fata, G., Gartner, A., Dominguez-Iturza, N., Dresselaers, T., Dawitz, J., Poorthuis, R. B., et al. (2014). FMRP regulates multipolar to bipolar transition affecting neuronal migration and cortical circuitry. *Nature Neuroscience, 17*(12), 1693–1700.
145. Boitard, M., Bocchi, R., Egervari, K., Petrenko, V., Viale, B., Gremaud, S., et al. (2015). Wnt signaling regulates multipolar-to-bipolar transition of migrating neurons in the cerebral cortex. *Cell Reports, 10*(8), 1349–1361.

146. Hori, K., & Hoshino, M. (2017). Neuronal migration and AUTS2 syndrome. *Brain Sciences, 7*(12), 54.
147. Hori, K., Nagai, T., Shan, W., Sakamoto, A., Taya, S., Hashimoto, R., et al. (2014). Cytoskeletal regulation by AUTS2 in neuronal migration and neuritogenesis. *Cell Reports, 9*(6), 2166–2179.
148. Yoo, H. (2015). Genetics of autism Spectrum disorder: Current status and possible clinical applications. *Exp Neurobiol., 24*(4), 257–272.
149. Buxbaum, J. D. (2009). Multiple rare variants in the etiology of autism spectrum disorders. *Dialogues in Clinical Neuroscience, 11*(1), 35–43.
150. Weiner, D. J., Wigdor, E. M., Ripke, S., Walters, R. K., Kosmicki, J. A., Grove, J., et al. (2017). Polygenic transmission disequilibrium confirms that common and rare variation act additively to create risk for autism spectrum disorders. *Nature Genetics, 49*(7), 978–985.
151. Bray, N. (2017). Neurodevelopmental disorders: Converging on autism spectrum disorder. *Nature Reviews. Neuroscience, 18*(2), 67.
152. Pinto, D., Delaby, E., Merico, D., Barbosa, M., Merikangas, A., Klei, L., et al. (2014). Convergence of genes and cellular pathways dysregulated in autism spectrum disorders. *American Journal of Human Genetics, 94*(5), 677–694.
153. Berg, J. M., & Geschwind, D. H. (2012). Autism genetics: Searching for specificity and convergence. *Genome Biology, 13*(7), 247.
154. Gupta, S., Ellis, S. E., Ashar, F. N., Moes, A., Bader, J. S., Zhan, J., et al. (2014). Transcriptome analysis reveals dysregulation of innate immune response genes and neuronal activity-dependent genes in autism. *Nature Communications, 5*, 5748.
155. Gokoolparsadh, A., Sutton, G. J., Charamko, A., Green, N. F., Pardy, C. J., & Voineagu, I. (2016). Searching for convergent pathways in autism spectrum disorders: Insights from human brain transcriptome studies. *Cellular and Molecular Life Sciences, 73*(23), 4517–4530.
156. Voineagu, I., & Eapen, V. (2013). Converging pathways in autism spectrum disorders: Interplay between synaptic dysfunction and immune responses. *Frontiers in Human Neuroscience, 7*, 738.
157. Voineagu, I., Wang, X., Johnston, P., Lowe, J. K., Tian, Y., Horvath, S., et al. (2011). Transcriptomic analysis of autistic brain reveals convergent molecular pathology. *Nature, 474*(7351), 380–384.
158. Wen, Y., Alshikho, M. J., & Herbert, M. R. (2016). Pathway network analyses for autism reveal multisystem involvement, major overlaps with other diseases and convergence upon MAPK and calcium signaling. *PLoS One, 11*(4), e0153329.
159. Luo, W., Zhang, C., Jiang, Y. H., & Brouwer, C. R. (2018). Systematic reconstruction of autism biology from massive genetic mutation profiles. *Science Advances, 4*(4), e1701799.
160. Sanders, S. J. (2015). First glimpses of the neurobiology of autism spectrum disorder. *Current Opinion in Genetics & Development, 33*, 80–92.
161. Ernst, C. (2016). Proliferation and differentiation deficits are a major convergence point for neurodevelopmental disorders. *Trends in Neurosciences, 39*(5), 290–299.
162. Stevens, H. E., Smith, K. M., Rash, B. G., & Vaccarino, F. M. (2010). Neural stem cell regulation, fibroblast growth factors, and the developmental origins of neuropsychiatric disorders. *Frontiers in Neuroscience, 4*, 59.
163. Sacco, R., Cacci, E., & Novarino, G. (2018). Neural stem cells in neuropsychiatric disorders. *Current Opinion in Neurobiology, 48*, 131–138.
164. Willsey, A. J., Sanders, S. J., Li, M., Dong, S., Tebbenkamp, A. T., Muhle, R. A., et al. (2013). Coexpression networks implicate human midfetal deep cortical projection neurons in the pathogenesis of autism. *Cell, 155*(5), 997–1007.
165. Parikshak, N. N., Luo, R., Zhang, A., Won, H., Lowe, J. K., Chandran, V., et al. (2013). Integrative functional genomic analyses implicate specific molecular pathways and circuits in autism. *Cell, 155*(5), 1008–1021.
166. Sara Ballouz, Paul Pavlidis, Jesse Gillis, Using predictive specificity to determine when gene set analysis is biologically meaningful. Nucleic Acids Research:gkw957.

167. Satterstrom F. K., Kosmicki J A., Wang J, Breen M S., De Rubeis S, Joon-Yong An, et al. (2020) Large-Scale Exome Sequencing Study Implicates Both Developmental and Functional Changes in the Neurobiology of Autism. *Cell 180*(3):568–584.e23
168. Gandal, M. J., Haney, J. R., Parikshak, N. N., Leppa, V., Ramaswami, G., Hartl, C., et al. (2018). Shared molecular neuropathology across major psychiatric disorders parallels polygenic overlap. *Science, 359*(6376), 693–697.
169. Hoeffer, C. A., Sanchez, E., Hagerman, R. J., Mu, Y., Nguyen, D. V., Wong, H., et al. (2012). Altered mTOR signaling and enhanced CYFIP2 expression levels in subjects with fragile X syndrome. *Genes, Brain, and Behavior, 11*(3), 332–341.
170. Olson, C. O., Pejhan, S., Kroft, D., Sheikholeslami, K., Fuss, D., Buist, M., et al. (2018). MECP2 mutation interrupts nucleolin-mTOR-P70S6K Signaling in Rett syndrome patients. *Frontiers in Genetics, 9*, 635.
171. Ricciardi, S., Boggio, E. M., Grosso, S., Lonetti, G., Forlani, G., Stefanelli, G., et al. (2011). Reduced AKT/mTOR signaling and protein synthesis dysregulation in a Rett syndrome animal model. *Human Molecular Genetics, 20*(6), 1182–1196.
172. Xing, X., Zhang, J., Wu, K., Cao, B., Li, X., Jiang, F., et al. (2019). Suppression of Akt-mTOR pathway rescued the social behavior in Cntnap2-deficient mice. *Scientific Reports, 9*(1), 3041.
173. Rosina, E., Battan, B., Siracusano, M., Di Criscio, L., Hollis, F., Pacini, L., et al. (2019). Disruption of mTOR and MAPK pathways correlates with severity in idiopathic autism. *Translational Psychiatry, 9*(1), 50.
174. Thomson, J. A., Itskovitz-Eldor, J., Shapiro, S. S., Waknitz, M. A., Swiergiel, J. J., Marshall, V. S., et al. (1998). Embryonic stem cell lines derived from human blastocysts. *Science, 282*(5391), 1145–1147.
175. Takahashi, K., & Yamanaka, S. (2006). Induction of pluripotent stem cells from mouse embryonic and adult fibroblast cultures by defined factors. *Cell, 126*(4), 663–676.
176. Yamanaka, S. (2006). Molecular mechanisms underlying pluripotency of embryonic stem cells. *Seikagaku, 78*(1), 27–33.
177. Okita, K., & Yamanaka, S. (2006). Intracellular signaling pathways regulating pluripotency of embryonic stem cells. *Current Stem Cell Research & Therapy, 1*(1), 103–111.
178. Nakagawa, M., Koyanagi, M., Tanabe, K., Takahashi, K., Ichisaka, T., Aoi, T., et al. (2008). Generation of induced pluripotent stem cells without Myc from mouse and human fibroblasts. *Nature Biotechnology, 26*(1), 101–106.
179. Takahashi, K., Tanabe, K., Ohnuki, M., Narita, M., Ichisaka, T., Tomoda, K., et al. (2007). Induction of pluripotent stem cells from adult human fibroblasts by defined factors. *Cell, 131*(5), 861–872.
180. Marchetto, M. C., Carromeu, C., Acab, A., Yu, D., Yeo, G. W., Mu, Y., et al. (2010). A model for neural development and treatment of Rett syndrome using human induced pluripotent stem cells. *Cell, 143*(4), 527–539.
181. Pasca, S. P., Portmann, T., Voineagu, I., Yazawa, M., Shcheglovitov, A., Pasca, A. M., et al. (2011). Using iPSC-derived neurons to uncover cellular phenotypes associated with Timothy syndrome. *Nature Medicine, 17*(12), 1657–1662.
182. Urbach, A., Bar-Nur, O., Daley, G. Q., & Benvenisty, N. (2010). Differential modeling of fragile X syndrome by human embryonic stem cells and induced pluripotent stem cells. *Cell Stem Cell, 6*(5), 407–411.
183. Krey, J. F., Pasca, S. P., Shcheglovitov, A., Yazawa, M., Schwemberger, R., Rasmusson, R., et al. (2013). Timothy syndrome is associated with activity-dependent dendritic retraction in rodent and human neurons. *Nature Neuroscience, 16*(2), 201–209.
184. Tian, Y., Voineagu, I., Pasca, S. P., Won, H., Chandran, V., Horvath, S., et al. (2014). Alteration in basal and depolarization induced transcriptional network in iPSC derived neurons from Timothy syndrome. *Genome Medicine, 6*(10), 75.
185. Mor-Shaked, H., & Eiges, R. (2016). Modeling fragile X syndrome using human pluripotent stem cells. *Genes, 7*(10), 77.

186. Li, M., Zhao, H., Ananiev, G. E., Musser, M. T., Ness, K. H., Maglaque, D. L., et al. (2017). Establishment of reporter lines for detecting fragile X mental retardation (FMR1) gene reactivation in human neural cells. *Stem Cells, 35*(1), 158–169.
187. Bhattacharyya, A., & Zhao, X. (2016). Human pluripotent stem cell models of Fragile X syndrome. *Molecular and Cellular Neurosciences, 73*, 43–51.
188. Doers, M. E., Musser, M. T., Nichol, R., Berndt, E. R., Baker, M., Gomez, T. M., et al. (2014). iPSC-derived forebrain neurons from FXS individuals show defects in initial neurite outgrowth. *Stem Cells and Development, 23*(15), 1777–1787.
189. Shcheglovitov, A., Shcheglovitova, O., Yazawa, M., Portmann, T., Shu, R., Sebastiano, V., et al. (2013). SHANK3 and IGF1 restore synaptic deficits in neurons from 22q13 deletion syndrome patients. *Nature, 503*(7475), 267–271.
190. Yi, F., Danko, T., Botelho, S. C., Patzke, C., Pak, C., Wernig, M., et al. (2016). Autism-associated SHANK3 haploinsufficiency causes Ih channelopathy in human neurons. *Science, 352*(6286), aaf2669.
191. Kathuria, A., Nowosiad, P., Jagasia, R., Aigner, S., Taylor, R. D., Andreae, L. C., et al. (2018). Stem cell-derived neurons from autistic individuals with SHANK3 mutation show morphogenetic abnormalities during early development. *Molecular Psychiatry, 23*(3), 735–746.
192. Deshpande, A., Yadav, S., Dao, D. Q., Wu, Z. Y., Hokanson, K. C., Cahill, M. K., et al. (2017). Cellular phenotypes in human iPSC-derived neurons from a genetic model of autism spectrum disorder. *Cell Reports, 21*(10), 2678–2687.
193. Flaherty, E., Deranieh, R. M., Artimovich, E., Lee, I. S., Siegel, A. J., Levy, D. L., et al. (2017). Patient-derived hiPSC neurons with heterozygous CNTNAP2 deletions display altered neuronal gene expression and network activity. *NPJ Schizophrenia, 3*, 35.
194. Brennand, K. J., Simone, A., Jou, J., Gelboin-Burkhart, C., Tran, N., Sangar, S., et al. (2011). Modelling schizophrenia using human induced pluripotent stem cells. *Nature, 473*(7346), 221–225.
195. Brennand, K., Savas, J. N., Kim, Y., Tran, N., Simone, A., Hashimoto-Torii, K., et al. (2015). Phenotypic differences in hiPSC NPCs derived from patients with schizophrenia. *Molecular Psychiatry, 20*(3), 361–368.
196. Griesi-Oliveira, K., Acab, A., Gupta, A. R., Sunaga, D. Y., Chailangkarn, T., Nicol, X., et al. (2015). Modeling non-syndromic autism and the impact of TRPC6 disruption in human neurons. *Molecular Psychiatry, 20*(11), 1350–1365.
197. Mariani, J., Coppola, G., Zhang, P., Abyzov, A., Provini, L., Tomasini, L., et al. (2015). FOXG1-dependent dysregulation of GABA/glutamate neuron differentiation in autism spectrum disorders. *Cell, 162*(2), 375–390.
198. Marchetto, M. C., Belinson, H., Tian, Y., Freitas, B. C., Fu, C., Vadodaria, K., et al. (2017). Altered proliferation and networks in neural cells derived from idiopathic autistic individuals. *Molecular Psychiatry, 22*(6), 820–835.
199. Liu, X., Campanac, E., Cheung, H. H., Ziats, M. N., Canterel-Thouennon, L., Raygada, M., et al. (2017). Idiopathic autism: Cellular and molecular phenotypes in pluripotent stem cell-derived neurons. *Molecular Neurobiology, 54*(6), 4507–4523.
200. Stadtfeld, M., & Hochedlinger, K. (2010). Induced pluripotency: History, mechanisms, and applications. *Genes & Development, 24*(20), 2239–2263.
201. Schwartzentruber, J., Foskolou, S., Kilpinen, H., Rodrigues, J., Alasoo, K., Knights, A. J., et al. (2018). Molecular and functional variation in iPSC-derived sensory neurons. *Nature Genetics, 50*(1), 54–61.
202. Vitale, A. M., Matigian, N. A., Ravishankar, S., Bellette, B., Wood, S. A., Wolvetang, E. J., et al. (2012). Variability in the generation of induced pluripotent stem cells: Importance for disease modeling. *Stem Cells Translational Medicine, 1*(9), 641–650.
203. Vigilante, A., Laddach, A., Moens, N., Meleckyte, R., Leha, A., Ghahramani, A., et al. (2019). Identifying extrinsic versus intrinsic drivers of variation in cell behavior in human iPSC lines from healthy donors. *Cell Reports, 26*(8), 2078–2087. e3.

204. Carcamo-Orive, I., Hoffman, G. E., Cundiff, P., Beckmann, N. D., D'Souza, S. L., Knowles, J. W., et al. (2017). Analysis of transcriptional variability in a large human iPSC library reveals genetic and non-genetic determinants of heterogeneity. *Cell Stem Cell, 20*(4), 518–532. e9.
205. Volpato, V., Smith, J., Sandor, C., Ried, J. S., Baud, A., Handel, A., et al. (2018). Reproducibility of molecular phenotypes after long-term differentiation to human iPSC-derived neurons: A multi-site omics study. *Stem Cell Reports, 11*(4), 897–911.
206. Deng, J., Shoemaker, R., Xie, B., Gore, A., LeProust, E. M., Antosiewicz-Bourget, J., et al. (2009). Targeted bisulfite sequencing reveals changes in DNA methylation associated with nuclear reprogramming. *Nature Biotechnology, 27*(4), 353–360.
207. Doi, A., Park, I. H., Wen, B., Murakami, P., Aryee, M. J., Irizarry, R., et al. (2009). Differential methylation of tissue- and cancer-specific CpG island shores distinguishes human induced pluripotent stem cells, embryonic stem cells and fibroblasts. *Nature Genetics, 41*(12), 1350–1353.
208. Lister, R., Pelizzola, M., Kida, Y. S., Hawkins, R. D., Nery, J. R., Hon, G., et al. (2011). Hotspots of aberrant epigenomic reprogramming in human induced pluripotent stem cells. *Nature, 471*(7336), 68–73.
209. Kim, K., Doi, A., Wen, B., Ng, K., Zhao, R., Cahan, P., et al. (2010). Epigenetic memory in induced pluripotent stem cells. *Nature, 467*(7313), 285–290.
210. Polo, J. M., Liu, S., Figueroa, M. E., Kulalert, W., Eminli, S., Tan, K. Y., et al. (2010). Cell type of origin influences the molecular and functional properties of mouse induced pluripotent stem cells. *Nature Biotechnology, 28*(8), 848–855.
211. Falk, A., Heine, V. M., Harwood, A. J., Sullivan, P. F., Peitz, M., Brüstle, O., Shen, S., Sun, Y-M., Glover, J. C., Posthuma, D., Djurovic, S. (2016) Modeling psychiatric disorders: from genomic findings to cellular phenotypes. *Molecular Psychiatry 21*(9):1167–1179.
212. Halevy, T., & Urbach, A. (2014). Comparing ESC and iPSC-based models for human genetic disorders. *Journal of Clinical Medicine, 3*(4), 1146–1162.
213. Williams, M., Prem, S., Zhou, X., Matteson, P., Yeung, P. L., & Lu, C. W., et al. (2018). Rapid detection of neurodevelopmental phenotypes in human neural precursor cells (NPCs). *Journal of Visualized Experiments* (133). https://doi.org/10.3791/56628
214. Rossman, I. T., Lin, L., Morgan, K. M., Digiovine, M., Van Buskirk, E. K., Kamdar, S., et al. (2014). Engrailed2 modulates cerebellar granule neuron precursor proliferation, differentiation and insulin-like growth factor 1 signaling during postnatal development. *Molecular Autism, 5*(1), 9.

Investigation of Schizophrenia with Human Induced Pluripotent Stem Cells

Samuel K. Powell, Callan P. O'Shea, Sara Rose Shannon, Schahram Akbarian, and Kristen J. Brennand

1 Introduction

The sheer complexity of the nervous system, the clinical and biological heterogeneity of affected patients, and limited approaches to access relevant tissues for disease modeling have together made the investigation of schizophrenia persistently daunting. New hope has been generated by the still recent advent of human induced pluripotent stem cell (hiPSC) models of neuropsychiatric conditions. The reprogramming of hiPSCs into defined cell types is a technical advancement that has made available living neural tissues derived from patients and controls for studying neurobiological abnormalities driving disease pathology, modeling the impact of genetic and non-genetic risk factors, and testing interventions—whether genetic or pharmacologic—to prevent or reverse disease-associated phenotypes. As a result, cell reprogramming enables both patient-specific study of genetic disease and in vitro modeling of the complex genetic risk factors underlying neurological disorders such as schizophrenia.

S. K. Powell
Medical Scientist Training Program, Icahn School of Medicine at Mount Sinai, New York, NY, USA

Friedman Brain Institute, Icahn School of Medicine at Mount Sinai, New York, NY, USA

Department of Genetics and Genomics, Icahn School of Medicine at Mount Sinai, New York, NY, USA

Department of Neuroscience, Icahn School of Medicine at Mount Sinai, New York, NY, USA

Department of Psychiatry, Icahn School of Medicine at Mount Sinai, New York, NY, USA

C. P. O'Shea · S. R. Shannon
Friedman Brain Institute, Icahn School of Medicine at Mount Sinai, New York, NY, USA

Department of Genetics and Genomics, Icahn School of Medicine at Mount Sinai, New York, NY, USA

E. DiCicco-Bloom, J. H. Millonig (eds.), *Neurodevelopmental Disorders*, Advances in Neurobiology 25, https://doi.org/10.1007/978-3-030-45493-7_6

In this chapter, we preview the syndromal presentations of schizophrenia and its treatment to provide the reader with clinically relevant information regarding both the severity of the disorder and the challenges associated with its management. As needed for integrating findings discussed later in the chapter, we highlight key genomic and biological processes disrupted in schizophrenia, with an emphasis on the essential role of neurodevelopmental perturbations in disease etiology. After previewing relevant methodological considerations, we discuss many of the studies that have employed hiPSC-based models to provide fundamental insights into mechanisms of disease in schizophrenia. We end with an overview of innovations that will facilitate the continued usefulness of hiPSCs for studying schizophrenia and for ultimately improving the lives of those affected by it.

2 Overview of Schizophrenia

2.1 Clinical Presentations of Schizophrenia

Schizophrenia is a neuropsychiatric condition that can have devastating impacts on the lives of patients and those around them. As is the case with all psychiatric disorders, diagnosis of schizophrenia is made based upon the presence of specific signs and symptoms that have persisted for a specified period of time, are not attributable to organic disease or substance use, and that cause significant distress, disability, and/or impairment in functioning. Unfortunately, there are no biologically informed tests or criteria that can be used to confirm the presence of schizophrenia, and there is notable heterogeneity among patients despite sharing an identical diagnosis.

Phenomenological descriptions of schizophrenia have varied throughout history and by clinical assessment methodologies. The development of symptom-rating scales and their widespread application across numerous patient samples and contexts served as key empirical foundations for disease description. Factor analyses [1–4] originally clustered the clinical features of schizophrenia into three subsyndromes,

S. Akbarian
Friedman Brain Institute, Icahn School of Medicine at Mount Sinai, New York, NY, USA

Department of Neuroscience, Icahn School of Medicine at Mount Sinai, New York, NY, USA

Department of Psychiatry, Icahn School of Medicine at Mount Sinai, New York, NY, USA

K. J. Brennand (✉)
Friedman Brain Institute, Icahn School of Medicine at Mount Sinai, New York, NY, USA

Department of Genetics and Genomics, Icahn School of Medicine at Mount Sinai,
New York, NY, USA

Department of Neuroscience, Icahn School of Medicine at Mount Sinai, New York, NY, USA

Department of Psychiatry, Icahn School of Medicine at Mount Sinai, New York, NY, USA
e-mail: kristen.brennand@mssm.edu

each relating to a specific group of defined symptoms: (1) psychotic symptoms (e.g., delusions and hallucinations), (2) negative symptoms (e.g., anhedonia, alogia, social withdrawal, blunted or flat affect), and (3) disorganized symptoms (e.g., thought disorder, bizarre behaviors). In recent years, particularly with the advent and extensive application of the Positive and Negative Syndrome Scale (PANSS) [5], robust support has been provided for an expanded *Five Factor Model* [6, 7] of schizophrenia that includes the addition of two more dimensions: (1) depression and anxiety and (2) agitation. These two different syndromal models of schizophrenia and their constituent signs and symptoms are listed and defined in Tables 1 and 2.

Table 1 Definition of factor item on the positive and negative syndrome scale (PANSS)

Factor item	Definition
Delusions	Beliefs which are unfounded, unrealistic, and idiosyncratic
Hallucinations	Verbal report or behavior indicating perceptions which are not generated by external stimuli. May occur in the auditory, visual, olfactory, or somatic realms
Grandiosity	Exaggerated self-opinion and unrealistic convictions of superiority, including delusions of extraordinary abilities, wealth, knowledge, fame, power, and moral righteousness
Suspiciousness/persecution	Unrealistic or exaggerated ideas of persecution, as reflected in guardedness, a distrustful attitude, suspicious hypervigilance, or frank delusions that others mean one harm
Unusual thought content	Thinking characterized by strange, fantastic, or bizarre ideas, ranging from those which are remote or atypical to those which are distorted, illogical, and patently absurd
Blunted affect	Diminished emotional responsiveness as characterized by a reduction in facial expression, modulation of feelings, and communicative gestures
Emotional withdrawal	Lack of interest in, involvement with, and affective commitment to life's events
Poor rapport	Lack of interpersonal empathy, openness in conversation, and sense of closeness, interest, or involvement with the interviewer
Passive/apathetic social withdrawal	Diminished interest or initiative in social interactions due to passivity, apathy, anergy, or avolition. This leads to reduced interpersonal involvement and neglect of activities of daily living
Lack of spontaneity and flow of conversation	Reduction in the normal flow of communication associated with apathy, avolition, defensiveness, or cognitive deficit. This is manifested by diminished fluidity and productivity of the verbal-interactional process
Motor retardation	Reduction in motor activity as reflected in slowing or lessening of movements and speech, diminished responsiveness to stimuli, and reduced body tone
Conceptual disorganization	Disorganized process of thinking characterized by disruption of goal-directed sequencing, e.g., circumstantiality, tangentiality, loose associations, non-sequiturs, gross illogicality, or thought block
Difficulty in abstract thinking	Impairment in the use of the abstract-symbolic mode of thinking, as evidenced by difficulty in classification, forming generalizations, and proceeding beyond concrete or egocentric thinking in problem-solving tasks

(continued)

Table 1 (continued)

Factor item	Definition
Abnormal mannerisms and posturing	Unnatural movements or posture as characterized by an awkward, stilted, disorganized, or bizarre appearance
Poor attention	Failure in focused alertness manifested by poor concentration, distractibility from internal and external stimuli, and difficulty in harnessing, sustaining, or shifting focus to new stimuli
Excitement	Hyperactivity as reflected in accelerated motor behavior, heightened responsivity to stimuli, hypervigilance, or excessive mood lability
Hostility	Verbal and nonverbal expressions of anger and resentment, including sarcasm, passive-aggressive behavior, verbal abuse, and assaultiveness
Uncooperativeness	Active refusal to comply with the will of significant others, including the interviewer, hospital staff, or family, which may be associated with distrust, defensiveness, stubbornness, negativism, rejection of authority, hostility, or belligerence
Poor impulse control	Disordered regulation and control of action on inner urges resulting in sudden, unmodulated, arbitrary, or misdirected discharge of tension and emotions without concern about consequences
Anxiety	Subjective experience of nervousness, worry, apprehension, or restlessness, ranging from excessive concern about the present or future to feelings of panic
Feelings of guilt	Sense or remorse or self-blame for real or imagined misdeeds in the past

Table 2 Dimensional models of schizophrenia: the three factor and five factor models

Three factor model

Positive	*Negative*	*Disorganized*
Delusions	Blunted affect	Conceptual disorganization
Hallucinations	Emotional withdrawal	Difficulty in abstract thinking
Grandiosity	Poor rapport	Abnormal mannerisms and posturing
Suspiciousness/ persecution	Passive/apathetic social withdrawal	Poor attention
Unusual thought content	Lack of spontaneity and flow of conversation	
	Motor retardation	

Five factor model

Positive	*Negative*	*Disorganized*	*Agitation and hostility*	*Depression and anxiety*
Delusions	Blunted affect	Conceptual disorganization	Excitement	Anxiety
Hallucinations	Emotional withdrawal	Difficulty in abstract thinking	Hostility	Feelings of guilt
Grandiosity	Poor rapport	Poor attention	Uncooperativeness	Depression
Unusual thought content	Passive/apathetic social withdrawal		Poor impulse control	
	Lack of spontaneity and flow of conversation			

Notwithstanding the recent advances in understanding the full spectrum of symptom dimensions in schizophrenia, criteria for the disorder employ a categorical approach for diagnosis. According to the criteria outlined in the *Diagnostic and Statistical Manual of Mental Disorders, Fifth Edition* (DSM-5) [8], the patient must present with two or more of the following for a majority of the time for a period of at least 1 month: delusions, hallucinations, disorganized speech, grossly disorganized or catatonic behavior, and negative symptoms. Furthermore, the DSM-5 specifies that one of those two (or more) must include one of the first three (i.e., delusions, hallucinations, disorganized speech). Overall symptoms, including the month of those outlined above along with prodromal and/or residual signs and symptoms, must persist for at least 6 months and cause significant functional impairment in one or more domains of life such as work, education, or interpersonal relationships. Finally, schizoaffective disorder, bipolar disorder, and major depressive disorder with psychotic features must be ruled out, and the symptoms cannot be attributable to substance abuse or medical illness (Table 3).

Table 3 Diagnostic criteria for schizophrenia in the DSM-5

1. Two or more of the following for at least a one-month (or longer) period of time, and at least one of them must be 1, 2, or 3:
• Delusions
• Hallucinations
• Disorganized speech
• Grossly disorganized or catatonic behavior
• Negative symptoms, such as diminished emotional expression
2. Impairment in one of the major areas of functioning for a significant period of time since the onset of the disturbance: Work, interpersonal relations, or self-care
3. Some signs of the disorder must last for a continuous period of at least 6 months. This six-month period must include at least 1 month of symptoms (or less if treated) that meet criterion A (active phase symptoms) and may include periods of residual symptoms. During residual periods, only negative symptoms may be present
4. Schizoaffective disorder and bipolar or depressive disorder with psychotic features have been ruled out:
• No major depressive or manic episodes occurred concurrently with active phase symptoms
• If mood episodes (depressive or manic) have occurred during active phase symptoms, they have been present for a minority of the total duration of the active and residual phases of the illness
5. The disturbance is not caused by the effects of a substance or another medical condition
6. If there is a history of autism spectrum disorder or a communication disorder (childhood onset), the diagnosis of schizophrenia is only made if prominent delusions or hallucinations, along with other symptoms, are present for at least 1 month

2.2 Strategies for the Treatment of Schizophrenia

Since the mid-twentieth century the mainstay of pharmacologic treatment of schizophrenia has been dopaminergic receptor antagonism with antipsychotic medication. The incidental discovery by French surgeon Henri Laborit that chlorpromazine had strikingly calming effects on patients preparing to undergo surgical procedures was the impetus to the medication's widespread use in psychotic patients beginning in 1952 [9]. Soon after, several other antipsychotics were synthesized, and it was discovered that the potency of these agents was mediated by blockade of dopamine 2 receptors [10] and that there was a direct, positive correlation between an antipsychotic's efficacy and its affinity for the D2 receptor [11, 12]. In the decades that followed, another type of antipsychotic was discovered that had affinity for both D2 and serotonergic receptors [13], leading to the widespread, although over-simplified and partially inaccurate, grouping of these medications into "first-" versus "second-generation antipsychotics." Large comparative studies have demonstrated that first- and second-generation antipsychotics have statistically indistinguishable effects on positive symptoms; instead, the two medication classes vary in their side effect profiles, with first-generation antipsychotics causing more extra-pyramidal symptoms and second-generation antipsychotics causing more notable metabolic side effects [14, 15]. One medication that stood out in its effectiveness for treatment-resistant schizophrenia was clozapine, but its substantial burden of side effects, particularly with the relatively common occurrence of agranulocytosis, has limited its use to refractory patients and to those with schizophrenia complicated by suicidality [16]. While the numerous antipsychotic medications available have clear benefit in the amelioration of the positive symptoms of schizophrenia, both first- and second-generation antipsychotics show, at most, only slight improvement in the negative and disorganization symptom dimensions [17, 18], and adverse effects from medications are a leading cause of non-adherence to them [19]. Furthermore, efforts to develop effective antipsychotics acting via other neurotransmitter systems have been disappointing (e.g., the mGluR2/3 agonist pomaglumetad methionil (LY2140023) [20]). These painful realities, along with the enormous burden of schizophrenia for patients, their families, and society, drive the urgent demand for novel approaches to understanding disease biology and therapeutic innovation.

2.3 Outcomes in Schizophrenia

Schizophrenia is associated with an expansive array of adverse outcomes. Despite treatment, at least one-third of patients in developed countries and about 60% of patients in developing countries do not achieve a satisfactory level of remission [21]. As a group, patients with schizophrenia have a life expectancy that is 15–20 years shorter than average [22] due to a multitude of factors. Rates of suicide range from 5 [23] to 13 [24] percent, and in a large study of Chinese patients, individuals with schizophrenia were 23 times more likely to die by suicide [25].

Accordingly, suicide has been shown to account for 28% of the excess mortality in patients with schizophrenia [26]. Additional likely contributing factors are adverse effects of chronic medication use [27], increased risk of several communicable diseases [28], cardiovascular disease and mortality [29], comorbid substance abuse [30], and markedly elevated rates of abuse and assault [31]. Overall, patients with schizophrenia carry a remarkable burden across domains of life, and advances in the understanding of the disease and its treatment are imperative.

3 Neurobiological Considerations of Schizophrenia

Having provided an overview of the clinical aspects of schizophrenia, we will now turn to relevant neurobiological considerations of the disorder. We begin with a discussion of the neurodevelopmental theory of schizophrenia, drawing upon numerous lines of evidence that support this line of thinking. Then, we focus on risk factors for schizophrenia, mentioning in brief those environmental or non-genetic factors, and then focusing more in-depth on the genetics of schizophrenia.

3.1 Schizophrenia as a Disorder of Neurodevelopment

Numerous lines of evidence implicate several abnormalities in neurodevelopment as contributing to schizophrenia (for review, see: [32, 33]). In addition to the impact of risk factors mentioned below that affect in utero development, birth, and early childhood, much additional data point to a "Neurodevelopmental Hypothesis of Schizophrenia." In this section, we provide a concise review of studies that have generated convincing support of this broader hypothesis.

Findings in Patients During First Episode of Psychosis

Brain development and maturation is a dynamic process beginning in early fetal development and extending up to the third decade of life [34]. Abnormalities documented in the brains of patients with schizophrenia relate strongly to biological processes most active during neurodevelopment, such as neuronal migration [35], proliferation [36], specification [37], maturation [38], and pruning [39]. These facts, coupled with the longitudinal course of brain development, make it exceedingly unlikely that the factors that ultimately contribute to schizophrenia take place in close temporal proximity to disease onset by current diagnostic formulations [38].

In patients presenting with their "First Episode of Psychosis," extensive reports of widespread brain abnormalities are consistent with a prolonged disease process that ultimately culminates in the manifestation of symptoms warranting acute clinical attention. First episode patients show volumetric abnormalities in the temporal lobe [40]; enlargement of CSF spaces [41]; smaller thalamic nuclei [42]; decreased grey matter in the prefrontal and temporal cortices [43, 44]; diminished grey matter in the frontal and hippocampal gyri [45]; progressively worsening reduction in overall cortical grey matter [46, 47] and in the cingulate and insular cortices [48–50]; abnormal functional connectivity [51]; and finally, disrupted maturation in several white matter tracts [52–55]. Importantly, many of these findings correlate positively with disease progression [47, 50, 52] and symptom severity [43, 45, 51], further supporting the relevance and longitudinal nature of these abnormalities. In sum, these data highly suggest that biological processes underlying schizophrenia begin many years prior to diagnosis, and as shown below, clinical studies of patients prior to diagnosis reveal many corresponding changes in psychological functioning.

Diagnosis of Schizophrenia Is Preceded by a Significant Prodromal Syndrome

Although schizophrenia is most commonly diagnosed in the late adolescent to early adult years, decades of clinical research have described the presence of substantial but non-specific psychopathology prior to disease diagnosis [56]. Regardless of prior risk determination, prodromal patients show markedly high rates of sleep alterations, anxious symptoms, suspiciousness, and non-hallucinogenic perceptual disturbances [57], as well as behavioral changes including decline in academic performance, impaired concentration, and social withdrawal [58–61], accompanied by early neurocognitive deficits (e.g., [62, 63]) that often precede illness diagnosis by many years. Of note, the duration of the prodromal syndrome correlates positively with the magnitude of grey matter volume reduction in several brain regions [64].

The heterogeneity and non-specificity of these signs and symptoms have prompted substantial research efforts into the prediction of later disease onset, particularly among children deemed "high risk" or "ultra-high risk" for schizophrenia based upon family history of the illness and/or early prodromal symptoms [65]. Presymptomatic "high-risk" individuals frequently show social, motor, and cognitive deficits from early age. In infants, delayed attainment of developmental milestones significantly increases risk of later development of schizophrenia in a dose-dependent manner and with additive effects when combined with obstetric complications [66]. Abnormalities beginning as early as age 4 in social behavior, affect, and motor development predicted later diagnosis of schizophrenia [67]. Consistent with this, IQ decline from ages 4 to 7 [68], poorer performance on intelligence and memory tasks at later developmental stages [69, 70] and decreased performance on IQ measures by age 13 predicted also predict schizophrenia [71]. A study of children aged 7–12 years who later developed schizophrenia documented substantial differences in several standardized behavioral metrics compared to children who did not develop the disease, and demonstrated that among these differences, performance in attention, memory, and motor skills predicted later onset of schizophrenia [72]. Moreover, higher rates of social maladjustment are observed in those who later received a diagnosis of schizophrenia [73]. Social and cognitive functioning deficits predict schizophrenia in adolescent males (aged 16–17) up to 10 years following initial testing [74]. Using extant data, investigators [75] have developed risk prediction tools that have high sensitivity (98%), but with less specificity (59%), for later conversion to schizophrenia, with a mean time of 4.3 years for females and 6.7 years for males between age of prodrome onset and diagnosis of schizophrenia [75]. Taken together, the data from these abundant reports show that formal diagnosis of schizophrenia is preceded by deficits across functional domains that often occur several years before onset of the classical disorder. Further strengthening the relevance of these findings are reports of alterations in neuronal functioning that concur with early psychological phenotypes, a topic to which we now turn.

Prodromal Symptoms of Schizophrenia Occur Concomitantly with Brain Abnormalities

Longitudinal neuroimaging studies of patients have documented several abnormalities that precede the diagnosis of schizophrenia. In high-risk persons, smaller grey matter volumes in several brain areas predicted onset of psychosis 1–2 years later [76]. Similar approaches have replicated and expanded on these findings, further implicating differences in subregions of the frontal, temporal, and parietal lobes [77]; thinning of the anterior cingulate cortex [78]; reduced grey matter volume in the insular cortex, [79], the superior temporal gyrus [80], and parahippocampal gyrus [81]; lower prefrontal cortex activation during a working memory task [82]; and generalized cortical thinning [83]. Strikingly, volumetric brain abnormalities have even been documented in the offspring of mothers with schizophrenia at the prenatal stage [84], providing a disease-relevant example of the high degree of heritability of several brain volumes [85]. Reduction in white matter tracts is also seen in patients showing prodromal symptoms, and the severity of the reduction predicts later diagnosis of schizophrenia [86]. Finally, high-risk patients who later transition to psychosis display abnormal white matter integrity up to 2 years prior to diagnosis [87].

Widespread reports of both psychological and biologic phenotypes preceding schizophrenia onset suggest strongly that processes driving disease production occur long before its formal diagnosis. When viewed within the context of the longitudinal course of brain development—spanning about three decades in length—the data discussed here constitute the foundation of a theory in which abnormal neurodevelopmental processes both predate and contribute to the likelihood of schizophrenia manifestation, particularly in those with elevated genetic risk. Although the early emergence of phenotypes associated with future disease incidence is well documented, the extent to which innate (i.e., genetic) and environmental factors contribute causally to their occurrence awaits clarification. This is highly non-trivial distinction to make: while variation in disease states, as well as normal traits, is almost always driven by variable combinations of genetic and non-genetic factors, an understanding of the relative etiologic contributions of each is a critical informant of appropriate and meaningful strategies to prevent or alleviate disease.

For those risk factors driven by a primary environmental insult, which of course may interact with genetic predisposition, there is ample opportunity—and, indeed, an imperative need—for early intervention to remove the risk factor or diminish its impact. Conversely, a better understanding of how innate, genetic pathways contribute to the development of schizophrenia may yield innovative treatment approaches that correct or mitigate the consequences of these inborn processes in a manner that improves the chances of averting disease presentation. While expanded and aggressive efforts to improve treatment strategies of bona fide schizophrenia remain imperative, the potential benefits of preventing its occurrence in the first place are unbounded.

3.2 *Epidemiology and Risk Factors for Schizophrenia*

Risk Factors for Schizophrenia: Environmental Factors

Highly consistent with the neurodevelopmental theory of schizophrenia are reports of numerous non-heritable factors can impact schizophrenia risk, particularly during early childhood and adolescent developmental periods. Several pregnancy- and birth-related complications increase disease risk, including maternal infection [88, 89], hypoxic events [90], maternal nutritional deficiency [91, 92], pre-term birth [93], and both low [94, 95] and high [96, 97] birth weight. In the Northern hemisphere, birth in the winter months is consistently associated with a small but significant increase in schizophrenia risk [98]. Additionally, early-life stressors and childhood traumas also increase risk [99, 100]. Although suggested causal relationships remain speculative, adolescent use of cannabis, especially early and heavy abuse, has been repeatedly associated with earlier-onset and/or increased severity of schizophrenia symptoms [101, 102]. Less well understood are reports that living in more urban than rural environments can increase schizophrenia risk [103]. The extent to which these environmental effects differentially impact individuals with high versus low genetic risk for schizophrenia remains unclear, and so there is a critical need to clarify these potential "gene x environment" interactions.

Risk Factors for Schizophrenia: Heritability

Schizophrenia has an estimated lifetime prevalence of 0.7% [104]. Family studies comparing the probability of having the disorder in those with an affected relative to those without have found an odds ratio about 10 [105]. Concordance rates among monozygotic twins are estimated in the range of 41–65% [106], indicating a substantial contribution of genetics to disease risk. In line with these findings are results from meta-analyses that are able to derive a "heritability estimate," the percent of variation in disease frequency attributable to heritable factors; as of this writing, the most recent assessment generated a heritability estimate of 79% for schizophrenia

[107]. In the subsequent section, we explore the structure and putative impacts of genetic variants that drive this high level of heritability.

3.3 *The Form and Function of Genetic Variants Driving Schizophrenia Heritability*

In this section of this chapter, we focus on the underlying biological nature of the genetic factors driving the substantial heritability of schizophrenia. Understanding both the types of variants contributing to disease transmission and the degree to which they influence longitudinal outcomes has essential relevance to knowledge of the neuropathology of the disorder and towards development of biologically informed treatment interventions. Broadly, the genetic contribution to schizophrenia risk consists of a combination of both rare and common variants [108]. Below, we provide an overview of current understanding of both types of variants and their relative contributions to disease risk.

Rare Variants Account for a Portion of Schizophrenia Heritability

A Role for De Novo Mutations

Reduced fecundity in patients with schizophrenia [109] poses a challenge to those studying the genetics of the disorder: How is this reality compatible with the high estimates of heritability? Because genetic variants leading to decreased reproductive fitness are subject to negative selection, the lack of evidence suggesting decreased disease prevalence renders genetic transmission alone an insufficient explanation for disease occurrence [110]. Indeed, partial explanation of this phenomenon lies in the importance of de novo mutations (DNMs). Consistent with a role of DNMs in the genetic etiology of schizophrenia are observations that risk of schizophrenia correlates positively with increasing paternal age [111], that transmission of DNMs to offspring increases with paternal age in general [112] and in schizophrenia in particular [113, 114], and that almost 80% of DNMs documented in exome-sequencing studies of schizophrenia are found on the paternal chromosome [115].

Studying trios of schizophrenia patients and their unaffected parents is one approach to identifying DNMs associated with incident cases. The first report using this design described increased DNMs in schizophrenia cases among a panel of several hundred synaptic genes [116]; targeted exome sequencing of a small number of such trios confirmed an increased rate of DNMs in schizophrenia cases [117]. Subsequent studies with expanded sample sizes found similarly increased rates of putatively damaging DNMs, including a report of substantial enrichment of the schizophrenia-associated DNMs in genes preferentially expressed in fetal but not post-natal brain [113, 114] as well as postsynaptic protein complexes and binding targets of the *Fragile X Mental Retardation Protein* (*FMRP*) [115].

Exome-sequencing studies may also be conducted using a more standard case-control sampling plan. While variants in any single gene were not associated with schizophrenia exome-wide, combined gene sets showed increased rates of rare variants in cases versus controls, with highest effects sizes observed for gene sets directly involved in *PSD-95* synaptic signaling and calcium channels [118]. While rates of disruptive, "ultra-rare" variants affecting protein-coding sequence appear to be increased in schizophrenia, [119], no single variant alone with exome-wide significance has been identified, possibly due to inadequate sample sizes, the rarity of such mutations in the general population, and the lack of transmission of exonic mutations of large effect sizes. Nevertheless, gene sets specific to brain tissue and neurons in general were highly enriched in such ultra-rare variants, while those in other tissues and non-neuronal cell types were not [120]. For these reasons, there is undoubtedly an increased burden of DNMs in single genes among patients with schizophrenia.

Rates of Copy Number Variants Are Increased in Patients with Schizophrenia

In addition to rare, de novo variants affecting a single gene, there is also an increased rate of large copy number variants (CNVs) in patients with schizophrenia compared to controls [121], and genes located within associated regions implicate disruption in synaptic pathways and neurodevelopment [122]. The largest and most recent analysis found eight CNVs that reached a strict genome-wide significance level, with an additional eight CNVs at a more relaxed threshold [123]. Strikingly, psychotic disorders are present in about 40% of adults carrying the 22q11.2 deletion [124], and this CNV is the most common genetic lesion associated with schizophrenia [125]. Although further research is needed to clarify this relationship, there appears to be an interaction between disease-associated CNVs and common variants in mediating risk variability and resultant phenotypes [126, 127]. In sum, whether they involve a single gene or multiple, there is clearly an important contribution of rare variants to schizophrenia heritability, and additional work to determine the utility of targeting implicated pathways for therapeutic intervention will be essential to the overall efforts to improve outcomes among affected patients.

Common Variation Contributes Substantially to Schizophrenia Heritability

Although rare variants constitute a portion of schizophrenia heritability, most of the genetic basis of the disease lies in the transmission of numerous (and presumably interacting) common variants that each confers a relatively small risk of schizophrenia on their own. In this section, we highlight the significant findings on the role of common variants contributing to schizophrenia risk as well as an exploration of their potential etiologic mechanisms.

The Polygenic Nature of Schizophrenia Heritability Points to Common Variants of Small Effect Size

Long before the advent of genome-wide association studies (GWAS), investigators posited that schizophrenia genetic risk was fundamentally polygenic in nature [128]). Pilot GWAS in several thousands of cases and controls demonstrated that substantial genetic risk for schizophrenia was conferred by single-nucleotide polymorphisms (SNPs), including several in the *Major Histocompatibility Complex* (MHC) region [129], that were both common and of small effect sizes individually [130–132]. Subsequent GWAS further expanded sample sizes and identified additional loci, including one with variants in *MIR-137* and its predicted targets [133]; several loci containing genes implicated in calcium signaling and numerous others containing long non-coding intergenic RNAs [134]; regions enriched in synaptic genes and genes involved in dopaminergic and glutamatergic neurotransmission [135]; and loci implicated in abnormal behavioral phenotypes, long-term potentiation, and targets of *FMRP* [136]. As seen in Fig. 1, the number of loci discovered is directly proportional to the sample size of the study. Importantly, these findings have been extensively replicated in cohorts of Han Chinese ancestry [138, 139] and among putatively more

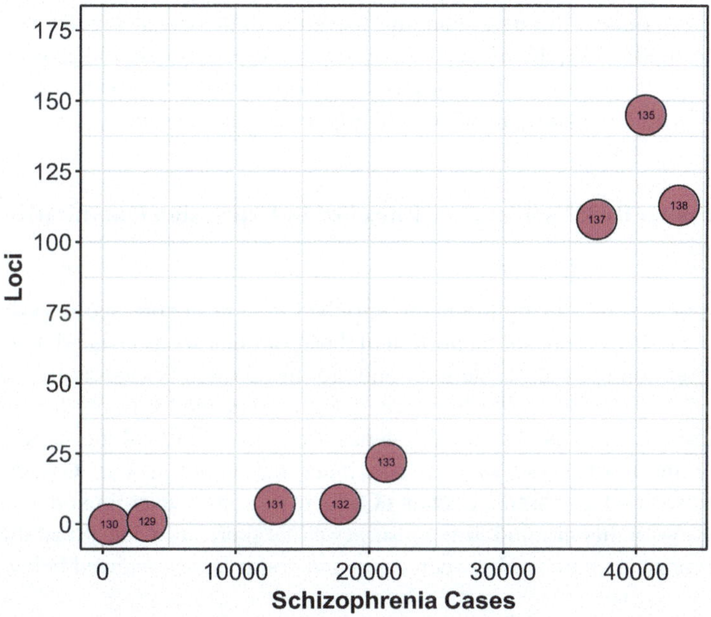

Fig. 1 The number of loci associated with schizophrenia through GWAS has dramatically expanded in the past decade, largely as a function of increasing sample sizes. Numbers overlapping each bubble correspond to the reference number: O'Donovan et al. [131] The International Schizophrenia Consortium [130] Stefansson et al. [132] The Schizophrenia Psychiatric Genome-Wide Association Study (GWAS) Consortium [133]; Ripke et al. [134] Schizophrenia Working Group of the Psychiatrics Genomics Consortium [137]; Li et al. [138] Pardanas et al. [136]

homogenous cohorts of schizophrenia patients taking clozapine [134, 136]. To date, the contribution of common variants to schizophrenia heritability is estimated to be about one-third [136], thus confirming the value of GWAS in assessing the genetic architecture of schizophrenia. Despite the utility of this approach in identifying disease-associated variants, the demonstration of a causal relationship between any given variant and occurrence of schizophrenia is not addressed by GWAS, let alone the mechanisms by which loci negatively impact normal human biology. In the following section, we explore nascent but promising avenues to assess the potential causal impacts of genic variants to schizophrenia development.

3.4 *Functional Genomics of Schizophrenia*

The common variants associated with schizophrenia risk are predicted to affect patterns of gene expression [137]. More generally, disease-associated variants are enriched in regions predicted to have gene-regulatory functions [140, 141]. Such variants may alter gene expression through differential affinities of proteins facilitating transcription [142], alterations of various DNA and histone post-translational modifications [143], differential splicing events leading to changes in isoform abundances [144], and/or through changing the three-dimensional regulatory architecture of chromatin [145]. In this section, we review findings of altered gene expression profiles in schizophrenia and studies using various approaches to identify the mechanisms by which risk variants adversely affect gene expression.

Schizophrenia Risk Variants Are Enriched in Expression Quantitative Trait Loci

SNPs implicated in schizophrenia are enriched in genes expressed in post-mortem brain tissue [146]. A substantial portion of these variants are associated with changes in the expression of at least one gene and are thus termed "expression Quantitative Trait Loci" (eQTL) [147]. Furthermore, disease-associated variants are enriched in chromatin-regulatory peaks among genes that drive cortical neogenesis [148]. Strikingly, the most recent analysis concluded that about 42% of GWAS variants associated with schizophrenia contain eQTLs in regions converging on gene regulation [149]. With this in mind, it is no surprise that there are widespread differences in gene expression patterns between cases and controls, as reviewed below.

Post-Mortem Gene Expression Analyses Highlight Neurodevelopmental Pathways and Specific Brain Regions and Cell Types

Numerous investigators have sought to characterize abnormalities in nervous system gene expression in schizophrenia in order to begin yielding mechanistic insights on molecular perturbations driving disease pathologies. With the discovery that common variants associated with schizophrenia are preferentially enriched in putative gene-regulatory regions and the advent of more high-throughput technologies for measuring gene expression, these efforts have expanded substantially. Here, we highlight some of the earlier reports of altered brain gene expression with a focus on highly replicated findings with enduring relevance to the mechanisms of schizophrenia. We then turn to an assessment of large datasets generated via modern RNA-sequencing and computational platforms. As shown below, gene expression abnormalities have begun to converge on several key neuronal processes, especially those critically regulating neurodevelopment and cell type and regional specification. With further developments in computational techniques and increasing sample sizes, studies of disrupted gene expression patterns in schizophrenia will continue to generate key insights connecting disease-associated genetic variants to etiologic molecular perturbations driving symptom manifestation.

Regardless of disease status, comparative evolutionary genomics studies have revealed that brain-related genes most recently acquired by *Homo sapiens* are preferentially expressed in fetal and infant but not adult neocortex [150], and data from large-scale genomics approaches indicate that variants associated with schizophrenia risk preferentially impact early neurodevelopmental pathways. Initial studies of patient post-mortem brain tissue documented cellular and anatomical defects that suggest disrupted cortical neuron development and migration [151, 152]. Expression analyses of genes driving GABAergic neuron specification found that the most dramatic differences in patient brain tissue were in transcripts whose expression undergoes the most significant changes in early rather than later life [153], and that patient brains exhibit more immature and under-developed expression patterns in GABAergic genes [154]. Meta-analysis of post-mortem microarray studies implicated a variety of cell types and gene sets involved in synaptic transmission [155], as well as disruption in the expression of genes most predominantly regulated during neurodevelopment [156]. Subsequent review of findings that specifically assessed data from post-mortem prefrontal cortex found enrichment in pathways related to synaptic transmission [157], as well as immune-related processes, myelination, and oxidative phosphorylation [158]. Consistent with the reports described below [159–161], similar analysis that included other brain regions found enrichment in modules specific to certain cell types and neurotransmitter systems (i.e., glutamatergic and inhibitory) and furthermore discovered diminished expression in genes that distinguish separate brain regions [162].

RNA-sequencing platforms and increasingly sophisticated statistical techniques enabled broader assessments of disease-associated changes in gene expression. Data from the *Common Mind Consortium* revealed subtle alterations in mRNA levels among hundreds of genes in dorsal lateral prefrontal cortex, and network analysis highlighted a gene "node" characterized by enrichment in genes importantly involved in synaptic transmission, neuronal subtype markers, and targets of *FMRP* [147]. An expanded analysis found that genes differentially expressed in schizophrenia were those enriched in neurodevelopmental regulation of neuronal cell type and discovered a dramatic shift in developmentally regulated isoform expression among genes involved in dopaminergic and glutamatergic synaptic functions [149]. Hierarchical clustering of gene expression in a different case-control sample revealed similar enrichment in modules defined by neurodevelopmental regulated genes, synaptic function, and cell type-specific markers [163]. In the largest transcriptomic study of psychiatric disorders to date, investigators discovered expansive alterations in isoform expression, particularly among non-coding RNAs and gene sets involved in synaptic function, cell-type markers, and the immune system [144].

Presently, the substantial cost of RNA-sequencing and the limited number of post-mortem samples, coupled with small effect sizes of individual loci, hinder the identification of robust gene expression changes associating with a particular trait or disease. As an alternative approach, transcriptome-wide association studies (TWAS) generate predictive models relating specific genotypes to gene expression phenotypes through imputation of available datasets describing gene expression changes associated with a disease risk variant [164]. In this way, leveraging large eQTL datasets with post-mortem RNA-sequencing from case-control studies enables the generation of models that predict alterations in gene expression from given disease-associated variants. Thus far, schizophrenia TWAS efforts have identified over 150 genes whose expression differs between cases and controls [160] [165]). Of note, sets of variants discovered through TWAS are overrepresented in neurodevelopmentally regulated genes [160] and impact the expression of several hundred of them in a remarkably site-specific and temporally regulated manner across many brain regions [165]. Interestingly, pathway analysis of the implicated genes documented strong enrichment in processes related to porphyric disorders and hexosaminidase-A deficiency [165]. A different study similarly documented over one-hundred TWAS hits; among the most significant were chromatin regulators and long intergenic non-coding RNAs [144]. By integrating gene expression taxonomies derived from single-cell RNA-sequencing datasets with schizophrenia GWAS signatures, specific neuronal cell types, including interneurons, pyramidal cells, and striatal medium spiny neurons, have been emphasized in driving disease risk [161]. In a related approach, analysis of the overlap between trait- and disease-associated loci and genes exhibiting expression specificity in numerous tissues and cell types found that schizophrenia loci were overrepresented in genes expressed in neurons, particularly glutamatergic neurons, but not in non-neural tissue [159]. In sum, these studies reveal a cell-type specific disruption of gene expression, particularly among sets key to neurodevelopment.

DNA and Histone Post-Translational Modifications Are Altered in Schizophrenia

Efforts to understand the mechanisms by which schizophrenia-associated risk variants alter gene expression have discovered key roles for several transcriptional regulatory processes. Assessment of the excess association of GWAS SNPs for several psychiatric disorders in particular biological pathways found that, for the combined datasets of schizophrenia, bipolar disorder, and depression data sets, histone H3K3me methylation showed the strongest enrichment overall, while the top pathways among schizophrenia SNPs were postsynaptic density and membrane, dendritic spine, H3K4 methylation, and axon part [166]. Across diseases and traits, associated SNPs are globally enriched in regions marked by histone post-translational modifications (PTMs), such as H3K4me3 and H3K9ac, indicating active enhancer, in a cell-type specific manner [167]. Using publicly available post-mortem ChIP-sequencing datasets from, Roussos et al. [168] found that this is also the case for SNPs associated with schizophrenia in both fetal and adult brain tissue. Based upon the growing consensus that epigenomic processes drive molecular phenotypes in numerous diseases, the PsychENCODE Project was developed to study the role of DNA regulatory elements in various psychiatric disorders [169]. A recent landmark study [170] assessed the enrichment of SNPs for several brain- and non-brain-related diseases and traits among specific regulatory chromatin marks in both brain tissue homogenate and neuron-enriched and neuron-depleted populations of cells. SNPs associated with schizophrenia demonstrated the strongest enrichment in open-chromatin peaks over all traits, including other brain-related traits and diseases; importantly, this enrichment was strengthened substantially in neuron-enriched samples over brain homogenate and non-neuronal samples, and the variable driving the most variation in histone peaks was cell-type identity (i.e., neuron-enriched versus non-depleted versus homogenate) [170]. Furthermore, many of the strongest histone QTLs (hQTLs) associated with schizophrenia were found in neuronal but not non-neuronal tissues [170]. This study highlighted, among other findings, both the strong enrichment of schizophrenia SNPs in regulatory chromatin regions and the importance of assessing potential functional roles of disease- and trait-associated variants in refined biological samples in order to more adequately capture meaningful signals. In reports of differential DNA post-translational modifications, schizophrenia SNPs are also enriched in sites of DNA methylation (meQTLs), which differ between cases and controls at genes that are strongly enriched in neuronal differentiation and neurodevelopment in frontal cortex [171], hippocampus [172], and prefrontal cortex [173].

Taken together, these data provide a mechanistic link between genetic variation and altered gene expression. The extent to which differential DNA and histone PTMs are driven by primary effects of genetic sequence variation or numerous other secondary processes that ultimately converge on alteration of these marks remains an area of significant uncertainty, but future studies will integrate the contributions of innate genetic variation and non-genetic factors in pathologically disrupting both the regulatory states and expression patterns of causal gene sets. We now turn to a budding field that extends the functional impact of schizophrenia risk loci on gene expression to include the pivotal role of three-dimensional chromatin-regulatory structures.

The Disruption of Three-Dimensional Chromatin Dynamics by Schizophrenia Risk Variants

A critical way in which DNA and histone PTMs change gene expression is through the alteration of three-dimensional (3D) chromatin structures. If variants associated with a disorder are enriched in putative participants in such structures, it follows that schizophrenia risk variants may cause pathogenic alterations in 3D chromatin architecture, thereby leading to disrupted patterns of gene expression [174]. While still a nascent field, the role of 3D genomics in schizophrenia disease biology has increasingly become come into focus for investigators. In 2013, Barhdwaj et al. identified a GABAergic neuron-specific, activity-regulated 3D chromatin interaction involving *GAD1* and an upstream transcriptional regulator, likely an enhancer, that was decreased in post-mortem samples of schizophrenia patients who also had decreased *GAD1* transcript levels [175]. In the following year, an activity-regulated distal regulatory element of the NMDA receptor subunit *GRIN2B* was identified using chromosome conformation capture 3C); this distal regulatory region was found to contain a SNP associated with schizophrenia risk, and post-mortem prefrontal cortex from patients with that SNP had lower levels of *GRIN2B* mRNA transcript compared to controls as well as patients without the risk allele [176]. Furthermore, studies in post-mortem brain tissue identified a remarkable overlap between regions of open chromatin and variants associated with schizophrenia risk [177–179]. Adapting 3C assays genome-wide, HiC analysis in human fetal brain tissue found enrichment of schizophrenia SNPs in 3D contacts of gene sets involved in neurogenesis, postsynaptic density, and chromatin remodeling proteins, among others [180], consistent with TWAS findings of a high degree of overlap between brain chromatin loops and signal for TWAS genes [160]. Recently, investigators demonstrated directly that risk variants for schizophrenia are enriched in 3D chromatin loops that are specific to neurons and neural progenitor cells (NPCs) [174].

Overall, genomic and functional genomic studies of schizophrenia heritability point to a model in which both rare and common genetic variants contribute to disease risk, and that common variants exert their effects in large part through altering gene-regulatory processes and thus disturbing normal neurodevelopment in specific cell types highly implicated in schizophrenia. In the next section, we begin our consideration of the role of hiPSC models in exploring schizophrenia with relevant technical considerations, and then dive into the numerous reports that have used hiPSCs to further expound disease biology.

4 Technical Considerations in Human Induced Pluripotent Stem Cells

A landmark screen of transcription factors determined that *OCT4, KLF4, SOX2*, and *c-MYC*, the now termed "Yamanaka Factors," were sufficient to reprogram somatic cells into induced pluripotent stem cells (iPSCs), thus bypassing the need for embryonic stem cells [181]. Critical for model validity, iPSCs and ESCs have been shown to be comparable across several features [182]. Advancements in hiPSC models and related technologies have fueled a new approach to studying disease-relevant biological pathways in living neural tissue from individuals with and without a particular trait or disease (for review see [183]).

4.1 Technical Challenges and Considerations in hiPSCs Models

hiPSC-based models offer as their primary advantage the ability to analyze living neuronal tissue in a way that preserves donor gene background. Nonetheless, both genetic [184] and epigenetic [185] errors are known to occur during the reprogramming process. Moreover, certain chromosomal abnormalities identified in patient somatic cells reprogrammed into hiPSCs may not retain the abnormality through traditional differentiation methods [186, 187]. While there is indeed variation in gene expression between hiPSCs derived from different donors (and to a lesser extent between hiPSCs from the same donor), genetic heritability also impacts gene expression patterns and differentiation potential [188]. Further variability is introduced during the neuronal differentiation process: RNA-seq analysis of hiPSC-derived neurons demonstrated that inter- and intra-donor variability could be decreased by correcting for variation in cell-type composition [189]. While hiPSC-based studies remain dramatically underpowered for the study of idiopathic disease, there is a minimal yet highly significant concordance between gene expression signal in hiPSC-derived neurons and RNA-seq datasets generated from the two largest post-mortem studies [189]. If applied appropriately, hiPSC nevertheless serves as a key modeling platform.

Brief Overview of Techniques for Generating Disease-Relevant Tissues from hiPSCs

In this section, we provide a targeted overview of techniques for generating the types of neural cell types relevant for the discussions throughout the chapter. Broadly speaking, techniques for producing neural tissue from hiPSCs may be classified as "directed differentiation" or "induction-based" approaches. Directed differentiations are those techniques that use a combination of small molecules, proteins, and other chemical factors to modulate intracellular signaling in order to recapitulate in vivo developmental pathways that give rise to the target cell type; these approaches often use varying combinations of agents at different steps of the protocol to mirror sequential phases of neurodevelopment. Approaches based upon inductions, on the other hand, employ transgenes, often packaged into viral vectors, to ectopically express transcription factors that are known to be necessary and/or sufficient to driving the hiPSC towards a specified neurodevelopmental pathway to produce the desired cell type. Each general approach has its own advantages and disadvantages, and an increasing number of techniques are employing combined strategies to improve cell type yields and enhance specific phenotypes of interest (e.g., [190]). We turn now to an overview of the protocols used to produce relevant cell types, with particular emphasis on those employed in the studies discussed herein. In both sections, we mention first the techniques for producing excitatory and inhibitory neurons, and then turn towards protocols that yield cells of other neurotransmitter system identities. For excellent and more thorough review, see Mertens et al. [191].

Directed Differentiation Approaches

A frequently used directed differentiation technique for deriving forebrain neural progenitor cells (NPCs) from hiPSCs involves the dual inhibition of SMAD signaling. The application of SMAD inhibitors *Noggin* and *SB431542* is known to reliably produce cultures that can be specialized to several neuron phenotypes [192]. To generate a mixed population of forebrain neurons consisting primarily of excitatory neurons, stepwise patterning of embryoid bodies into neural rosettes and NPCs using developmental signals and then differentiating NPCs with defined neuronal growth factors are used [193]. Cortical interneurons can be efficiently generated from hPSCs with a combination of dual-SMAD inhibition and small-molecule application [194]. In the most frequently used directed differentiation approach for producing dopaminergic neurons, hiPSCs are first patterned into ventral midbrain floor plate progenitors and then matured into dopaminergic neurons with a standard mixture of neurotrophic factors and small molecules [195]. Serotonergic neurons may be differentiated from hiPSCs through sequential dual-SMAD inhibition and modulation of *WNT*, *Sonic Hedgehog* (*SHH*), and *FGF4*signaling [196]. Using sequential application of small-molecule inhibitors and morphogens, hippocampal progenitor cells and dentate granule neurons can be produced [197], and a related technique enriches for CA3 hippocampal neurons [198]. Combinatorial small-molecule approaches are also capable of producing sensory neurons [199, 200]. Often, co-culturing developing neurons with astrocytes during directed differentiation yields more mature, functional neurons. A study conducted by Kuijlaars and colleagues demonstrated enhanced synchronized synaptic activity in a population of GABAergic and glutamatergic neurons differentiated in a co-culture of astrocytes via a small-molecule technique [201]. In a simplified approach, astrocytes and neurons are produced together from the same starting hiPSCs with modified differentiation strategies [202].

Induction Approaches

Fibroblasts can be reprogrammed into functional induced neuronal (iN) cells via overexpression of the transcription factors *ASCL1*, *BRN2,* and *MYTL1* [203]; the same three factors yield human neurons when combined with the transcription factor *NEUROD1*[204]. Greatly simplifying the ability of investigators to produce cortical excitatory neurons was the approached developed by Zhang et al. [205] involving the overexpression of *NGN2* alone in antibiotic-selected hiPSCs. This technique works similarly when starting from NPCs [206] and was recently modified to include the addition of small molecules in order to improve functional maturation [190]. Several cell-type specific induction techniques have now been

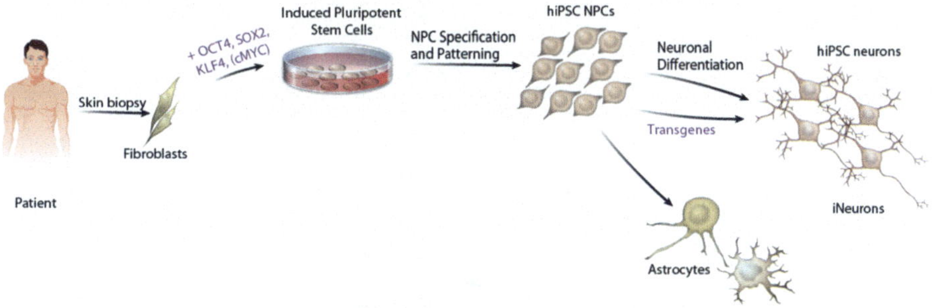

Fig. 2 Generalized schematic of hiPSC-based models

developed. GABAergic neurons of varying subtype identities and maturities can be produced through the overexpression of different combinations of transcription factors [207–209]. In order to produce midbrain dopaminergic neurons, hiPSCs can be transduced with a combination of *ASCL1, LMX1a/b,* and *NURR1* [210, 211], among others. Production of serotonergic neurons with induction approaches has been achieved from fibroblasts using a combination of either *LMX1b*, *FOXA2*, *ASCL1*, and *FEV* [212] or *LMX1b, FEV, NKX2.2, GATA2, ASCL1, and NGN2* [213]. See Fig. 2 for an overview of these techniques.

Overall, these approaches enable the production of numerous cell types relevant to schizophrenia. Continued work to improve neuronal maturity and subtype specificity will only bolster the field, and efforts to render approaches more scalable and reproducible across donors and laboratories will increase the attractiveness of hiPSC models of neuropsychiatric disease to many investigators (Fig. 3).

5 Investigation of Schizophrenia with Human Induced Pluripotent Stem Cells

We turn now to the primary focus of this chapter. Here, we discuss studies that have explored various aspects of schizophrenia biology using human induced pluripotent stem cells (hiPSCs). We begin with pioneering reports that ignited the field and highlight cellular and molecular phenotypes that were documented in patient cells. Then, we evaluate studies that have assessed the impact of both common and rare genetic variants associated with schizophrenia neuronal phenotypes. Afterwards, we share results of approaches that have used hiPSC models to explore the impact of non-genetic factors implicated in schizophrenia. We end with a discussion of recent advances in using hiPSC platforms for drug screening purposes and applications towards improving clinical outcomes.

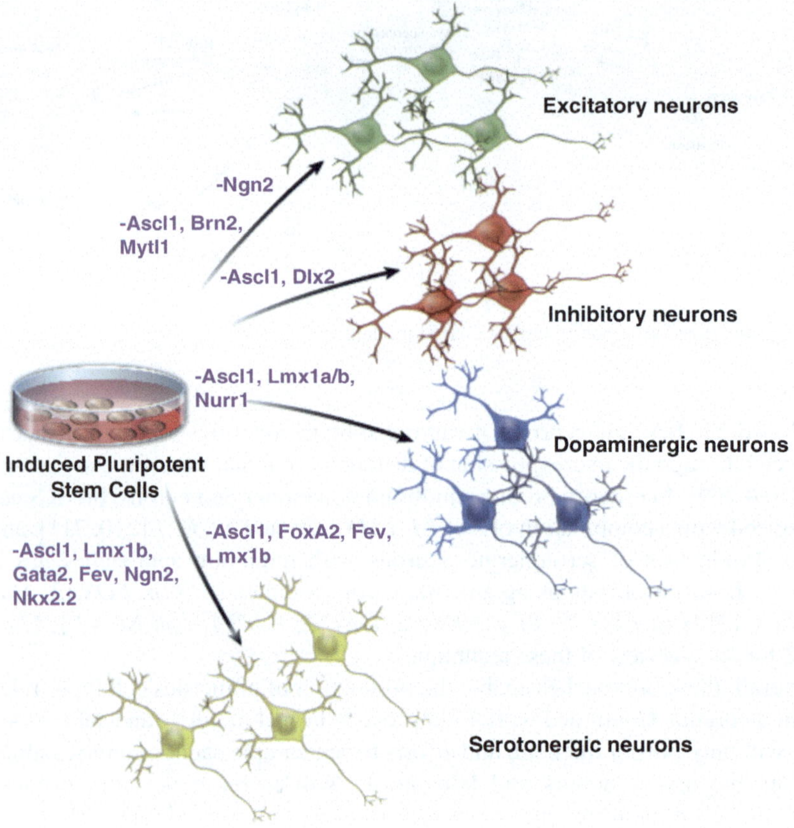

Fig. 3 Transduction approaches to generate samples of defined neurotransmitter system identities

5.1 Assessment of Phenotypic Differences in Schizophrenia Across Neurodevelopment with hiPSCs

The first report assessing phenotypic differences in hiPSC-neurons from patients and controls found that patient-derived forebrain neurons exhibited decreased connectivity, fewer neurites, decreased expression of the synaptic protein *PSD95*, and altered gene expression profiles in pathways important to *WNT* signaling, glutamatergic neurotransmission, and cAMP-related processes. Strikingly, the antipsychotic loxapine improved connectivity and partially reversed abnormal gene expression in patient neurons [214]. Two follow-up studies from these same hiPSCs revealed reduced excitatory synaptic activity [197] and altered dopamine release [215]. In an independent study, hiPSC-derived dopaminergic and glutamatergic neurons showed diminished ability to develop into morphologically mature neurons and displayed several mitochondrial abnormalities [216]. Experiments in schizophrenia hiPSC-derived forebrain and *NGN2* neurons were coupled with two mouse genetic models of schizophrenia to provide robust, cross-model validation of perturbed expression and activity of the striatal-enriched tyrosine phosphatase isoform 61 ($STEP_{61}$) [217], a postsynaptic density-enriched enzyme implicated in animal and pharmacologic models of schizophrenia [218].

Whereas *forebrain* neural progenitor cells (NPCs) from patients with schizophrenia exhibited reduced neural migration [219], altered *WNT* signaling [220] and increased abundance of translation machinery [221], *hippocampal* NPCs from this same hiPSC cohort revealed decreased expression of several key markers of hippocampal neurogenesis [197]. Moreover, hippocampal CA3 neurons derived from these patients had several electrophysiological abnormalities and altered network connectivity when co-cultured with human dentate granule neurons [198]. An independent group confirmed defective migration patterns in patient NPCs, and further reported depressed expression of several angiogenic proteins and reduced angiogenic capacity [222]. These findings are intriguing in light of reports of altered angiogenesis protein expression post-mortem [223] and in living patients [224], as well as diminished vasculature in patients with schizophrenia [225], highlighting the importance of studying non-neuronal cell types in schizophrenia [226].

Evidence is gradually accumulating that abnormalities in mitochondria and reactive oxygen species (ROS) may contribute to schizophrenia etiology [227–229]. A study of ROS production in neural cells derived from a single patient with schizophrenia showed increased ROS, a phenotype that could be alleviated by exposure to the mood-stabilizer and anti-epileptic valproic acid [230]. Subsequently, proteomic analysis revealed alterations in oxidative stress pathways [219] and mitochondrial abnormalities [216] in schizophrenia hiPSC NPCs and neurons, respectively. Intriguingly, transfer of healthy control mitochondria to differentiating hiPSCs derived from schizophrenia patients ameliorated abnormalities in mitochondrial bioenergetics and neuronal differentiation [231].

A pivotal role for abnormal expression of microRNAs and their targets in schizophrenia pathology has garnered increasing levels of support [232]. Studying gene expression across dopaminergic differentiation of hiPSCs, Shi et al. [233] confirmed findings from post-mortem data on an inverse correlation between expression of the dopamine receptor 2 (*DRD2*) and a regulatory microRNA, *miRNA-326*. After establishing a key role for *miR-19* in the regulation of NPC proliferation and migration, Han et al. [234] documented increased *miR-19* expression and a corresponding decrease in the RNA and protein levels of one of its key regulatory targets. A broad examination of microRNA expression levels in NPCs derived from schizophrenia patients and controls identified *miR-9* as the most substantially decreased transcript [235]. Subsequent analysis correlated decreased *miR-9* expression levels with defective migration of patient NPCs in a neurosphere migration assay; abnormalities in NPC migration and gene expression could be partially rescued by overexpression of *miR-9*, and knockdown in control NPCs resulted in the production of a "SCZ NPC" gene expression and migration phenotype [235]. Separately, GWAS studies identified SNPs associated with *miR-9* targets to be enriched in schizophrenia [236], independently validating this hiPSC-based discovery.

Taken together, these reports indicate the pivotal role of hiPSC-based models in parsing altered molecular pathways in living neurons from patients. As shown, the generation of isogenic neural tissue from patients and appropriate controls has been a fruitful approach to further assessing cell and molecular phenotypes observed in schizophrenia. Next, we focus on studies that have used hiPSC-derived tissue to test specific hypotheses regarding putative roles of genetic variants in disease biology.

5.2 Application of hiPSC Models to Study the Effects of Common Variants

Analyses of the contribution of common genetic variation to schizophrenia indicate that hundreds of single-nucleotide polymorphisms (SNPs) may be involved in disease risk [136] and that common variation may account for one-third to a half of the genetic risk for schizophrenia [130, 134]. Individually, each implicated SNP has an incredibly small effect size, with odds ratios (OR) typically ranging from about 1.05 to 1.20 [137]. The remarkably polygenic nature of common variants in schizophrenia, coupled with their low effect sizes, indicates that studying any single, individual SNP in a case-control fashion is highly unlikely to yield meaningful results. While the sample sizes required to detect phenotypic differences attributable to common variants in hiPSC-based models remain undetermined, it is noteworthy that a supplementary analysis conducted by members of the Common Mind Consortium [147] found that the median sample size needed to detect a genome-wide significant difference in the expression of a gene (among 10,000 genes and assuming a mean allele frequency similar to that found in existing data) [147] to be 28,500. Despite this sobering reality, hiPSCs models may still be used to test specific hypotheses on the impacts of a given genetic variant on gene expression, chromatin biology, and cell phenotypes, and they offer the advantage of exploring these effects in cells of a defined genetic background. Below, we discuss promising findings that have used this approach to make important contributions to understanding of the neurobiology of schizophrenia.

Rather than causing damaging mutations in protein-coding genes, common variants for schizophrenia are thought to contribute to disease risk through alteration of gene expression [137]. Increasingly, investigations into the biological effects of disease-associated quantitative trait loci on gene expression, alternative splicing, and chromatin features are employing hiPSC-based models [200, 237]. For example, hiPSC platforms have been used to assess activity-dependent differences in gene expression in patient and control lines [238]. At the present time, hiPSCs models are beginning to serve as a viable platform to study the actual mechanisms of gene expression alteration by schizophrenia risk loci. Ascertainment of hiPSC lines from individuals homozygous for a schizophrenia risk allele or the protective allele, as well as heterozygotes, for a voltage-gated calcium channel (*CACNA1C*) found altered *CACNA1C* gene expression and electrophysiological properties upon conversion to induced neurons [239]. Additionally, hiPSC-derived NPCs were employed to show that knockdown of *FURIN*, a gene for which CMC analysis found a highly significant eQTL, depresses normal migration in neurospheres [147]. Further supporting the relevance of microRNAs to disease biology, longitudinal ATAC-Seq analysis of neurons differentiating from hiPSCs found enrichment of schizophrenia risk loci in open chromatin regions in neurons, and genetic alteration of that a risk allele within an open chromatin region at *miRNA137* led to altered dendritic morphology and synaptic maturity [240]. Additionally, CRISPR/Cas9 approaches enable the manipulation of gene sequence or expression with exquisite precision, and applications of CRISPR-based tools are beginning to make fundamental contributions to the study of schizophrenia genetic risk variants in hiPSC models [241]. In a landmark report, Ho et al. [242] provided the first demonstration of altering expression of specific schizophrenia risk genes with CRISPR-dCas9 in hiPSC-derived NPCs, astrocytes, and neurons. Continued development of this approach and related techniques to improve reproducibility, scalability, and bidirectionality will likely yield a key platform for exploring the functional impacts of disease-associated gene expression abnormalities.

In investigations of the potential role of specific chromatin structures in mediating schizophrenia susceptibility, hiPSC-derived NPCs and neurons have been a key model system to further validate findings in post-mortem and animal model studies and to test the effects of manipulating such structures in living human brain tissue [243]. Intriguingly, chromosome conformation capture (3C) assessment of patient-derived neurons revealed increased contact frequency between a schizophrenia SNP predicted to affect gene expression of *CACNA1C*, confirming a finding that had also seen in post-mortem samples [168]. In a recent report, Rajarajan et al. [244] profiled global patterns of three-dimensional chromatin architecture in hiPSC-derived excitatory neurons, NPCs, and glial cells. Among their findings included the discovery of about double the localization of schizophrenia risk loci in chromatin loops specific to neurons and NPCs over those specific to glial cells [244], thus indicating the pivotal role of cell-type identity in assessing the participation of risk loci gene-regulatory regions. Of note, CRISPR/dCas9-mediated targeting of transcriptional effectors to putative gene-regulatory structures, often separated by several hundred kilobases of DNA, confirmed functional capabilities of selected loops [244]. Future studies will further expand these findings to additional cell types relevant to schizophrenia pathophysiology to provide a comprehensive assessment of the potential disruption of 3D gene-regulatory structures by disease-associated loci.

5.3 Abnormalities in hiPSC-Derived Neurons Harboring Rare Schizophrenia Risk Variants

The higher penetrance typically seen with rare variants associated with schizophrenia makes them feasible contexts to analyze in a case-control fashion. In this section, we discuss reports highlighting such differences in lines containing rare variants implicated in schizophrenia.

(I) Copy Number Variants

Copy number variants (CNVs) are often duplications or deletions ranging from 50 bp to several hundred kilobases in length [245]. The Psychiatric Genomics Consortium has performed the largest genome-wide analysis of CNVs in schizophrenia to date and identified eight loci associated with disease risk [123]. Furthermore, there is a particularly high incidence of CNVs associated with cases of COS [246]. In the text that follows, we discuss investigations of the impact of schizophrenia-associated CNVs on neuronal phenotype using hiPSCs (reviewed in [247].

22q11 Deletion

The 22q11.2 microdeletion is the strongest genetic risk factor for schizophrenia [123], and several prospective studies have shown that at least one-third of such patients develop some sort of psychotic disorder [248–250]. Derivation of hiPSCs from patients with the 22q11.2 deletion was one of the first published reports on in vitro models of schizophrenia [251]. Transcriptomic profiling of neurons derived from 22q11.2 deletion hiPSCs provided a list of hundreds of genes that were differentially expressed between patients and controls; GO analysis of top hits strongly implicated molecular pathways regulating apoptosis, the cell cycle, and neural proliferation [252]. Importantly, assessment of these processes in NPCs generated from the same lines revealed diminished proliferation [252]. 22q11.2 hiPSCs generated from SCZ patients demonstrate reductions in neurosphere size upon differentiation without a decrease in the total number of neurospheres and several morphological abnormalities [252]. Given the location of *DCG8*, a microRNA-regulating protein, in the 22q11.2 deletion band, Zhao et al. [253] sought to assess the impact of this CNV on microRNA expression in lines derived from patients carrying the deletion; they found that several microRNAs and their targets were perturbed by the deletion [253]. Furthermore, gene expression analysis from 22q11.2 neural tissue confirmed substantial alterations in microRNA levels across numerous genes [254]. These data confirm an important role for disruption in microRNA pathways, particularly those involved in neurodevelopment [255], in 22q11.2 deletion carriers with schizophrenia. Future studies will further dissect the molecular pathways by which 22q11.2 deletion contributes to cell-type specific abnormalities to further inform knowledge of disease mechanisms and reveal plausible targets of therapeutic intervention.

15q11.2

An additional CNV implicated in schizophrenia is 15q11.2 [123]. Neural rosettes generated from hiPSC lines harboring a 15q11.2 heterozygous microdeletion displayed abnormal polarity and adherens junction distributions, and complementation experiments in rosettes and NPCs demonstrated that haploinsufficiency of *CYFIP1*, located within the 15q11.2 region, drove these alterations in a process dependent on *WAVE* [256], a member of a complex regulating actin cytoskeletal organization. Importantly, the in vitro and rodent in vivo findings in this study led the authors to perform a targeted eQTL analysis to examine interactions between components of the WAVE pathway, even though individual variants by themselves did not show GWAS-level association with schizophrenia risk [256], and thus demonstrated the power of observations in pre-clinical models to inform larger-scale human genomics studies to discover findings not observed from a relatively unbiased—"omics" approach.

16p11.2

Copy number variations in the 16p11.2 band are also associated with schizophrenia [257]. In a remarkable demonstration of the utility of hiPSC-derived neuron models in recapitulating clinical phenotypes and revealing corresponding disease mechanisms, Deshpande et al. reported opposing cell morphological abnormalities in 16p11.2 duplication and deletion forebrain neurons that mirrored clinical phenotypes of micro- and macrocephaly, respectively. Subsequent functional studies revealed a mechanism connecting altered cell morphologies to specific electrophysiological and synaptic abnormalities in these neurons [258].

(II) Other Variants

Neurexins and Their Loci

Mutations in *Neurexin1* have been implicated in several neurodevelopmental disorders, including schizophrenia [259], and RNA-seq analysis on neural stem cells was employed to investigate the effects of *NRXN1* knockdown on the expression of several genes potentially important in disease processes [260]. Generation of human ESC lines with conditional knockout of *NRXN1* and their subsequent conversion to *NGN2* neurons found impaired functional synaptic activity but not structure [261]. By using lines from patients with childhood-onset schizophrenia with specific deletions in *NRNX1*, Flaherty et al. (2019) [262] discovered alterations in neurexin isoform expression and neuronal phenotypes associated with defined deletions and demonstrated the remarkable utility of hiPSC platforms to elucidate disease biology.

Assessment of the Role of *DISC1* in Schizophrenia with hiPSCs

The origin of investigations into the so-called disrupted in schizophrenia I (*DISC1*) gene lies in the identification of a balanced t(1;11) (q42;q14) translocation in a Scottish adolescent boy discovered as part of a cytogenetic survey of detainees in an juvenile delinquent center [263]. The proband had a diagnosis of conduct disorder, and subsequent assessment of family members found exceptionally high rates of several mental illnesses among carriers of the translocation, a minority of which met criteria for schizophrenia [264]. Although only five of the 34 members who carried the translocation (77 in pedigree as a whole) had a diagnosis of schizophrenia or schizoaffective disorder, the protein-coding gene altered by the translocation was named disrupted in schizophrenia 1 (*DISC1*) [265].

After initial publication of the original *DISC1* translocation found in the Scottish pedigree, a subsequent report documented a four base-pair deletion in *DISC1* in a proband with schizophrenia that co-segregated with other family members who had the same condition or schizoaffective disorder [266]. Although this mutation was not found to associated with risk for schizophrenia in a larger follow-up study [267], it served as a valuable platform for modeling disease biology in patients with defined genetic lesions. In fact, the first report of the generation of hiPSCs from patients with schizophrenia was the publication of integration-free hiPSC from two patients with this particular *DISC1* mutation [268]. In a landmark study, investigators [269] derived hiPSCs from four members of the same family. Forebrain neurons generated from mutant patient lines expressed substantially less wild type *DISC1* protein and had defective glutamatergic synapses [269]. Strikingly, TALEN-mediated introduction of the mutant form of *DISC1* in isogenic lines from unaffected family members as well as correction of the mutation in an affected cell line confirmed a causal role for the *DISC1* mutation in production of abnormal synapses. Finally, global transcription profiling confirmed robust alteration in the expression of genes essential in synaptic processes and neural development, as well as numerous genes previously implicated in schizophrenia and other psychiatric disorders [269]. In a follow-up report, neural stem cells derived from the same *DISC1* mutation and isogenic control lines were used to provide evidence for a relationship between altered *DISC1* expression and a microRNA pathway that was shown to regulate neural stem cell proliferation; both patient and mutant-edited isogenic displayed alterations in this pathway and differentiation abnormalities [270]. Most recently, Yalla et al. [271] explored the role of the ubiquitin-proteasome system in regulating levels of *DISC1* protein and identified a crucial regulator of *DISC1* turnover whose disruption could increase the abnormally low levels of *DISC1* in mutant NPC lines. Generation of medium spiny neuron-like cells [272] from hiPSC lines was used as a platform to functionally validate the role of a schizophrenia-associated protein and *DISC1* interaction partner, *TRAX1*, in neuroprotection and DNA damage repair [273]. On the other hand, generation of hiPSCs with TALEN-mediated deletion of *DISC1* exons relevant to the Scottish translocation and comparison with isogenic controls revealed altered expression of cell fate markers, neurodevelopment, Wnt signaling, and schizophrenia-related genes by mutant *DISC1* [274].

Evidence suggests that *DISC1* exerts many of its neuronal effects through its protein-binding partners [275–278]. In a post-mortem gene expression analysis of cases and controls, a small group of SNPs in *DISC1* were associated with altered expression of *DISC1* isoforms in brain tissue of schizophrenia patients [279]. A recent report [280] documented generation of an hiPSC line with CRISPR/

Cas9-mediated insertion of a 3X FLAG tag downstream of endogenous *DISC1* and characterized *DISC1*-interaction proteomes across multiple cell types; NPCs and astrocytes showed cell-type specificity in *DISC1* binding patterns, and the "*DISC1* interactome" of NPCs was enriched in genes previously identified as candidate de novo variants associated with SCZ [281]. Elegant biochemical experiments confirmed an interaction between a specific residue of *DISC1* protein and *NDEL1*, a component of the dynein complex that is highly important in the brain [282], and disruption of this interaction altered cell cycle progression in radial glial cells in developing murine cortex [283]. Furthermore, the importance of the *DISC1-NDEL1* interaction was confirmed when its disruption reduced neural stem cell proliferation in a human cerebral organoid model; strikingly, cerebral organoids derived from a schizophrenia patient carrying the 4 bp *DISC1* deletion in the *NDEL1*-interacting residue exhibited delayed cell-cycle progression in radial glial cells and confirmed the key role of *DISC1-NDEL1* interaction in regulation of neural proliferation [283].

In total, hiPSC-based models of *DISC1* mutation have generated important insights about the functions of *DISC1* protein in normal neuronal biology and have certainly yielded data on mechanistic pathways contributing to disease in those patients harboring lesions in this gene. The number of patients carrying the defined lesions highlighted, however, is thus far limited to the few studies in which they were originally identified. The importance of *DISC1* variants to the vast majority of "idiopathic" cases of schizophrenia overall is at best unclear and at worst increasingly doubtful. Genetic studies of both common [136, 137, 284] and rare [119, 123, 285] variants have all failed to detect genome-wide significant associations between *DISC1* and risk of schizophrenia. As aptly pointed out already [286, 287], the fact that repeated studies evidence neuronal functions for *DISC1* is both unsurprising given its effects in the (very) small number of people carrying *DISC1* mutations, and insufficient justification alone to assert its relevance to schizophrenia as a whole, as a substantial proportion of all genes could plausibly lead to neuronal phenotypes upon their disruption.

We turn now to mention remaining reports on hiPSC models of selected genetic mutations. Studies of NPCs, oligodendrocyte precursor cells (OPCs), and neurons from a trio containing a patient with an exonic deletion in contactin-associated protein-like 2 (*CNTNAP2*) revealed altered migration patterns [288] and expression patterns in synaptic genes [289]. Derivation of hiPSCs from a family with a missense mutation in *CSPG4*, a gene highly enriched in OPCs [290], revealed that OPCs from carriers of the mutation demonstrated altered patterns of *CSPG4* protein subcellular localization and ratios of modified versus unmodified protein as well as reduced viability and oligodendrogenesis [290]. Remarkably, diffusion tensor imaging of affected patients compared to sibling controls demonstrated white matter abnormalities, connecting cellular phenotypes observed in vitro to abnormalities in vivo [290]. Taken together, investigations seeking to uncover the impact of genetic variants associated with schizophrenia have benefited tremendously from hiPSC models, and we expect substantial expansion of this field going forward.

5.4 Exploring the Impact of Non-Genetic Risk Factors with hiPSCs

As enumerated above, several non-genetic factors have been implicated in schizophrenia risk. In several cases, risk factors that are thought to be proximal causes of disease initiation and/or exacerbation can be isolated and studied upon the exposure of hiPSCs to them throughout experimental time points. In this section, we highlight studies that have assessed the impact of environmental risk factors for schizophrenia using hiPSCs-based modes.

Cannabis Exposure

Meta-analyses have found an increased risk of psychotic disorders in anyone who has ever used cannabis (OR = 1.4) and in frequent users (OR = 2.1) [102]. hiPSC-derived forebrain neurons exposed to THC showed transcriptional alterations of numerous genes implicated in schizophrenia [291], including *WNT* signaling pathways and mitochondrial processes, and expressed a blunted response to KCl-mediated depolarization similar to that seen in patient-derived forebrain neurons in previous studies [238]. In a related report, Obiorah et al. [292] showed that THC exposure reduced the expression of several glutamatergic receptor subunits in hiPSC-derived neurons. Future studies will integrate the ability of hiPSCs to model both typical neurodevelopment and disease phenotypes to continue shedding light on the mechanisms by which cannabis and other drugs of abuse contribute to schizophrenia risk.

Immunologic Processes and Schizophrenia

Another non-genetic riskfactor implicated in schizophrenia is maternal infection [293]. However, data in humans is largely based upon cohort and case-control studies [294], and such study designs do not allow the establishment of causation. While animal models may serve as important tools in exploring the effects of maternal infections on brain development and disease, many infectious agents provoke immune responses that differ among host organisms [295]. For these reasons, hiPSC models of neurodevelopment are an attractive approach for mechanistic studies on the effects of infectious agents on brain development and disease-associated phenotypes.

The increased risk of schizophrenia associated with maternal infection may be due at least in part to the elevation of maternal cytokines and other stress-related cellular processes [296]. Exposure of in vitro generated neural aggregates to heat shock altered the expression of several genes implicated in schizophrenia and autism [297]. After demonstrating a role for heat shock protein (HSP)-mediated processes in protection of developing cortex in response to subthreshold exposure to environmental toxins, investigators [298] also documented increased variance in *HSP* expression levels in response to the same toxins in NPCs derived from patient hiPSC lines. Follow-up analysis confirmed in both animal models and NPCs that exposure to environmental toxins induces *HSP* signaling in a probabilistic manner, generating significant mosaicism among affected cells [299]. These studies serve as notable examples of combining animal in vivo approaches with human in vitro neural models to provide robust exploration of non-genetic riskfactors for schizophrenia [135].

6 Future Directions

6.1 Further Elucidation of Risk Factor Biology in Schizophrenia

While studies of cannabis use and risk of schizophrenia are numerous, similar approaches to assessing the impact of other drugs of abuse on schizophrenia risk are comparatively sparse. Several drugs of abuse can produce a psychotic state during acute intoxication and may also increase risk for later onset of schizophrenia. Psychosis was documented in 6.3% (n = 329) of 5529 hospital drug intoxication cases, particularly among those intoxicated with cannabis (25.9%), amphetamine (25.0%), and cocaine (16.1%) [300]. Patients admitted for substance use disorder to methamphetamine had the highest risk of schizophrenia (hazard ratio = 9.37), followed by cannabis (Hazard ratio = 8.16), cocaine (HR = 5.84), alcohol (HR = 5.56), and opioids (HR = 3.60) [301]. Overall, diagnosis of any substance abuse disorder significantly elevated risk of schizophrenia (HR = 6.04) [302], but most strongly alcohol or cannabis use disorders (HR = 3.38 and 5.20, respectively) [302]. Similar findings have been reported for tobacco smoking (HR of 5.9 [303]. Because cannabis alone is not the only substance associated with schizophrenia risk, and in light of recent genomic findings suggesting a reverse direction of affect [304], hiPSC models should be used to explore the impact of other drugs of abuse on the nervous system in general and on potentially increasing risk of schizophrenia in particular. The differentiation of NPCs to forebrain neurons was used to model the effects of nicotine exposure on neurodevelopment [305], and we anticipate that future studies will assess the effects of other drugs of abuse as well.

6.2 Screening for New Schizophrenia Therapeutics

Progress in understanding disease biology must be coupled with therapeutic innovations derived from emerging discoveries. Considerable interest has been generated in using hiPSC-derived tissues to assess responses of disease-relevant cells to interventions seeking to alleviate disease processes the so-called precision medicine. hiPSC-neurons from cohorts of patients with a given psychiatric diagnosis who exhibit differential responses to treatment will serve as an invaluable strategy to correlate patient genotypes with phenotypic differences in in vitro neural tissue; data derived from such cohorts may be employed to predict treatment responses of psychiatric conditions to specific medications, a longstanding but nevertheless unobtained goal of psychiatric medicine [306]. Comparisons of gene sets impacted by different pharmacologic agents confirmed overlap between genes enriched in schizophrenia SNPs and those involved in antipsychotic response [307], and early-stage studies are now investigating the potential utility of schizophrenia genetic risk variants to predict response to specific antipsychotic medications (e.g., [308]). Further supporting the relationship between schizophrenia risk variants and treatment response, RNA-sequencing analysis of brain tissue of mice chronically treated with the antipsychotic Haloperidol found that differentially expressed genes (DEGs) were enriched in schizophrenia risk loci and biological pathways implicated in the disease [309]. Recently, a screen of over 100 compounds predicted to have relevance to schizophrenia treatment and biology in NPCs and cancer cell lines (CCLs) demonstrated the advantages of using disease-relevant cell types in drug screens and identified compounds that reversed abnormal gene expression signature documented in post-mortem brain samples [310]. Further demonstrating the future utility of hiPSC-based drug screening, high-throughput screening has identified hit compounds that inhibit Zika virus replication [311], infection [312] and reverse Zika-induced neurodevelopmental phenotypes [313]. Taken together, these studies garner much optimism to the prospect of using more disease-relevant and patient-specific approaches toward driving therapeutic innovation.

7 Conclusion

hiPSC enables a broader "ex vivo" study of normal human development and disease mechanisms. Of particular importance to neuropsychiatry, hiPSC-based models produce living, functional, human brain cells accessible for characterization, manipulation, and disease modeling. Insofar as hiPSCs remain isogenic to their donors, they provide a platform for studying the impact of genetics and associated non-genetic factors on neurodevelopment, schizophrenia-associated phenotypes, and response to various interventions both genetic and pharmacological. For these reasons, the further application of hiPSC-based models and the advancement of current technologies should provide a level of hope for investigators striving to understand neuropsychiatric conditions like schizophrenia as well for those affected by them.

References

1. Carpenter Jr., W. T., Strauss, J. S., & Bartko, J. J. (1974). An approach to the diagnosis and understanding of schizophrenia. Introduction. *Schizophrenia Bulletin* (11), 35–36. https://doi.org/10.1093/schbul/1.11.35
2. Crow, T. J. (1985). The two-syndrome concept: origins and current status. *Schizophrenia Bulletin, 11*(3), 471–486.

3. Sartorius, N., Shapiro, R., Kimura, M., & Barrett, K. (1972). WHO international pilot study of schizophrenia. *Psychological Medicine, 2*(4), 422–425.
4. Strauss, J. S., Carpenter Jr., W. T., & Bartko, J. J. (1974). The diagnosis and understanding of schizophrenia. Summary and conclusions. *Schizophrenia Bulletin* (11), 70–80.
5. Kay, S. R., Opler, L. A., & Lindenmayer, J. P. (1988). Reliability and validity of the positive and negative syndrome scale for schizophrenics. *Psychiatry Research, 23*(1), 99–110.
6. Lindenmayer, J. P., Bernstein-Hyman, R., & Grochowski, S. (1994). A new five factor model of schizophrenia. *Psychiatric Quarterly, 65*(4), 299–322.
7. Wallwork, R. S., Fortgang, R., Hashimoto, R., Weinberger, D. R., & Dickinson, D. (2012). Searching for a consensus five-factor model of the Positive and Negative Syndrome Scale for schizophrenia. *Schizophrenia Research, 137*(1–3), 246–250. https://doi.org/10.1016/j.schres.2012.01.031
8. American Psychiatric Association. (2013). *Diagnostic and statistical manual of mental disorders* (5th ed.). Arlington, VA: Author.
9. Staehelin, J. E., & Kielholz, P. (1953). Largactil, a new vegetative damping agent in mental disorders. *Schweizerische Medizinische Wochenschrift, 83*(25), 581–586.
10. Carlsson, A., & Lindqvist, M. (1963). Effect of chlorpromazine or haloperidol on formation of 3methoxytyramine and normetanephrine in mouse brain. *Acta Pharmacologica et Toxicologica, 20*, 140–144.
11. Creese, I., Burt, D. R., & Snyder, S. H. (1976). Dopamine receptor binding predicts clinical and pharmacological potencies of antischizophrenic drugs. *Science, 192*(4238), 481–483.
12. Seeman, P., & Lee, T. (1975). Antipsychotic drugs: direct correlation between clinical potency and presynaptic action on dopamine neurons. *Science, 188*(4194), 1217–1219.
13. Borison, R. L., Pathiraja, A. P., Diamond, B. I., & Meibach, R. C. (1992). Risperidone: clinical safety and efficacy in schizophrenia. *Psychopharmacology Bulletin, 28*(2), 213–218.
14. Jones, P. B., Barnes, T. R., Davies, L., Dunn, G., Lloyd, H., Hayhurst, K. P., et al. (2006). Randomized controlled trial of the effect on quality of life of second- vs first-generation antipsychotic drugs in schizophrenia: cost utility of the latest antipsychotic drugs in schizophrenia study (CUtLASS 1). *Archives of General Psychiatry, 63*(10), 1079–1087. https://doi.org/10.1001/archpsyc.63.10.1079
15. Lieberman, J. A., Stroup, T. S., McEvoy, J. P., Swartz, M. S., Rosenheck, R. A., Perkins, D. O., et al. (2005). Effectiveness of antipsychotic drugs in patients with chronic schizophrenia. *The New England Journal of Medicine, 353*(12), 1209–1223. https://doi.org/10.1056/NEJMoa051688
16. Kane, J., Honigfeld, G., Singer, J., & Meltzer, H. (1988). Clozapine for the treatment-resistant schizophrenic. A double-blind comparison with chlorpromazine. *Archives of General Psychiatry, 45*(9), 789–796.
17. Fusar-Poli, P., Papanastasiou, E., Stahl, D., Rocchetti, M., Carpenter, W., Shergill, S., et al. (2015). Treatments of negative symptoms in schizophrenia: meta-analysis of 168 randomized placebo-controlled trials. *Schizophrenia Bulletin, 41*(4), 892–899. https://doi.org/10.1093/schbul/sbu170
18. Leucht, S., Cipriani, A., Spineli, L., Mavridis, D., Orey, D., Richter, F., et al. (2013). Comparative efficacy and tolerability of 15 antipsychotic drugs in schizophrenia: a multiple-treatments meta-analysis. *Lancet, 382*(9896), 951–962. https://doi.org/10.1016/S0140-6736(13)60733-3
19. Naber, D., & Lambert, M. (2009). The CATIE and CUtLASS studies in schizophrenia: results and implications for clinicians. *CNS Drugs, 23*(8), 649–659. https://doi.org/10.2165/00023210-200923080-00002
20. Downing, A. M., Kinon, B. J., Millen, B. A., Zhang, L., Liu, L., Morozova, M. A., et al. (2014). A double-blind, placebo-controlled comparator study of LY2140023 monohydrate in patients with schizophrenia. *BMC Psychiatry, 14*, 351. https://doi.org/10.1186/s12888-014-0351-3
21. Jablensky, A., Sartorius, N., Ernberg, G., Anker, M., Korten, A., Cooper, J. E., et al. (1992). Schizophrenia: manifestations, incidence and course in different cultures. A World Health Organization ten-country study. *Psychological Medicine. Monograph Supplement, 20*, 1–97.

22. Hjorthoj, C., Sturup, A. E., McGrath, J. J., & Nordentoft, M. (2017). Years of potential life lost and life expectancy in schizophrenia: a systematic review and meta-analysis. *Lancet Psychiatry, 4*(4), 295–301. https://doi.org/10.1016/S2215-0366(17)30078-0
23. Palmer, B. A., Pankratz, V. S., & Bostwick, J. M. (2005). The lifetime risk of suicide in schizophrenia: a reexamination. *Archives of General Psychiatry, 62*(3), 247–253. https://doi.org/10.1001/archpsyc.62.3.247
24. Caldwell, C. B., & Gottesman, I. I. (1990). Schizophrenics kill themselves too: a review of risk factors for suicide. *Schizophrenia Bulletin, 16*(4), 571–589.
25. Phillips, M. R., Yang, G., Li, S., & Li, Y. (2004). Suicide and the unique prevalence pattern of schizophrenia in mainland China: a retrospective observational study. *Lancet, 364*(9439), 1062–1068. https://doi.org/10.1016/S0140-6736(04)17061-X
26. Brown, S. (1997). Excess mortality of schizophrenia. A meta-analysis. *The British Journal of Psychiatry, 171*, 502–508.
27. Weinmann, S., Read, J., & Aderhold, V. (2009). Influence of antipsychotics on mortality in schizophrenia: systematic review. *Schizophrenia Research, 113*(1), 1–11. https://doi.org/10.1016/j.schres.2009.05.018
28. Nielsen, P. R., Laursen, T. M., & Agerbo, E. (2016). Comorbidity of schizophrenia and infection: a population-based cohort study. *Social Psychiatry and Psychiatric Epidemiology, 51*(12), 1581–1589. https://doi.org/10.1007/s00127-016-1297-1
29. Goff, D. C., Cather, C., Evins, A. E., Henderson, D. C., Freudenreich, O., Copeland, P. M., et al. (2005). Medical morbidity and mortality in schizophrenia: guidelines for psychiatrists. *Journal of Clinical Psychiatry, 66*(2), 183–194; quiz 147, 273-184.
30. Winklbaur, B., Ebner, N., Sachs, G., Thau, K., & Fischer, G. (2006). Substance abuse in patients with schizophrenia. *Dialogues in Clinical Neuroscience, 8*(1), 37–43.
31. Brekke, J. S., Prindle, C., Bae, S. W., & Long, J. D. (2001). Risks for individuals with schizophrenia who are living in the community. *Psychiatric Services, 52*(10), 1358–1366. https://doi.org/10.1176/appi.ps.52.10.1358
32. Rapoport, J. L., Addington, A. M., Frangou, S., & Psych, M. R. (2005). The neurodevelopmental model of schizophrenia: update 2005. *Molecular Psychiatry, 10*(5), 434–449. https://doi.org/10.1038/sj.mp.4001642
33. Rapoport, J. L., Giedd, J. N., & Gogtay, N. (2012). Neurodevelopmental model of schizophrenia: update 2012. *Molecular Psychiatry, 17*(12), 1228–1238. https://doi.org/10.1038/mp.2012.23
34. Stiles, J., & Jernigan, T. L. (2010). The basics of brain development. *Neuropsychology Review, 20*(4), 327–348. https://doi.org/10.1007/s11065-010-9148-4
35. Muraki, K., & Tanigaki, K. (2015). Neuronal migration abnormalities and its possible implications for schizophrenia. *Frontiers in Neuroscience, 9*, 74. https://doi.org/10.3389/fnins.2015.00074
36. Schoenfeld, T. J., & Cameron, H. A. (2015). Adult neurogenesis and mental illness. *Neuropsychopharmacology, 40*(1), 113–128. https://doi.org/10.1038/npp.2014.230
37. Schmidt, M. J., & Mirnics, K. (2015). Neurodevelopment, GABA system dysfunction, and schizophrenia. *Neuropsychopharmacology, 40*(1), 190–206. https://doi.org/10.1038/npp.2014.95
38. Bartzokis, G. (2002). Schizophrenia: breakdown in the well-regulated lifelong process of brain development and maturation. *Neuropsychopharmacology, 27*(4), 672–683. https://doi.org/10.1016/S0893-133X(02)00364-0
39. Forsyth, J. K., & Lewis, D. A. (2017). Mapping the consequences of impaired synaptic plasticity in schizophrenia through development: an integrative model for diverse clinical features. *Trends in Cognitive Sciences, 21*(10), 760–778. https://doi.org/10.1016/j.tics.2017.06.006
40. Hirayasu, Y., Shenton, M. E., Salisbury, D. F., Dickey, C. C., Fischer, I. A., Mazzoni, P., et al. (1998). Lower left temporal lobe MRI volumes in patients with first-episode schizophrenia compared with psychotic patients with first-episode affective disorder and normal

subjects. *The American Journal of Psychiatry, 155*(10), 1384–1391. https://doi.org/10.1176/ajp.155.10.1384
41. Wilke, M., Kaufmann, C., Grabner, A., Putz, B., Wetter, T. C., & Auer, D. P. (2001). Gray matter-changes and correlates of disease severity in schizophrenia: a statistical parametric mapping study. *NeuroImage, 13*(5), 814–824. https://doi.org/10.1006/nimg.2001.0751
42. Salgado-Pineda, P., Baeza, I., Perez-Gomez, M., Vendrell, P., Junque, C., Bargallo, N., et al. (2003). Sustained attention impairment correlates to gray matter decreases in first episode neuroleptic-naive schizophrenic patients. *NeuroImage, 19*(2 Pt 1), 365–375.
43. Berge, D., Carmona, S., Rovira, M., Bulbena, A., Salgado, P., & Vilarroya, O. (2011). Gray matter volume deficits and correlation with insight and negative symptoms in first-psychotic-episode subjects. *Acta Psychiatrica Scandinavica, 123*(6), 431–439. https://doi.org/10.1111/j.1600-0447.2010.01635.x
44. Hirayasu, Y., Tanaka, S., Shenton, M. E., Salisbury, D. F., DeSantis, M. A., Levitt, J. J., et al. (2001). Prefrontal gray matter volume reduction in first episode schizophrenia. *Cerebral Cortex, 11*(4), 374–381.
45. Paillere-Martinot, M., Caclin, A., Artiges, E., Poline, J. B., Joliot, M., Mallet, L., et al. (2001). Cerebral gray and white matter reductions and clinical correlates in patients with early onset schizophrenia. *Schizophrenia Research, 50*(1–2), 19–26.
46. Crespo-Facorro, B., Roiz-Santianez, R., Perez-Iglesias, R., Rodriguez-Sanchez, J. M., Mata, I., Tordesillas-Gutierrez, D., et al. (2011). Global and regional cortical thinning in first-episode psychosis patients: relationships with clinical and cognitive features. *Psychological Medicine, 41*(7), 1449–1460. https://doi.org/10.1017/S003329171000200X
47. Whitford, T. J., Grieve, S. M., Farrow, T. F., Gomes, L., Brennan, J., Harris, A. W., et al. (2006). Progressive grey matter atrophy over the first 2–3 years of illness in first-episode schizophrenia: a tensor-based morphometry study. *NeuroImage, 32*(2), 511–519. https://doi.org/10.1016/j.neuroimage.2006.03.041
48. Hirayasu, Y., Shenton, M. E., Salisbury, D. F., Kwon, J. S., Wible, C. G., Fischer, I. A., et al. (1999). Subgenual cingulate cortex volume in first-episode psychosis. *The American Journal of Psychiatry, 156*(7), 1091–1093. https://doi.org/10.1176/ajp.156.7.1091
49. Kasai, K., Shenton, M. E., Salisbury, D. F., Onitsuka, T., Toner, S. K., Yurgelun-Todd, D., et al. (2003). Differences and similarities in insular and temporal pole MRI gray matter volume abnormalities in first-episode schizophrenia and affective psychosis. *Archives of General Psychiatry, 60*(11), 1069–1077. https://doi.org/10.1001/archpsyc.60.11.1069
50. Rothlisberger, M., Riecher-Rossler, A., Aston, J., Fusar-Poli, P., Radu, E. W., & Borgwardt, S. (2012). Cingulate volume abnormalities in emerging psychosis. *Current Pharmaceutical Design, 18*(4), 495–504.
51. Liu, J., Pearlson, G., Windemuth, A., Ruano, G., Perrone-Bizzozero, N. I., & Calhoun, V. (2009). Combining fMRI and SNP data to investigate connections between brain function and genetics using parallel ICA. *Human Brain Mapping, 30*(1), 241–255. https://doi.org/10.1002/hbm.20508
52. Carpenter, D. M., Tang, C. Y., Friedman, J. I., Hof, P. R., Stewart, D. G., Buchsbaum, M. S., et al. (2008). Temporal characteristics of tract-specific anisotropy abnormalities in schizophrenia. *Neuroreport, 19*(14), 1369–1372. https://doi.org/10.1097/WNR.0b013e32830abc35
53. Karlsgodt, K. H., van Erp, T. G., Poldrack, R. A., Bearden, C. E., Nuechterlein, K. H., & Cannon, T. D. (2008). Diffusion tensor imaging of the superior longitudinal fasciculus and working memory in recent-onset schizophrenia. *Biological Psychiatry, 63*(5), 512–518. https://doi.org/10.1016/j.biopsych.2007.06.017
54. Perez-Iglesias, R., Tordesillas-Gutierrez, D., Barker, G. J., McGuire, P. K., Roiz-Santianez, R., Mata, I., et al. (2010). White matter defects in first episode psychosis patients: a voxelwise analysis of diffusion tensor imaging. *NeuroImage, 49*(1), 199–204. https://doi.org/10.1016/j.neuroimage.2009.07.016
55. Ruef, A., Curtis, L., Moy, G., Bessero, S., Badan Ba, M., Lazeyras, F., et al. (2012). Magnetic resonance imaging correlates of first-episode psychosis in young adult male patients: com-

bined analysis of grey and white matter. *Journal of Psychiatry & Neuroscience, 37*(5), 305–312. https://doi.org/10.1503/jpn.110057
56. White, T., Anjum, A., & Schulz, S. C. (2006). The schizophrenia prodrome. *The American Journal of Psychiatry, 163*(3), 376–380. https://doi.org/10.1176/appi.ajp.163.3.376
57. Yung, A. R., & McGorry, P. D. (1996a). The initial prodrome in psychosis: descriptive and qualitative aspects. *The Australian and New Zealand Journal of Psychiatry, 30*(5), 587–599. https://doi.org/10.3109/00048679609062654
58. Beiser, M., Erickson, D., Fleming, J. A., & Iacono, W. G. (1993). Establishing the onset of psychotic illness. *The American Journal of Psychiatry, 150*(9), 1349–1354. https://doi.org/10.1176/ajp.150.9.1349
59. Lencz, T., Cornblatt, B., & Bilder, R. M. (2001). Neurodevelopmental models of schizophrenia: pathophysiologic synthesis and directions for intervention research. *Psychopharmacology Bulletin, 35*(1), 95–125.
60. Tsuang, M. T., Faraone, S. V., Bingham, S., Young, K., Prabhudesai, S., Haverstock, S. L., et al. (2000). Department of Veterans Affairs Cooperative Studies Program genetic linkage study of schizophrenia: ascertainment methods and sample description. *American Journal of Medical Genetics, 96*(3), 342–347.
61. Yung, A. R., & McGorry, P. D. (1996b). The prodromal phase of first-episode psychosis: past and current conceptualizations. *Schizophrenia Bulletin, 22*(2), 353–370.
62. Cornblatt, B., Lencz, T., & Obuchowski, M. (2002). The schizophrenia prodrome: treatment and high-risk perspectives. *Schizophrenia Research, 54*(1–2), 177–186.
63. Cornblatt, B., Obuchowski, M., Roberts, S., Pollack, S., & Erlenmeyer-Kimling, L. (1999). Cognitive and behavioral precursors of schizophrenia. *Development and Psychopathology, 11*(3), 487–508.
64. Lappin, J. M., Dazzan, P., Morgan, K., Morgan, C., Chitnis, X., Suckling, J., et al. (2007). Duration of prodromal phase and severity of volumetric abnormalities in first-episode psychosis. *The British Journal of Psychiatry. Supplement, 51*, s123–s127. https://doi.org/10.1192/bjp.191.51.s123
65. Fusar-Poli, P., Tantardini, M., De Simone, S., Ramella-Cravaro, V., Oliver, D., Kingdon, J., et al. (2017). Deconstructing vulnerability for psychosis: meta-analysis of environmental risk factors for psychosis in subjects at ultra high-risk. *European Psychiatry, 40*, 65–75. https://doi.org/10.1016/j.eurpsy.2016.09.003
66. Clarke, M. C., Tanskanen, A., Huttunen, M., Leon, D. A., Murray, R. M., Jones, P. B., et al. (2011). Increased risk of schizophrenia from additive interaction between infant motor developmental delay and obstetric complications: evidence from a population-based longitudinal study. *The American Journal of Psychiatry, 168*(12), 1295–1302. https://doi.org/10.1176/appi.ajp.2011.11010011
67. Jones, P., Rodgers, B., Murray, R., & Marmot, M. (1994). Child development risk factors for adult schizophrenia in the British 1946 birth cohort. *Lancet, 344*(8934), 1398–1402.
68. Kremen, W. S., Buka, S. L., Seidman, L. J., Goldstein, J. M., Koren, D., & Tsuang, M. T. (1998). IQ decline during childhood and adult psychotic symptoms in a community sample: a 19-year longitudinal study. *The American Journal of Psychiatry, 155*(5), 672–677. https://doi.org/10.1176/ajp.155.5.672
69. Wood, S. J., Pantelis, C., Proffitt, T., Phillips, L. J., Stuart, G. W., Buchanan, J. A., et al. (2003). Spatial working memory ability is a marker of risk-for-psychosis. *Psychological Medicine, 33*(7), 1239–1247.
70. Brewer, W. J., Francey, S. M., Wood, S. J., Jackson, H. J., Pantelis, C., Phillips, L. J., et al. (2005). Memory impairments identified in people at ultra-high risk for psychosis who later develop first-episode psychosis. *The American Journal of Psychiatry, 162*(1), 71–78. https://doi.org/10.1176/appi.ajp.162.1.71
71. Dickson, H., Laurens, K. R., Cullen, A. E., & Hodgins, S. (2012). Meta-analyses of cognitive and motor function in youth aged 16 years and younger who subsequently develop schizophrenia. *Psychological Medicine, 42*(4), 743–755. https://doi.org/10.1017/S0033291711001693

72. Erlenmeyer-Kimling, L., Rock, D., Roberts, S. A., Janal, M., Kestenbaum, C., Cornblatt, B., et al. (2000). Attention, memory, and motor skills as childhood predictors of schizophrenia-related psychoses: the New York High-Risk Project. *The American Journal of Psychiatry, 157*(9), 1416–1422. https://doi.org/10.1176/appi.ajp.157.9.1416
73. Done, D. J., Crow, T. J., Johnstone, E. C., & Sacker, A. (1994). Childhood antecedents of schizophrenia and affective illness: social adjustment at ages 7 and 11. *BMJ, 309*(6956), 699–703.
74. Davidson, M., Reichenberg, A., Rabinowitz, J., Weiser, M., Kaplan, Z., & Mark, M. (1999). Behavioral and intellectual markers for schizophrenia in apparently healthy male adolescents. *The American Journal of Psychiatry, 156*(9), 1328–1335. https://doi.org/10.1176/ajp.156.9.1328
75. Klosterkotter, J., Hellmich, M., Steinmeyer, E. M., & Schultze-Lutter, F. (2001). Diagnosing schizophrenia in the initial prodromal phase. *Archives of General Psychiatry, 58*(2), 158–164.
76. Pantelis, C., Velakoulis, D., McGorry, P. D., Wood, S. J., Suckling, J., Phillips, L. J., et al. (2003). Neuroanatomical abnormalities before and after onset of psychosis: a cross-sectional and longitudinal MRI comparison. *Lancet, 361*(9354), 281–288. https://doi.org/10.1016/S0140-6736(03)12323-9
77. Borgwardt, S. J., McGuire, P. K., Aston, J., Berger, G., Dazzan, P., Gschwandtner, U., et al. (2007). Structural brain abnormalities in individuals with an at-risk mental state who later develop psychosis. *The British Journal of Psychiatry. Supplement, 51*, s69–s75. https://doi.org/10.1192/bjp.191.51.s69
78. Fornito, A., Yung, A. R., Wood, S. J., Phillips, L. J., Nelson, B., Cotton, S., et al. (2008). Anatomic abnormalities of the anterior cingulate cortex before psychosis onset: an MRI study of ultra-high-risk individuals. *Biological Psychiatry, 64*(9), 758–765. https://doi.org/10.1016/j.biopsych.2008.05.032
79. Takahashi, T., Wood, S. J., Soulsby, B., Kawasaki, Y., McGorry, P. D., Suzuki, M., et al. (2009a). An MRI study of the superior temporal subregions in first-episode patients with various psychotic disorders. *Schizophrenia Research, 113*(2–3), 158–166. https://doi.org/10.1016/j.schres.2009.06.016
80. Takahashi, T., Wood, S. J., Yung, A. R., Phillips, L. J., Soulsby, B., McGorry, P. D., et al. (2009b). Insular cortex gray matter changes in individuals at ultra-high-risk of developing psychosis. *Schizophrenia Research, 111*(1–3), 94–102. https://doi.org/10.1016/j.schres.2009.03.024
81. Mechelli, A., Riecher-Rossler, A., Meisenzahl, E. M., Tognin, S., Wood, S. J., Borgwardt, S. J., et al. (2011). Neuroanatomical abnormalities that predate the onset of psychosis: a multicenter study. *Archives of General Psychiatry, 68*(5), 489–495. https://doi.org/10.1001/archgenpsychiatry.2011.42
82. Fusar-Poli, P., Broome, M. R., Woolley, J. B., Johns, L. C., Tabraham, P., Bramon, E., et al. (2011). Altered brain function directly related to structural abnormalities in people at ultra high risk of psychosis: longitudinal VBM-fMRI study. *Journal of Psychiatric Research, 45*(2), 190–198. https://doi.org/10.1016/j.jpsychires.2010.05.012
83. Jung, W. H., Kim, J. S., Jang, J. H., Choi, J. S., Jung, M. H., Park, J. Y., et al. (2011). Cortical thickness reduction in individuals at ultra-high-risk for psychosis. *Schizophrenia Bulletin, 37*(4), 839–849. https://doi.org/10.1093/schbul/sbp151
84. Gilmore, J. H., Kang, C., Evans, D. D., Wolfe, H. M., Smith, J. K., Lieberman, J. A., et al. (2010a). Prenatal and neonatal brain structure and white matter maturation in children at high risk for schizophrenia. *The American Journal of Psychiatry, 167*(9), 1083–1091. https://doi.org/10.1176/appi.ajp.2010.09101492
85. Gilmore, J. H., Schmitt, J. E., Knickmeyer, R. C., Smith, J. K., Lin, W., Styner, M., et al. (2010b). Genetic and environmental contributions to neonatal brain structure: A twin study. *Human Brain Mapping, 31*(8), 1174–1182. https://doi.org/10.1002/hbm.20926

86. Walterfang, M., McGuire, P. K., Yung, A. R., Phillips, L. J., Velakoulis, D., Wood, S. J., et al. (2008). White matter volume changes in people who develop psychosis. *The British Journal of Psychiatry, 193*(3), 210–215. https://doi.org/10.1192/bjp.bp.107.043463
87. Bloemen, O. J., de Koning, M. B., Schmitz, N., Nieman, D. H., Becker, H. E., de Haan, L., et al. (2010). White-matter markers for psychosis in a prospective ultra-high-risk cohort. *Psychological Medicine, 40*(8), 1297–1304. https://doi.org/10.1017/S0033291709991711
88. Brown, A. S. (2006). Prenatal infection as a risk factor for schizophrenia. *Schizophrenia Bulletin, 32*(2), 200–202. https://doi.org/10.1093/schbul/sbj052
89. Brown, A. S. (2012). Epidemiologic studies of exposure to prenatal infection and risk of schizophrenia and autism. *Developmental Neurobiology, 72*(10), 1272–1276. https://doi.org/10.1002/dneu.22024
90. Cannon, M., Jones, P. B., & Murray, R. M. (2002). Obstetric complications and schizophrenia: historical and meta-analytic review. *The American Journal of Psychiatry, 159*(7), 1080–1092. https://doi.org/10.1176/appi.ajp.159.7.1080
91. Picker, J. D., & Coyle, J. T. (2005). Do maternal folate and homocysteine levels play a role in neurodevelopmental processes that increase risk for schizophrenia? *Harvard Review of Psychiatry, 13*(4), 197–205. https://doi.org/10.1080/10673220500243372
92. Roseboom, T. J., Painter, R. C., van Abeelen, A. F., Veenendaal, M. V., & de Rooij, S. R. (2011). Hungry in the womb: what are the consequences? Lessons from the Dutch famine. *Maturitas, 70*(2), 141–145. https://doi.org/10.1016/j.maturitas.2011.06.017
93. Knud Larsen, J., Bendsen, B. B., Foldager, L., & Munk-Jorgensen, P. (2010). Prematurity and low birth weight as risk factors for the development of affective disorder, especially depression and schizophrenia: a register study. *Acta Neuropsychiatrica, 22*(6), 284–291. https://doi.org/10.1111/j.1601-5215.2010.00498.x
94. Rifkin, L., Lewis, S., Jones, P., Toone, B., & Murray, R. (1994). Low birth weight and schizophrenia. *The British Journal of Psychiatry, 165*(3), 357–362.
95. Wahlbeck, K., Forsen, T., Osmond, C., Barker, D. J., & Eriksson, J. G. (2001). Association of schizophrenia with low maternal body mass index, small size at birth, and thinness during childhood. *Archives of General Psychiatry, 58*(1), 48–52.
96. Torniainen, M., Wegelius, A., Tuulio-Henriksson, A., Lonnqvist, J., & Suvisaari, J. (2013). Both low birthweight and high birthweight are associated with cognitive impairment in persons with schizophrenia and their first-degree relatives. *Psychological Medicine, 43*(11), 2361–2367. https://doi.org/10.1017/S0033291713000032
97. Moilanen, K., Jokelainen, J., Jones, P. B., Hartikainen, A. L., Jarvelin, M. R., & Isohanni, M. (2010). Deviant intrauterine growth and risk of schizophrenia: a 34-year follow-up of the Northern Finland 1966 Birth Cohort. *Schizophrenia Research, 124*(1–3), 223–230. https://doi.org/10.1016/j.schres.2010.09.006
98. Davies, G., Welham, J., Chant, D., Torrey, E. F., & McGrath, J. (2003). A systematic review and meta-analysis of Northern Hemisphere season of birth studies in schizophrenia. *Schizophrenia Bulletin, 29*(3), 587–593.
99. Frissen, A., Lieverse, R., Drukker, M., van Winkel, R., Delespaul, P., & Investigators, G. (2015). Childhood trauma and childhood urbanicity in relation to psychotic disorder. *Social Psychiatry and Psychiatric Epidemiology, 50*(10), 1481–1488. https://doi.org/10.1007/s00127-015-1049-7
100. Lataster, J., Myin-Germeys, I., Lieb, R., Wittchen, H. U., & van Os, J. (2012). Adversity and psychosis: a 10-year prospective study investigating synergism between early and recent adversity in psychosis. *Acta Psychiatrica Scandinavica, 125*(5), 388–399. https://doi.org/10.1111/j.1600-0447.2011.01805.x
101. Marconi, A., Di Forti, M., Lewis, C. M., Murray, R. M., & Vassos, E. (2016). Meta-analysis of the association between the level of cannabis use and risk of psychosis. *Schizophrenia Bulletin, 42*(5), 1262–1269. https://doi.org/10.1093/schbul/sbw003

102. Moore, T. H., Zammit, S., Lingford-Hughes, A., Barnes, T. R., Jones, P. B., Burke, M., et al. (2007). Cannabis use and risk of psychotic or affective mental health outcomes: a systematic review. *Lancet, 370*(9584), 319–328. https://doi.org/10.1016/S0140-6736(07)61162-3
103. Heinz, A., Deserno, L., & Reininghaus, U. (2013). Urbanicity, social adversity and psychosis. *World Psychiatry, 12*(3), 187–197. https://doi.org/10.1002/wps.20056
104. Lichtenstein, P., Yip, B. H., Bjork, C., Pawitan, Y., Cannon, T. D., Sullivan, P. F., et al. (2009). Common genetic determinants of schizophrenia and bipolar disorder in Swedish families: a population-based study. *Lancet, 373*(9659), 234–239. https://doi.org/10.1016/S0140-6736(09)60072-6
105. Lichtenstein, P., Bjork, C., Hultman, C. M., Scolnick, E., Sklar, P., & Sullivan, P. F. (2006). Recurrence risks for schizophrenia in a Swedish national cohort. *Psychological Medicine, 36*(10), 1417–1425. https://doi.org/10.1017/S0033291706008385
106. Cardno, A. G., & Gottesman, I. I. (2000). Twin studies of schizophrenia: from bow-and-arrow concordances to star wars Mx and functional genomics. *American Journal of Medical Genetics, 97*(1), 12–17.
107. Hilker, R., Helenius, D., Fagerlund, B., Skytthe, A., Christensen, K., Werge, T. M., et al. (2018). Heritability of schizophrenia and schizophrenia spectrum based on the Nationwide Danish Twin Register. *Biological Psychiatry, 83*(6), 492–498. https://doi.org/10.1016/j.biopsych.2017.08.017
108. Sullivan, P. F., Agrawal, A., Bulik, C. M., Andreassen, O. A., Borglum, A. D., Breen, G., et al. (2018). Psychiatric genomics: an update and an agenda. *The American Journal of Psychiatry, 175*(1), 15–27. https://doi.org/10.1176/appi.ajp.2017.17030283
109. Power, R. A., Kyaga, S., Uher, R., MacCabe, J. H., Langstrom, N., Landen, M., et al. (2013). Fecundity of patients with schizophrenia, autism, bipolar disorder, depression, anorexia nervosa, or substance abuse vs their unaffected siblings. *JAMA Psychiatry, 70*(1), 22–30. https://doi.org/10.1001/jamapsychiatry.2013.268
110. Gershon, E. S., Alliey-Rodriguez, N., & Liu, C. (2011). After GWAS: searching for genetic risk for schizophrenia and bipolar disorder. *The American Journal of Psychiatry, 168*(3), 253–256. https://doi.org/10.1176/appi.ajp.2010.10091340
111. Malaspina, D., Brown, A., Goetz, D., Alia-Klein, N., Harkavy-Friedman, J., Harlap, S., et al. (2002). Schizophrenia risk and paternal age: a potential role for de novo mutations in schizophrenia vulnerability genes. *CNS Spectrums, 7*(1), 26–29.
112. Kong, A., Frigge, M. L., Masson, G., Besenbacher, S., Sulem, P., Magnusson, G., et al. (2012). Rate of de novo mutations and the importance of father's age to disease risk. *Nature, 488*(7412), 471–475. https://doi.org/10.1038/nature11396
113. Gulsuner, S., Walsh, T., Watts, A. C., Lee, M. K., Thornton, A. M., Casadei, S., et al. (2013). Spatial and temporal mapping of de novo mutations in schizophrenia to a fetal prefrontal cortical network. *Cell, 154*(3), 518–529. https://doi.org/10.1016/j.cell.2013.06.049
114. Xu, B., Ionita-Laza, I., Roos, J. L., Boone, B., Woodrick, S., Sun, Y., et al. (2012). De novo gene mutations highlight patterns of genetic and neural complexity in schizophrenia. *Nature Genetics, 44*(12), 1365–1369. https://doi.org/10.1038/ng.2446
115. Fromer, M., Pocklington, A. J., Kavanagh, D. H., Williams, H. J., Dwyer, S., Gormley, P., et al. (2014). De novo mutations in schizophrenia implicate synaptic networks. *Nature, 506*(7487), 179–184. https://doi.org/10.1038/nature12929
116. Awadalla, P., Gauthier, J., Myers, R. A., Casals, F., Hamdan, F. F., Griffing, A. R., et al. (2010). Direct measure of the de novo mutation rate in autism and schizophrenia cohorts. *American Journal of Human Genetics, 87*(3), 316–324. https://doi.org/10.1016/j.ajhg.2010.07.019
117. Girard, S. L., Gauthier, J., Noreau, A., Xiong, L., Zhou, S., Jouan, L., et al. (2011). Increased exonic de novo mutation rate in individuals with schizophrenia. *Nature Genetics, 43*(9), 860–863. https://doi.org/10.1038/ng.886
118. Purcell, S. M., Moran, J. L., Fromer, M., Ruderfer, D., Solovieff, N., Roussos, P., et al. (2014). A polygenic burden of rare disruptive mutations in schizophrenia. *Nature, 506*(7487), 185–190. https://doi.org/10.1038/nature12975

119. Genovese, G., Fromer, M., Stahl, E. A., Ruderfer, D. M., Chambert, K., Landen, M., et al. (2016). Increased burden of ultra-rare protein-altering variants among 4,877 individuals with schizophrenia. *Nature Neuroscience, 19*(11), 1433–1441. https://doi.org/10.1038/nn.4402
120. Genovese G., Fromer M., Stahl E. A., Ruderfer D. M., Chambert K., Landén M., et al. (2016) Increased burden of ultra-rare protein-altering variants among 4,877 individuals with schizophrenia. *Nature Neuroscience 19*(11):1433–1441
121. Szatkiewicz, J. P., O'Dushlaine, C., Chen, G., Chambert, K., Moran, J. L., Neale, B. M., et al. (2014). Copy number variation in schizophrenia in Sweden. *Molecular Psychiatry, 19*(7), 762–773. https://doi.org/10.1038/mp.2014.40
122. Rees, E., Kirov, G., O'Donovan, M. C., & Owen, M. J. (2012). De novo mutation in schizophrenia. *Schizophrenia Bulletin, 38*(3), 377–381. https://doi.org/10.1093/schbul/sbs047
123. Marshall, C. R., Howrigan, D. P., Merico, D., Thiruvahindrapuram, B., Wu, W., Greer, D. S., et al. (2017). Contribution of copy number variants to schizophrenia from a genome-wide study of 41,321 subjects. *Nature Genetics, 49*(1), 27–35. https://doi.org/10.1038/ng.3725
124. Schneider, M., Debbane, M., Bassett, A. S., Chow, E. W., Fung, W. L., van den Bree, M., et al. (2014). Psychiatric disorders from childhood to adulthood in 22q11.2 deletion syndrome: results from the International Consortium on Brain and Behavior in 22q11.2 deletion syndrome. *The American Journal of Psychiatry, 171*(6), 627–639. https://doi.org/10.1176/appi.ajp.2013.13070864
125. Van, L., Boot, E., & Bassett, A. S. (2017). Update on the 22q11.2 deletion syndrome and its relevance to schizophrenia. *Current Opinion in Psychiatry, 30*(3), 191–196. https://doi.org/10.1097/YCO.0000000000000324
126. Bergen, S. E., Ploner, A., Howrigan, D., CNV Analysis Group and the Schizophrenia Working Group of the Psychiatric Genomics Consortium, O'Donovan, M. C., Smoller, J. W., et al. (2018). Joint contributions of rare copy number variants and common SNPs to risk for schizophrenia. *Am J Psychiatry, 176*, 29. https://doi.org/10.1176/appi.ajp.2018.17040467
127. Tansey, K. E., Rees, E., Linden, D. E., Ripke, S., Chambert, K. D., Moran, J. L., et al. (2016). Common alleles contribute to schizophrenia in CNV carriers. *Molecular Psychiatry, 21*(8), 1153. https://doi.org/10.1038/mp.2015.170
128. Gottesman, I. I., & Shields, J. (1967). A polygenic theory of schizophrenia. *Proceedings of the National Academy of Sciences of the United States of America, 58*(1), 199–205.
129. Shi, J., Levinson, D. F., Duan, J., Sanders, A. R., Zheng, Y., Pe'er, I., et al. (2009). Common variants on chromosome 6p22.1 are associated with schizophrenia. *Nature, 460*(7256), 753–757. https://doi.org/10.1038/nature08192
130. International Schizophrenia Consortium, Purcell, S. M., Wray, N. R., Stone, J. L., Visscher, P. M., O'Donovan, M. C., et al. (2009). Common polygenic variation contributes to risk of schizophrenia and bipolar disorder. *Nature, 460*(7256), 748–752. https://doi.org/10.1038/nature08185
131. O'Donovan, M. C., Craddock, N., Norton, N., Williams, H., Peirce, T., Moskvina, V., et al. (2008). Identification of loci associated with schizophrenia by genome-wide association and follow-up. *Nature Genetics, 40*(9), 1053–1055. https://doi.org/10.1038/ng.201
132. Stefansson, H., Ophoff, R. A., Steinberg, S., Andreassen, O. A., Cichon, S., Rujescu, D., et al. (2009). Common variants conferring risk of schizophrenia. *Nature, 460*(7256), 744–747. https://doi.org/10.1038/nature08186
133. Schizophrenia Psychiatric Genome-Wide Association Study (GWAS) Consortium, Ripke, S., Sanders, A. R., Kendler, K. S., Levinson, D. F., Sklar, P., et al. (2011). Genome-wide association study identifies five new schizophrenia loci. *Nature Genetics, 43*(10), 969–976. https://doi.org/10.1038/ng.940
134. Ripke, S., O'Dushlaine, C., Chambert, K., Moran, J. L., Kahler, A. K., Akterin, S., et al. (2013). Genome-wide association analysis identifies 13 new risk loci for schizophrenia. *Nature Genetics, 45*(10), 1150–1159. https://doi.org/10.1038/ng.2742

135. Schmitt, A., Malchow, B., Hasan, A., & Falkai, P. (2014). The impact of environmental factors in severe psychiatric disorders. *Frontiers in Neuroscience, 8*, 19. https://doi.org/10.3389/fnins.2014.00019
136. Pardinas, A. F., Holmans, P., Pocklington, A. J., Escott-Price, V., Ripke, S., Carrera, N., et al. (2018). Common schizophrenia alleles are enriched in mutation-intolerant genes and in regions under strong background selection. *Nature Genetics, 50*(3), 381–389. https://doi.org/10.1038/s41588-018-0059-2
137. Schizophrenia Working Group of the Psychiatric Genomics Consortium. (2014). Biological insights from 108 schizophrenia-associated genetic loci. *Nature, 511*(7510), 421–427. https://doi.org/10.1038/nature13595
138. Li, Z., Chen, J., Yu, H., He, L., Xu, Y., Zhang, D., et al. (2017). Genome-wide association analysis identifies 30 new susceptibility loci for schizophrenia. *Nature Genetics, 49*(11), 1576–1583. https://doi.org/10.1038/ng.3973
139. Shi, Y., Li, Z., Xu, Q., Wang, T., Li, T., Shen, J., et al. (2011). Common variants on 8p12 and 1q24.2 confer risk of schizophrenia. *Nature Genetics, 43*(12), 1224–1227. https://doi.org/10.1038/ng.980
140. GTEx Consortium, Laboratory, Data Analysis &Coordinating Center (LDACC)—Analysis Working Group, Statistical Methods groups—Analysis Working Group, Enhancing GTEx (eGTEx) groups, NIH Common Fund, NIH/NCI, et al. (2017). Genetic effects on gene expression across human tissues. *Nature, 550*(7675), 204–213. https://doi.org/10.1038/nature24277
141. Maurano, M. T., Humbert, R., Rynes, E., Thurman, R. E., Haugen, E., Wang, H., et al. (2012). Systematic localization of common disease-associated variation in regulatory DNA. *Science, 337*(6099), 1190–1195. https://doi.org/10.1126/science.1222794
142. Albert, F. W., & Kruglyak, L. (2015). The role of regulatory variation in complex traits and disease. *Nature Reviews Genetics, 16*(4), 197–212. https://doi.org/10.1038/nrg3891
143. Ng, B., White, C. C., Klein, H. U., Sieberts, S. K., McCabe, C., Patrick, E., et al. (2017). An xQTL map integrates the genetic architecture of the human brain's transcriptome and epigenome. *Nature Neuroscience, 20*(10), 1418–1426. https://doi.org/10.1038/nn.4632
144. Gandal, M. J., Zhang, P., Hadjimichael, E., Walker, R. L., Chen, C., Liu, S., et al. (2018). Transcriptome-wide isoform-level dysregulation in ASD, schizophrenia, and bipolar disorder. *Science, 362*(6420). https://doi.org/10.1126/science.aat8127
145. Rajarajan, P., Gil, S. E., Brennand, K. J., & Akbarian, S. (2016). Spatial genome organization and cognition. *Nature Reviews Neuroscience, 17*(11), 681–691. https://doi.org/10.1038/nrn.2016.124
146. Richards, A. L., Jones, L., Moskvina, V., Kirov, G., Gejman, P. V., Levinson, D. F., et al. (2012). Schizophrenia susceptibility alleles are enriched for alleles that affect gene expression in adult human brain. *Molecular Psychiatry, 17*(2), 193–201. https://doi.org/10.1038/mp.2011.11
147. Fromer, M., Roussos, P., Sieberts, S. K., Johnson, J. S., Kavanagh, D. H., Perumal, T. M., et al. (2016). Gene expression elucidates functional impact of polygenic risk for schizophrenia. *Nature Neuroscience, 19*(11), 1442–1453. https://doi.org/10.1038/nn.4399
148. de la Torre-Ubieta, L., Stein, J. L., Won, H., Opland, C. K., Liang, D., Lu, D., et al. (2018). The dynamic landscape of open chromatin during human cortical neurogenesis. *Cell2, 172*(1–2), 289–304, e218. https://doi.org/10.1016/j.cell.2017.12.014
149. Jaffe, A. E., Straub, R. E., Shin, J. H., Tao, R., Gao, Y., Collado-Torres, L., et al. (2018). Developmental and genetic regulation of the human cortex transcriptome illuminate schizophrenia pathogenesis. *Nature Neuroscience, 21*(8), 1117–1125. https://doi.org/10.1038/s41593-018-0197-y
150. Zhang, Y. E., Landback, P., Vibranovski, M. D., & Long, M. (2011). Accelerated recruitment of new brain development genes into the human genome. *PLoS Biology, 9*(10), e1001179. https://doi.org/10.1371/journal.pbio.1001179

151. Akbarian, S., Bunney Jr., W. E., Potkin, S. G., Wigal, S. B., Hagman, J. O., Sandman, C. A., et al. (1993). Altered distribution of nicotinamide-adenine dinucleotide phosphate-diaphorase cells in frontal lobe of schizophrenics implies disturbances of cortical development. *Archives of General Psychiatry, 50*(3), 169–177.
152. Jakob, H., & Beckmann, H. (1986). Prenatal developmental disturbances in the limbic allocortex in schizophrenics. *Journal of Neural Transmission, 65*(3–4), 303–326.
153. Fung, S. J., Webster, M. J., Sivagnanasundaram, S., Duncan, C., Elashoff, M., & Weickert, C. S. (2010). Expression of interneuron markers in the dorsolateral prefrontal cortex of the developing human and in schizophrenia. *The American Journal of Psychiatry, 167*(12), 1479–1488. https://doi.org/10.1176/appi.ajp.2010.09060784
154. Hyde, T. M., Lipska, B. K., Ali, T., Mathew, S. V., Law, A. J., Metitiri, O. E., et al. (2011). Expression of GABA signaling molecules KCC2, NKCC1, and GAD1 in cortical development and schizophrenia. *The Journal of Neuroscience, 31*(30), 11088–11095. https://doi.org/10.1523/JNEUROSCI.1234-11.2011
155. Horváth S., Janka Z., Mirnics K., (2011) Analyzing Schizophrenia by DNA Microarrays. *Biological Psychiatry 69*(2):157–162
156. Torkamani, A., Dean, B., Schork, N. J., & Thomas, E. A. (2010). Coexpression network analysis of neural tissue reveals perturbations in developmental processes in schizophrenia. *Genome Research, 20*(4), 403–412. https://doi.org/10.1101/gr.101956.109
157. Mistry, M., Gillis, J., & Pavlidis, P. (2013a). Genome-wide expression profiling of schizophrenia using a large combined cohort. *Molecular Psychiatry, 18*(2), 215–225. https://doi.org/10.1038/mp.2011.172
158. Mistry, M., Gillis, J., & Pavlidis, P. (2013b). Meta-analysis of gene coexpression networks in the post-mortem prefrontal cortex of patients with schizophrenia and unaffected controls. *BMC Neuroscience, 14*, 105. https://doi.org/10.1186/1471-2202-14-105
159. Finucane, H. K., Reshef, Y. A., Anttila, V., Slowikowski, K., Gusev, A., Byrnes, A., et al. (2018). Heritability enrichment of specifically expressed genes identifies disease-relevant tissues and cell types. *Nature Genetics, 50*(4), 621–629. https://doi.org/10.1038/s41588-018-0081-4
160. Gusev, A., Mancuso, N., Won, H., Kousi, M., Finucane, H. K., Reshef, Y., et al. (2018). Transcriptome-wide association study of schizophrenia and chromatin activity yields mechanistic disease insights. *Nature Genetics, 50*(4), 538–548. https://doi.org/10.1038/s41588-018-0092-1
161. Skene, N. G., Bryois, J., Bakken, T. E., Breen, G., Crowley, J. J., Gaspar, H. A., et al. (2018). Genetic identification of brain cell types underlying schizophrenia. *Nature Genetics, 50*(6), 825–833. https://doi.org/10.1038/s41588-018-0129-5
162. Roussos, P., Katsel, P., Davis, K. L., Siever, L. J., & Haroutunian, V. (2012). A system-level transcriptomic analysis of schizophrenia using postmortem brain tissue samples. *Archives of General Psychiatry, 69*(12), 1205–1213. https://doi.org/10.1001/archgenpsychiatry.2012.704
163. Radulescu, E., Jaffe, A. E., Straub, R. E., Chen, Q., Shin, J. H., Hyde, T. M., et al. (2018). Identification and prioritization of gene sets associated with schizophrenia risk by co-expression network analysis in human brain. *Molecular Psychiatry*. https://doi.org/10.1038/s41380-018-0304-1
164. Gusev, A., Ko, A., Shi, H., Bhatia, G., Chung, W., Penninx, B. W., et al. (2016). Integrative approaches for large-scale transcriptome-wide association studies. *Nature Genetics, 48*(3), 245–252. https://doi.org/10.1038/ng.3506
165. Huckins L. M., Dobbyn A., Ruderfer D. M., Hoffman G., Wang W., Pardiñas A. F., et al. (2019) Gene expression imputation across multiple brain regions provides insights into schizophrenia risk. *Nature Genetics 51*(4):659–674
166. The Network, O'Dushlaine, C., Rossin, L., Lee, P. H., Duncan, L., Parikshak, N. N., et al. (2015). Psychiatric genome-wide association study analyses implicate neuronal, immune and histone pathways. *Nature Neuroscience, 18*, 199. https://doi.org/10.1038/nn.3922. https://www.nature.com/articles/nn.3922#supplementary-information

167. Trynka, G., Sandor, C., Han, B., Xu, H., Stranger, B. E., Liu, X. S., et al. (2013). Chromatin marks identify critical cell types for fine mapping complex trait variants. *Nature Genetics, 45*(2), 124–130. https://doi.org/10.1038/ng.2504
168. Roussos, P., Mitchell, A. C., Voloudakis, G., Fullard, J. F., Pothula, V. M., Tsang, J., et al. (2014). A role for noncoding variation in schizophrenia. *Cell Reports, 9*(4), 1417–1429. https://doi.org/10.1016/j.celrep.2014.10.015
169. Psych, E. C., Akbarian, S., Liu, C., Knowles, J. A., Vaccarino, F. M., Farnham, P. J., et al. (2015). The PsychENCODE project. *Nature Neuroscience, 18*(12), 1707–1712. https://doi.org/10.1038/nn.4156
170. Girdhar, K., Hoffman, G. E., Jiang, Y., Brown, L., Kundakovic, M., Hauberg, M. E., et al. (2018). Cell-specific histone modification maps in the human frontal lobe link schizophrenia risk to the neuronal epigenome. *Nature Neuroscience, 21*(8), 1126–1136. https://doi.org/10.1038/s41593-018-0187-0
171. Jaffe, A. E., Gao, Y., Deep-Soboslay, A., Tao, R., Hyde, T. M., Weinberger, D. R., et al. (2016). Mapping DNA methylation across development, genotype and schizophrenia in the human frontal cortex. *Nature Neuroscience, 19*(1), 40–47. https://doi.org/10.1038/nn.4181
172. Schulz, H., Ruppert, A. K., Herms, S., Wolf, C., Mirza-Schreiber, N., Stegle, O., et al. (2017). Genome-wide mapping of genetic determinants influencing DNA methylation and gene expression in human hippocampus. *Nature Communications, 8*(1), 1511. https://doi.org/10.1038/s41467-017-01818-4
173. Dobbyn, A., Huckins, L. M., Boocock, J., Sloofman, L. G., Glicksberg, B. S., Giambartolomei, C., et al. (2018). Landscape of conditional eQTL in dorsolateral prefrontal cortex and co-localization with schizophrenia GWAS. *American Journal of Human Genetics, 102*(6), 1169–1184. https://doi.org/10.1016/j.ajhg.2018.04.011
174. Rajarajan, P., Borrman, T., Liao, W., Schrode, N., Flaherty, E., Casino, C., et al. (2018a). Neuron-specific signatures in the chromosomal connectome associated with schizophrenia risk. *Science, 362*(6420), eaat4311. https://doi.org/10.1126/science.aat4311
175. Bharadwaj, R., Jiang, Y., Mao, W., Jakovcevski, M., Dincer, A., Krueger, W., et al. (2013). Conserved chromosome 2q31 conformations are associated with transcriptional regulation of GAD1 GABA synthesis enzyme and altered in prefrontal cortex of subjects with schizophrenia. *The Journal of Neuroscience, 33*(29), 11839–11851. https://doi.org/10.1523/JNEUROSCI.1252-13.2013
176. Bharadwaj, R., Peter, C. J., Jiang, Y., Roussos, P., Vogel-Ciernia, A., Shen, E. Y., et al. (2014). Conserved higher-order chromatin regulates NMDA receptor gene expression and cognition. *Neuron, 84*(5), 997–1008. https://doi.org/10.1016/j.neuron.2014.10.032
177. Bryois, J., Garrett, M. E., Song, L., Safi, A., Giusti-Rodriguez, P., Johnson, G. D., et al. (2018). Evaluation of chromatin accessibility in prefrontal cortex of individuals with schizophrenia. *Nature Communications, 9*(1), 3121. https://doi.org/10.1038/s41467-018-05379-y
178. Fullard, J. F., Giambartolomei, C., Hauberg, M. E., Xu, K., Voloudakis, G., Shao, Z., et al. (2017). Open chromatin profiling of human postmortem brain infers functional roles for non-coding schizophrenia loci. *Human Molecular Genetics, 26*(10), 1942–1951. https://doi.org/10.1093/hmg/ddx103
179. Fullard, J. F., Hauberg, M. E., Bendl, J., Egervari, G., Cirnaru, M. D., Reach, S. M., et al. (2018). An atlas of chromatin accessibility in the adult human brain. *Genome Research, 28*(8), 1243–1252. https://doi.org/10.1101/gr.232488.117
180. Won, H., de la Torre-Ubieta, L., Stein, J. L., Parikshak, N. N., Huang, J., Opland, C. K., et al. (2016). Chromosome conformation elucidates regulatory relationships in developing human brain. *Nature, 538*(7626), 523–527. https://doi.org/10.1038/nature19847
181. Takahashi, K., & Yamanaka, S. (2006). Induction of pluripotent stem cells from mouse embryonic and adult fibroblast cultures by defined factors. *Cell, 126*(4), 663–676. https://doi.org/10.1016/j.cell.2006.07.024

182. Narsinh, K. H., Plews, J., & Wu, J. C. (2011). Comparison of human induced pluripotent and embryonic stem cells: fraternal or identical twins? *Molecular Therapy, 19*(4), 635–638. https://doi.org/10.1038/mt.2011.41
183. Hoffman, G. E., Schrode, N., Flaherty, E., & Brennand, K. J. (2018). New considerations for hiPSC-based models of neuropsychiatric disorders. *Molecular Psychiatry, 24*, 49. https://doi.org/10.1038/s41380-018-0029-1
184. Laurent, L. C., Ulitsky, I., Slavin, I., Tran, H., Schork, A., Morey, R., et al. (2011). Dynamic changes in the copy number of pluripotency and cell proliferation genes in human ESCs and iPSCs during reprogramming and time in culture. *Cell Stem Cell, 8*(1), 106–118. https://doi.org/10.1016/j.stem.2010.12.003
185. Lister, R., Pelizzola, M., Kida, Y. S., Hawkins, R. D., Nery, J. R., Hon, G., et al. (2011). Hotspots of aberrant epigenomic reprogramming in human induced pluripotent stem cells. *Nature, 471*(7336), 68–73. https://doi.org/10.1038/nature09798
186. Julia, T. C. W., Carvalho, C. M. B., Yuan, B., Gu, S., Altheimer, A. N., McCarthy, S., et al. (2017). Divergent levels of marker chromosomes in an hiPSC-based model of psychosis. *Stem Cell Reports, 8*(3), 519–528. https://doi.org/10.1016/j.stemcr.2017.01.010
187. Grochowski, C. M., Gu, S., Yuan, B., Tcw, J., Brennand, K. J., Sebat, J., et al. (2018). Marker chromosome genomic structure and temporal origin implicate a chromoanasynthesis event in a family with pleiotropic psychiatric phenotypes. *Human Mutation, 39*(7), 939–946. https://doi.org/10.1002/humu.23537
188. Kyttala, A., Moraghebi, R., Valensisi, C., Kettunen, J., Andrus, C., Pasumarthy, K. K., et al. (2016). Genetic variability overrides the impact of parental cell type and determines iPSC Differentiation Potential. *Stem Cell Reports, 6*(2), 200–212. https://doi.org/10.1016/j.stemcr.2015.12.009
189. Hoffman, G. E., Hartley, B. J., Flaherty, E., Ladran, I., Gochman, P., Ruderfer, D. M., et al. (2017). Transcriptional signatures of schizophrenia in hiPSC-derived NPCs and neurons are concordant with post-mortem adult brains. *Nature Communications, 8*(1), 2225. https://doi.org/10.1038/s41467-017-02330-5
190. Nehme, R., Zuccaro, E., Ghosh, S. D., Li, C., Sherwood, J. L., Pietilainen, O., et al. (2018). Combining NGN2 Programming with developmental patterning generates human excitatory neurons with NMDAR-mediated synaptic transmission. *Cell Reports, 23*(8), 2509–2523. https://doi.org/10.1016/j.celrep.2018.04.066
191. Mertens, J., Marchetto, M. C., Bardy, C., & Gage, F. H. (2016). Evaluating cell reprogramming, differentiation and conversion technologies in neuroscience. *Nature Reviews Neuroscience, 17*(7), 424–437. https://doi.org/10.1038/nrn.2016.46
192. Chambers, S. M., Fasano, C. A., Papapetrou, E. P., Tomishima, M., Sadelain, M., & Studer, L. (2009). Highly efficient neural conversion of human ES and iPS cells by dual inhibition of SMAD signaling. *Nature Biotechnology, 27*(3), 275–280. https://doi.org/10.1038/nbt.1529
193. Marchetto, M. C., Carromeu, C., Acab, A., Yu, D., Yeo, G. W., Mu, Y., et al. (2010). A model for neural development and treatment of Rett syndrome using human induced pluripotent stem cells. *Cell, 143*(4), 527–539. https://doi.org/10.1016/j.cell.2010.10.016
194. Maroof, A. M., Keros, S., Tyson, J. A., Ying, S. W., Ganat, Y. M., Merkle, F. T., et al. (2013). Directed differentiation and functional maturation of cortical interneurons from human embryonic stem cells. *Cell Stem Cell, 12*(5), 559–572. https://doi.org/10.1016/j.stem.2013.04.008
195. Kriks, S., Shim, J. W., Piao, J., Ganat, Y. M., Wakeman, D. R., Xie, Z., et al. (2011). Dopamine neurons derived from human ES cells efficiently engraft in animal models of Parkinson's disease. *Nature, 480*(7378), 547–551. https://doi.org/10.1038/nature10648
196. Lu J., Zhong X., Liu H., Hao L., Tzu-Ling Huang C., Sherafat M. A., et al. (2016) Generation of serotonin neurons from human pluripotent stem cells. *Nature Biotechnology 34*(1):89–94
197. Yu, D. X., Di Giorgio, F. P., Yao, J., Marchetto, M. C., Brennand, K., Wright, R., et al. (2014). Modeling hippocampal neurogenesis using human pluripotent stem cells. *Stem Cell Reports, 2*(3), 295–310. https://doi.org/10.1016/j.stemcr.2014.01.009

198. Sarkar, A., Mei, A., Paquola, A. C. M., Stern, S., Bardy, C., Klug, J. R., et al. (2018). Efficient generation of CA3 neurons from human pluripotent stem cells enables modeling of hippocampal connectivity in vitro. *Cell Stem Cell, 22*(5), 684–697. e689. https://doi.org/10.1016/j.stem.2018.04.009
199. Qi, Y., Zhang, X. J., Renier, N., Wu, Z., Atkin, T., Sun, Z., et al. (2017). Combined small-molecule inhibition accelerates the derivation of functional cortical neurons from human pluripotent stem cells. *Nature Biotechnology, 35*(2), 154–163. https://doi.org/10.1038/nbt.3777
200. Schwartzentruber, J., Foskolou, S., Kilpinen, H., Rodrigues, J., Alasoo, K., Knights, A. J., et al. (2018). Molecular and functional variation in iPSC-derived sensory neurons. *Nature Genetics, 50*(1), 54–61. https://doi.org/10.1038/s41588-017-0005-8
201. Kuijlaars, J., Oyelami, T., Diels, A., Rohrbacher, J., Versweyveld, S., Meneghello, G., et al. (2016). Sustained synchronized neuronal network activity in a human astrocyte co-culture system. *Scientific Reports, 6*, 36529. https://doi.org/10.1038/srep36529
202. Gunhanlar, N., Shpak, G., van der Kroeg, M., Gouty-Colomer, L. A., Munshi, S. T., Lendemeijer, B., et al. (2018). A simplified protocol for differentiation of electrophysiologically mature neuronal networks from human induced pluripotent stem cells. *Molecular Psychiatry, 23*(5), 1336–1344. https://doi.org/10.1038/mp.2017.56
203. Vierbuchen, T., Ostermeier, A., Pang, Z. P., Kokubu, Y., Sudhof, T. C., & Wernig, M. (2010). Direct conversion of fibroblasts to functional neurons by defined factors. *Nature, 463*(7284), 1035–1041. https://doi.org/10.1038/nature08797
204. Pang, Z. P., Yang, N., Vierbuchen, T., Ostermeier, A., Fuentes, D. R., Yang, T. Q., et al. (2011). Induction of human neuronal cells by defined transcription factors. *Nature, 476*(7359), 220–223. https://doi.org/10.1038/nature10202
205. Zhang, Y., Pak, C., Han, Y., Ahlenius, H., Zhang, Z., Chanda, S., et al. (2013). Rapid single-step induction of functional neurons from human pluripotent stem cells. *Neuron, 78*(5), 785–798. https://doi.org/10.1016/j.neuron.2013.05.029
206. Ho, S. M., Hartley, B. J., Tcw, J., Beaumont, M., Stafford, K., Slesinger, P. A., et al. (2016). Rapid Ngn2-induction of excitatory neurons from hiPSC-derived neural progenitor cells. *Methods, 101*, 113–124. https://doi.org/10.1016/j.ymeth.2015.11.019
207. Colasante, G., Lignani, G., Rubio, A., Medrihan, L., Yekhlef, L., Sessa, A., et al. (2015). Rapid conversion of fibroblasts into functional forebrain GABAergic interneurons by direct genetic reprogramming. *Cell Stem Cell, 17*(6), 719–734. https://doi.org/10.1016/j.stem.2015.09.002
208. Sun, A. X., Yuan, Q., Tan, S., Xiao, Y., Wang, D., Khoo, A. T., et al. (2016). Direct induction and functional maturation of forebrain GABAergic neurons from human pluripotent stem cells. *Cell Reports, 16*(7), 1942–1953. https://doi.org/10.1016/j.celrep.2016.07.035
209. Yang, N., Chanda, S., Marro, S., Ng, Y. H., Janas, J. A., Haag, D., et al. (2017). Generation of pure GABAergic neurons by transcription factor programming. *Nature Methods, 14*(6), 621–628. https://doi.org/10.1038/nmeth.4291
210. Caiazzo, M., Dell'Anno, M. T., Dvoretskova, E., Lazarevic, D., Taverna, S., Leo, D., et al. (2011). Direct generation of functional dopaminergic neurons from mouse and human fibroblasts. *Nature, 476*(7359), 224–227. https://doi.org/10.1038/nature10284
211. Theka, I., Caiazzo, M., Dvoretskova, E., Leo, D., Ungaro, F., Curreli, S., et al. (2013). Rapid generation of functional dopaminergic neurons from human induced pluripotent stem cells through a single-step procedure using cell lineage transcription factors. *Stem Cells Translational Medicine, 2*(6), 473–479. https://doi.org/10.5966/sctm.2012-0133
212. Lu, J., Zhong, X., Liu, H., Hao, L., Huang, C. T., Sherafat, M. A., et al. (2016). Generation of serotonin neurons from human pluripotent stem cells. *Nature Biotechnology, 34*(1), 89–94. https://doi.org/10.1038/nbt.3435
213. Vadodaria, K. C., Mertens, J., Paquola, A., Bardy, C., Li, X., Jappelli, R., et al. (2016). Generation of functional human serotonergic neurons from fibroblasts. *Molecular Psychiatry, 21*(1), 49–61. https://doi.org/10.1038/mp.2015.161

214. Brennand, K. J., Simone, A., Jou, J., Gelboin-Burkhart, C., Tran, N., Sangar, S., et al. (2011). Modelling schizophrenia using human induced pluripotent stem cells. *Nature, 473*(7346), 221–225. https://doi.org/10.1038/nature09915
215. Hook, V., Brennand, K. J., Kim, Y., Toneff, T., Funkelstein, L., Lee, K. C., et al. (2014). Human iPSC neurons display activity-dependent neurotransmitter secretion: aberrant catecholamine levels in schizophrenia neurons. *Stem Cell Reports, 3*(4), 531–538. https://doi.org/10.1016/j.stemcr.2014.08.001
216. Robicsek, O., Karry, R., Petit, I., Salman-Kesner, N., Muller, F. J., Klein, E., et al. (2013). Abnormal neuronal differentiation and mitochondrial dysfunction in hair follicle-derived induced pluripotent stem cells of schizophrenia patients. *Molecular Psychiatry, 18*(10), 1067–1076. https://doi.org/10.1038/mp.2013.67
217. Xu, J., Hartley, B. J., Kurup, P., Phillips, A., Topol, A., Xu, M., et al. (2018). Inhibition of STEP61 ameliorates deficits in mouse and hiPSC-based schizophrenia models. *Molecular Psychiatry, 23*(2), 271–281. https://doi.org/10.1038/mp.2016.163
218. Carty, N. C., Xu, J., Kurup, P., Brouillette, J., Goebel-Goody, S. M., Austin, D. R., et al. (2012). The tyrosine phosphatase STEP: implications in schizophrenia and the molecular mechanism underlying antipsychotic medications. *Translational Psychiatry, 2*, e137. https://doi.org/10.1038/tp.2012.63
219. Brennand, K., Savas, J. N., Kim, Y., Tran, N., Simone, A., Hashimoto-Torii, K., et al. (2015). Phenotypic differences in hiPSC NPCs derived from patients with schizophrenia. *Molecular Psychiatry, 20*(3), 361–368. https://doi.org/10.1038/mp.2014.22
220. Topol, A., English, J. A., Flaherty, E., Rajarajan, P., Hartley, B. J., Gupta, S., et al. (2015a). Increased abundance of translation machinery in stem cell-derived neural progenitor cells from four schizophrenia patients. *Translational Psychiatry, 5*, e662. https://doi.org/10.1038/tp.2015.118
221. Topol, A., Zhu, S., Tran, N., Simone, A., Fang, G., & Brennand, K. J. (2015b). Altered WNT signaling in human induced pluripotent stem cell neural progenitor cells derived from four schizophrenia patients. *Biological Psychiatry, 78*(6), e29–e34. https://doi.org/10.1016/j.biopsych.2014.12.028
222. Casas, B. S., Vitoria, G., do Costa, M. N., Madeiro da Costa, R., Trindade, P., Maciel, R., et al. (2018). hiPSC-derived neural stem cells from patients with schizophrenia induce an impaired angiogenesis. *Translational Psychiatry, 8*(1), 48. https://doi.org/10.1038/s41398-018-0095-9
223. Hino, M., Kunii, Y., Matsumoto, J., Wada, A., Nagaoka, A., Niwa, S., et al. (2016). Decreased VEGFR2 expression and increased phosphorylated Akt1 in the prefrontal cortex of individuals with schizophrenia. *Journal of Psychiatric Research, 82*, 100–108. https://doi.org/10.1016/j.jpsychires.2016.07.018
224. Lee, B. H., Hong, J. P., Hwang, J. A., Ham, B. J., Na, K. S., Kim, W. J., et al. (2015). Alterations in plasma vascular endothelial growth factor levels in patients with schizophrenia before and after treatment. *Psychiatry Research, 228*(1), 95–99. https://doi.org/10.1016/j.psychres.2015.04.020
225. Lopes, R., Soares, R., Coelho, R., & Figueiredo-Braga, M. (2015). Angiogenesis in the pathophysiology of schizophrenia - a comprehensive review and a conceptual hypothesis. *Life Sciences, 128*, 79–93. https://doi.org/10.1016/j.lfs.2015.02.010
226. Gonzalez, D. M., Gregory, J., & Brennand, K. J. (2017). The importance of non-neuronal cell types in hiPSC-based disease modeling and drug screening. *Frontiers in Cell and Development Biology, 5*, 117. https://doi.org/10.3389/fcell.2017.00117
227. Ben-Shachar, D. (2002). Mitochondrial dysfunction in schizophrenia: a possible linkage to dopamine. *Journal of Neurochemistry, 83*(6), 1241–1251.
228. Prabakaran, S., Swatton, J. E., Ryan, M. M., Huffaker, S. J., Huang, J. T., Griffin, J. L., et al. (2004). Mitochondrial dysfunction in schizophrenia: evidence for compromised brain metabolism and oxidative stress. *Molecular Psychiatry, 9*(7), 684–697, 643. https://doi.org/10.1038/sj.mp.4001511

229. Uguz, A. C., Demirci, K., & Espino, J. (2016). The importance of melatonin and mitochondria interaction in mood disorders and schizophrenia: a current assessment. *Current Medicinal Chemistry, 23*(20), 2146–2158.
230. Paulsen Bda, S., de Moraes Maciel, R., Galina, A., Souza da Silveira, M., dos Santos Souza, C., Drummond, H., et al. (2012). Altered oxygen metabolism associated to neurogenesis of induced pluripotent stem cells derived from a schizophrenic patient. *Cell Transplantation, 21*(7), 1547–1559. https://doi.org/10.3727/096368911X600957
231. Robicsek, O., Ene, H. M., Karry, R., Ytzhaki, O., Asor, E., McPhie, D., et al. (2018). Isolated mitochondria transfer improves neuronal differentiation of schizophrenia-derived induced pluripotent stem cells and rescues deficits in a rat model of the disorder. *Schizophrenia Bulletin, 44*(2), 432–442. https://doi.org/10.1093/schbul/sbx077
232. Caputo, V., Ciolfi, A., Macri, S., & Pizzuti, A. (2015). The emerging role of MicroRNA in schizophrenia. *CNS & Neurological Disorders Drug Targets, 14*(2), 208–221.
233. Shi, S., Leites, C., He, D., Schwartz, D., Moy, W., Shi, J., et al. (2014). MicroRNA-9 and microRNA-326 regulate human dopamine D2 receptor expression, and the microRNA-mediated expression regulation is altered by a genetic variant. *The Journal of Biological Chemistry, 289*(19), 13434–13444. https://doi.org/10.1074/jbc.M113.535203
234. Han, J., Kim, H. J., Schafer, S. T., Paquola, A., Clemenson, G. D., Toda, T., et al. (2016). Functional implications of miR-19 in the migration of newborn neurons in the adult brain. *Neuron, 91*(1), 79–89. https://doi.org/10.1016/j.neuron.2016.05.034
235. Topol, A., Zhu, S., Hartley, B. J., English, J., Hauberg, M. E., Tran, N., et al. (2016). Dysregulation of miRNA-9 in a subset of schizophrenia patient-derived neural progenitor cells. *Cell Reports, 15*(5), 1024–1036. https://doi.org/10.1016/j.celrep.2016.03.090
236. Hauberg, M. E., Roussos, P., Grove, J., Borglum, A. D., Mattheisen, M., & Schizophrenia Working Group of the Psychiatric Genomics Consortium. (2016). Analyzing the role of MicroRNAs in schizophrenia in the context of common genetic risk variants. *JAMA Psychiatry, 73*(4), 369–377. https://doi.org/10.1001/jamapsychiatry.2015.3018
237. Hoffman, G. E., & Brennand, K. J. (2018). Mapping regulatory variants in hiPSC models. *Nature Genetics, 50*(1), 1–2. https://doi.org/10.1038/s41588-017-0017-4
238. Roussos, P., Guennewig, B., Kaczorowski, D. C., Barry, G., & Brennand, K. J. (2016). Activity-dependent changes in gene expression in schizophrenia human-induced pluripotent stem cell neurons. *JAMA Psychiatry, 73*(11), 1180–1188. https://doi.org/10.1001/jamapsychiatry.2016.2575
239. Yoshimizu, T., Pan, J. Q., Mungenast, A. E., Madison, J. M., Su, S., Ketterman, J., et al. (2015). Functional implications of a psychiatric risk variant within CACNA1C in induced human neurons. *Molecular Psychiatry, 20*(2), 162–169. https://doi.org/10.1038/mp.2014.143
240. Forrest, M. P., Zhang, H., Moy, W., McGowan, H., Leites, C., Dionisio, L. E., et al. (2017). Open chromatin profiling in hiPSC-derived neurons prioritizes functional noncoding psychiatric risk variants and highlights neurodevelopmental loci. *Cell Stem Cell, 21*(3), 305–318. e308. https://doi.org/10.1016/j.stem.2017.07.008
241. Powell, S. K., Gregory, J., Akbarian, S., & Brennand, K. J. (2017). Application of CRISPR/Cas9 to the study of brain development and neuropsychiatric disease. *Molecular and Cellular Neurosciences, 82*, 157–166. https://doi.org/10.1016/j.mcn.2017.05.007
242. Ho, S. M., Hartley, B. J., Flaherty, E., Rajarajan, P., Abdelaal, R., Obiorah, I., et al. (2017). Evaluating synthetic activation and repression of neuropsychiatric-related genes in hiPSC-derived NPCs, neurons, and astrocytes. *Stem Cell Reports, 9*(2), 615–628. https://doi.org/10.1016/j.stemcr.2017.06.012
243. Jiang, Y., Loh, Y. E., Rajarajan, P., Hirayama, T., Liao, W., Kassim, B. S., et al. (2017). The methyltransferase SETDB1 regulates a large neuron-specific topological chromatin domain. *Nature Genetics, 49*(8), 1239–1250. https://doi.org/10.1038/ng.3906
244. Rajarajan, P., Jiang, Y., Kassim, B. S., & Akbarian, S. (2018b). Chromosomal conformations and epigenomic regulation in schizophrenia. *Progress in Molecular Biology and Translational Science, 157*, 21–40. https://doi.org/10.1016/bs.pmbts.2017.11.022

245. Zarrei, M., MacDonald, J. R., Merico, D., & Scherer, S. W. (2015). A copy number variation map of the human genome. *Nature Reviews Genetics, 16*(3), 172–183. https://doi.org/10.1038/nrg3871
246. Ahn, K., Gotay, N., Andersen, T. M., Anvari, A. A., Gochman, P., Lee, Y., et al. (2014). High rate of disease-related copy number variations in childhood onset schizophrenia. *Molecular Psychiatry, 19*(5), 568–572. https://doi.org/10.1038/mp.2013.59
247. Flaherty E. K., Brennand K. J., (2017) Using hiPSCs to model neuropsychiatric copy number variations (CNVs) has potential to reveal underlying disease mechanisms. *Brain Research* 1655:283–293
248. Gothelf, D., Eliez, S., Thompson, T., Hinard, C., Penniman, L., Feinstein, C., et al. (2005). COMT genotype predicts longitudinal cognitive decline and psychosis in 22q11.2 deletion syndrome. *Nature Neuroscience, 8*(11), 1500–1502. https://doi.org/10.1038/nn1572
249. Gothelf, D., Feinstein, C., Thompson, T., Gu, E., Penniman, L., Van Stone, E., et al. (2007). Risk factors for the emergence of psychotic disorders in adolescents with 22q11.2 deletion syndrome. *The American Journal of Psychiatry, 164*(4), 663–669. https://doi.org/10.1176/ajp.2007.164.4.663
250. Murphy, K. C., Jones, L. A., & Owen, M. J. (1999). High rates of schizophrenia in adults with velo-cardio-facial syndrome. *Archives of General Psychiatry, 56*(10), 940–945.
251. Pedrosa, E., Sandler, V., Shah, A., Carroll, R., Chang, C., Rockowitz, S., et al. (2011). Development of patient-specific neurons in schizophrenia using induced pluripotent stem cells. *Journal of Neurogenetics, 25*(3), 88–103. https://doi.org/10.3109/01677063.2011.597908
252. Lin, M., Pedrosa, E., Hrabovsky, A., Chen, J., Puliafito, B. R., Gilbert, S. R., et al. (2016). Integrative transcriptome network analysis of iPSC-derived neurons from schizophrenia and schizoaffective disorder patients with 22q11.2 deletion. *BMC Systems Biology, 10*(1), 105. https://doi.org/10.1186/s12918-016-0366-0
253. Zhao, D., Lin, M., Chen, J., Pedrosa, E., Hrabovsky, A., Fourcade, H. M., et al. (2015). MicroRNA profiling of neurons generated using induced pluripotent stem cells derived from patients with schizophrenia and schizoaffective disorder, and 22q11.2 Del. *PLoS One, 10*(7), e0132387. https://doi.org/10.1371/journal.pone.0132387
254. Toyoshima, M., Akamatsu, W., Okada, Y., Ohnishi, T., Balan, S., Hisano, Y., et al. (2016). Analysis of induced pluripotent stem cells carrying 22q11.2 deletion. *Translational Psychiatry, 6*(11), e934. https://doi.org/10.1038/tp.2016.206
255. Warnica, W., Merico, D., Costain, G., Alfred, S. E., Wei, J., Marshall, C. R., et al. (2015). Copy number variable microRNAs in schizophrenia and their neurodevelopmental gene targets. *Biological Psychiatry, 77*(2), 158–166. https://doi.org/10.1016/j.biopsych.2014.05.011
256. Yoon, K. J., Nguyen, H. N., Ursini, G., Zhang, F., Kim, N. S., Wen, Z., et al. (2014). Modeling a genetic risk for schizophrenia in iPSCs and mice reveals neural stem cell deficits associated with adherens junctions and polarity. *Cell Stem Cell, 15*(1), 79–91. https://doi.org/10.1016/j.stem.2014.05.003
257. McCarthy S. E., Makarov V., Kirov G., Addington A. M., McClellan J., Yoon S., et al. (2009) Microduplications of 16p11.2 are associated with schizophrenia. *Nature Genetics 41* (11):1223–1227
258. Deshpande, A., Yadav, S., Dao, D. Q., Wu, Z. Y., Hokanson, K. C., Cahill, M. K., et al. (2017). Cellular phenotypes in human iPSC-derived neurons from a genetic model of autism spectrum disorder. *Cell Reports, 21*(10), 2678–2687. https://doi.org/10.1016/j.celrep.2017.11.037
259. Rujescu, D., Ingason, A., Cichon, S., Pietilainen, O. P., Barnes, M. R., Toulopoulou, T., et al. (2009). Disruption of the neurexin 1 gene is associated with schizophrenia. *Human Molecular Genetics, 18*(5), 988–996. https://doi.org/10.1093/hmg/ddn351
260. Zeng, L., Zhang, P., Shi, L., Yamamoto, V., Lu, W., & Wang, K. (2013). Functional impacts of NRXN1 knockdown on neurodevelopment in stem cell models. *PLoS One, 8*(3), e59685. https://doi.org/10.1371/journal.pone.0059685

261. Pak, C., Danko, T., Zhang, Y., Aoto, J., Anderson, G., Maxeiner, S., et al. (2015). Human neuropsychiatric disease modeling using conditional deletion reveals synaptic transmission defects caused by heterozygous mutations in NRXN1. *Cell Stem Cell, 17*(3), 316–328. https://doi.org/10.1016/j.stem.2015.07.017
262. Flaherty E., Zhu S., Barretto N., Cheng E., Michael Deans P. J., Fernando M. B., et al. (2019) Neuronal impact of patient-specific aberrant NRXN1α splicing. *Nature Genetics 51* (12):1679–1690
263. Jacobs, P., Brunton, M., Frackiewicz, A., Newton, M., Cook, P., & Robson, E. (1970). Studies on a family with three cytogenetic markers. *Annals of Human Genetics, 33*, 325–336.
264. St Clair, D., Blackwood, D., Muir, W., Carothers, A., Walker, M., Spowart, G., et al. (1990). Association within a family of a balanced autosomal translocation with major mental illness. *Lancet, 336*(8706), 13–16.
265. Millar, J. K., Wilson-Annan, J. C., Anderson, S., Christie, S., Taylor, M. S., Semple, C. A., et al. (2000). Disruption of two novel genes by a translocation co-segregating with schizophrenia. *Human Molecular Genetics, 9*(9), 1415–1423.
266. Sachs, N. A., Sawa, A., Holmes, S. E., Ross, C. A., DeLisi, L. E., & Margolis, R. L. (2005). A frameshift mutation in Disrupted in Schizophrenia 1 in an American family with schizophrenia and schizoaffective disorder. *Molecular Psychiatry, 10*(8), 758–764. https://doi.org/10.1038/sj.mp.4001667
267. Green, E. K., Norton, N., Peirce, T., Grozeva, D., Kirov, G., Owen, M. J., et al. (2006). Evidence that a DISC1 frame-shift deletion associated with psychosis in a single family may not be a pathogenic mutation. *Molecular Psychiatry, 11*(9), 798–799. https://doi.org/10.1038/sj.mp.4001853
268. Chiang, C. H., Su, Y., Wen, Z., Yoritomo, N., Ross, C. A., Margolis, R. L., et al. (2011). Integration-free induced pluripotent stem cells derived from schizophrenia patients with a DISC1 mutation. *Molecular Psychiatry, 16*(4), 358–360. https://doi.org/10.1038/mp.2011.13
269. Wen, Z., Nguyen, H. N., Guo, Z., Lalli, M. A., Wang, X., Su, Y., et al. (2014). Synaptic dysregulation in a human iPS cell model of mental disorders. *Nature, 515*(7527), 414–418. https://doi.org/10.1038/nature13716
270. Murai, K., Sun, G., Ye, P., Tian, E., Yang, S., Cui, Q., et al. (2016). The TLX-miR-219 cascade regulates neural stem cell proliferation in neurodevelopment and schizophrenia iPSC model. *Nature Communications, 7*, 10965. https://doi.org/10.1038/ncomms10965
271. Yalla, K., Elliott, C., Day, J. P., Findlay, J., Barratt, S., Hughes, Z. A., et al. (2018). FBXW7 regulates DISC1 stability via the ubiquitin-proteosome system. *Molecular Psychiatry, 23*(5), 1278–1286. https://doi.org/10.1038/mp.2017.138
272. Chiu, F. L., Lin, J. T., Chuang, C. Y., Chien, T., Chen, C. M., Chen, K. H., et al. (2015). Elucidating the role of the A2A adenosine receptor in neurodegeneration using neurons derived from Huntington's disease iPSCs. *Human Molecular Genetics, 24*(21), 6066–6079. https://doi.org/10.1093/hmg/ddv318
273. Chien, T., Weng, Y. T., Chang, S. Y., Lai, H. L., Chiu, F. L., Kuo, H. C., et al. (2018). GSK3beta negatively regulates TRAX, a scaffold protein implicated in mental disorders, for NHEJ-mediated DNA repair in neurons. *Molecular Psychiatry*. https://doi.org/10.1038/s41380-017-0007-z
274. Srikanth, P., Han, K., Callahan, D. G., Makovkina, E., Muratore, C. R., Lalli, M. A., et al. (2015). Genomic DISC1 disruption in hiPSCs alters Wnt signaling and neural cell fate. *Cell Reports, 12*(9), 1414–1429. https://doi.org/10.1016/j.celrep.2015.07.061
275. Bradshaw, N. J., & Porteous, D. J. (2012). DISC1-binding proteins in neural development, signalling and schizophrenia. *Neuropharmacology, 62*(3), 1230–1241. https://doi.org/10.1016/j.neuropharm.2010.12.027
276. Camargo, L. M., Collura, V., Rain, J. C., Mizuguchi, K., Hermjakob, H., Kerrien, S., et al. (2007). Disrupted in schizophrenia 1 interactome: evidence for the close connectivity of risk genes and a potential synaptic basis for schizophrenia. *Molecular Psychiatry, 12*(1), 74–86. https://doi.org/10.1038/sj.mp.4001880

277. Camargo, L. M., Wang, Q., & Brandon, N. J. (2008). What can we learn from the disrupted in schizophrenia 1 interactome: lessons for target identification and disease biology? *Novartis Foundation Symposium, 289*, 208–216; discussion 216-221, 238-240.
278. Teng, S., Thomson, P. A., McCarthy, S., Kramer, M., Muller, S., & Lihm, J. (2018). Rare disruptive variants in the DISC1 Interactome and Regulome: association with cognitive ability and schizophrenia. *Molecular Psychiatry, 23*(5), 1270–1277. https://doi.org/10.1038/mp.2017.115
279. Nakata, K., Lipska, B. K., Hyde, T. M., Ye, T., Newburn, E. N., Morita, Y., et al. (2009). DISC1 splice variants are upregulated in schizophrenia and associated with risk polymorphisms. *Proceedings of the National Academy of Sciences of the United States of America, 106*(37), 15873–15878. https://doi.org/10.1073/pnas.0903413106
280. Wilkinson, B., Evgrafov, O. V., Zheng, D., Hartel, N., Knowles, J. A., Graham, N. A., et al. (2018). Endogenous cell type-specific disrupted in schizophrenia 1 interactomes reveal protein networks associated with neurodevelopmental disorders. *Biological Psychiatry, 85*, 305. https://doi.org/10.1016/j.biopsych.2018.05.009
281. Turner, T. N., Yi, Q., Krumm, N., Huddleston, J., Hoekzema, K., Stessman, H. A., et al. (2017). denovo-db: a compendium of human de novo variants. *Nucleic Acids Research, 45*(D1), D804–D811. https://doi.org/10.1093/nar/gkw865
282. Bakircioglu, M., Carvalho, O. P., Khurshid, M., Cox, J. J., Tuysuz, B., Barak, T., et al. (2011). The essential role of centrosomal NDE1 in human cerebral cortex neurogenesis. *American Journal of Human Genetics, 88*(5), 523–535. https://doi.org/10.1016/j.ajhg.2011.03.019
283. Ye, F., Kang, E., Yu, C., Qian, X., Jacob, F., Yu, C., et al. (2017). DISC1 regulates neurogenesis via modulating kinetochore attachment of Ndel1/Nde1 during mitosis. *Neuron, 96*(5), 1041–1054. e1045. https://doi.org/10.1016/j.neuron.2017.10.010
284. Mathieson, I., Munafo, M. R., & Flint, J. (2012). Meta-analysis indicates that common variants at the DISC1 locus are not associated with schizophrenia. *Molecular Psychiatry, 17*(6), 634–641. https://doi.org/10.1038/mp.2011.41
285. Richards, A. L., Leonenko, G., Walters, J. T., Kavanagh, D. H., Rees, E. G., Evans, A., et al. (2016). Exome arrays capture polygenic rare variant contributions to schizophrenia. *Human Molecular Genetics, 25*(5), 1001–1007. https://doi.org/10.1093/hmg/ddv620
286. Farrell, M. S., Werge, T., Sklar, P., Owen, M. J., Ophoff, R. A., O'Donovan, M. C., et al. (2015). Evaluating historical candidate genes for schizophrenia. *Molecular Psychiatry, 20*(5), 555–562. https://doi.org/10.1038/mp.2015.16
287. Sullivan, P. F. (2013). Questions about DISC1 as a genetic risk factor for schizophrenia. *Molecular Psychiatry, 18*(10), 1050–1052. https://doi.org/10.1038/mp.2012.182
288. Lee I. S., Carvalho C. M. B., Douvaras P., Ho S-M, Hartley B. J., Zuccherato L. W., et al. (2015) Characterization of molecular and cellular phenotypes associated with a heterozygous CNTNAP2 deletion using patient-derived hiPSC neural cells. npj Schizophrenia 1 (1)
289. Flaherty, E., Deranieh, R. M., Artimovich, E., Lee, I. S., Siegel, A. J., Levy, D. L., et al. (2017). Patient-derived hiPSC neurons with heterozygous CNTNAP2 deletions display altered neuronal gene expression and network activity. *NPJ Schizophrenia, 3*, 35. https://doi.org/10.1038/s41537-017-0033-5
290. de Vrij, F. M., Bouwkamp, C. G., Gunhanlar, N., Shpak, G., Lendemeijer, B., Baghdadi, M., et al. (2018). Candidate CSPG4 mutations and induced pluripotent stem cell modeling implicate oligodendrocyte progenitor cell dysfunction in familial schizophrenia. *Molecular Psychiatry, 24*, 757. https://doi.org/10.1038/s41380-017-0004-2
291. Guennewig, B., Bitar, M., Obiorah, I., Hanks, J., O'Brien, E. A., Kaczorowski, D. C., et al. (2018). THC exposure of human iPSC neurons impacts genes associated with neuropsychiatric disorders. *Translational Psychiatry, 8*(1), 89. https://doi.org/10.1038/s41398-018-0137-3
292. Obiorah, I. V., Muhammad, H., Stafford, K., Flaherty, E. K., & Brennand, K. J. (2017). THC treatment alters glutamate receptor gene expression in human stem cell-derived neurons. *Molecular Neuropsychiatry, 3*(2), 73–84. https://doi.org/10.1159/000477762

293. Khandaker, G. M., Zimbron, J., Lewis, G., & Jones, P. B. (2013). Prenatal maternal infection, neurodevelopment and adult schizophrenia: a systematic review of population-based studies. *Psychological Medicine, 43*(2), 239–257. https://doi.org/10.1017/S0033291712000736
294. Kahn, R. S., Sommer, I. E., Murray, R. M., Meyer-Lindenberg, A., Weinberger, D. R., Cannon, T. D., et al. (2015). Schizophrenia. *Nature Reviews Disease Primers, 1*, 15067. https://doi.org/10.1038/nrdp.2015.67
295. Walsh, N. C., Kenney, L. L., Jangalwe, S., Aryee, K. E., Greiner, D. L., Brehm, M. A., et al. (2017). Humanized mouse models of clinical disease. *Annual Review of Pathology, 12*, 187–215. https://doi.org/10.1146/annurev-pathol-052016-100332
296. Allswede, D. M., Buka, S. L., Yolken, R. H., Torrey, E. F., & Cannon, T. D. (2016). Elevated maternal cytokine levels at birth and risk for psychosis in adult offspring. *Schizophrenia Research, 172*(1–3), 41–45. https://doi.org/10.1016/j.schres.2016.02.022
297. Lin, M., Zhao, D., Hrabovsky, A., Pedrosa, E., Zheng, D., & Lachman, H. M. (2014). Heat shock alters the expression of schizophrenia and autism candidate genes in an induced pluripotent stem cell model of the human telencephalon. *PLoS One, 9*(4), e94968. https://doi.org/10.1371/journal.pone.0094968
298. Hashimoto-Torii, K., Torii, M., Fujimoto, M., Nakai, A., El Fatimy, R., Mezger, V., et al. (2014). Roles of heat shock factor 1 in neuronal response to fetal environmental risks and its relevance to brain disorders. *Neuron, 82*(3), 560–572. https://doi.org/10.1016/j.neuron.2014.03.002
299. Ishii, S., Torii, M., Son, A. I., Rajendraprasad, M., Morozov, Y. M., Kawasawa, Y. I., et al. (2017). Variations in brain defects result from cellular mosaicism in the activation of heat shock signalling. *Nature Communications, 8*, 15157. https://doi.org/10.1038/ncomms15157
300. Vallersnes, O. M., Dines, A. M., Wood, D. M., Yates, C., Heyerdahl, F., Hovda, K. E., et al. (2016). Psychosis associated with acute recreational drug toxicity: a European case series. *BMC Psychiatry, 16*, 293. https://doi.org/10.1186/s12888-016-1002-7
301. Callaghan, R. C., Cunningham, J. K., Allebeck, P., Arenovich, T., Sajeev, G., Remington, G., et al. (2012). Methamphetamine use and schizophrenia: a population-based cohort study in California. *The American Journal of Psychiatry, 169*(4), 389–396. https://doi.org/10.1176/appi.ajp.2011.10070937
302. Nielsen, S. M., Toftdahl, N. G., Nordentoft, M., & Hjorthoj, C. (2017). Association between alcohol, cannabis, and other illicit substance abuse and risk of developing schizophrenia: a nationwide population based register study. *Psychological Medicine, 47*(9), 1668–1677. https://doi.org/10.1017/S0033291717000162
303. de Leon, J., & Diaz, F. J. (2005). A meta-analysis of worldwide studies demonstrates an association between schizophrenia and tobacco smoking behaviors. *Schizophrenia Research, 76*(2–3), 135–157. https://doi.org/10.1016/j.schres.2005.02.010
304. Pasman, J. A., Verweij, K. J. H., Gerring, Z., Stringer, S., Sanchez-Roige, S., Treur, J. L., et al. (2018). GWAS of lifetime cannabis use reveals new risk loci, genetic overlap with psychiatric traits, and a causal influence of schizophrenia. *Nature Neuroscience, 21*(9), 1161–1170. https://doi.org/10.1038/s41593-018-0206-1
305. Chatterton, Z., Hartley, B. J., Seok, M. H., Mendelev, N., Chen, S., Milekic, M., et al. (2017). In utero exposure to maternal smoking is associated with DNA methylation alterations and reduced neuronal content in the developing fetal brain. *Epigenetics & Chromatin, 10*, 4. https://doi.org/10.1186/s13072-017-0111-y
306. Oedegaard, K. J., Alda, M., Anand, A., Andreassen, O. A., Balaraman, Y., Berrettini, W. H., et al. (2016). The pharmacogenomics of bipolar disorder study (PGBD): identification of genes for lithium response in a prospective sample. *BMC Psychiatry, 16*, 129. https://doi.org/10.1186/s12888-016-0732-x
307. Ruderfer, D. M., Charney, A. W., Readhead, B., Kidd, B. A., Kahler, A. K., Kenny, P. J., et al. (2016). Polygenic overlap between schizophrenia risk and antipsychotic response: a genomic medicine approach. *Lancet Psychiatry, 3*(4), 350–357. https://doi.org/10.1016/S2215-0366(15)00553-2

308. Li, J., Yoshikawa, A., Brennan, M. D., Ramsey, T. L., & Meltzer, H. Y. (2018). Genetic predictors of antipsychotic response to lurasidone identified in a genome wide association study and by schizophrenia risk genes. *Schizophrenia Research, 192*, 194–204. https://doi.org/10.1016/j.schres.2017.04.009
309. Kim, Y., Giusti-Rodriguez, P., Crowley, J. J., Bryois, J., Nonneman, R. J., Ryan, A. K., et al. (2018). Comparative genomic evidence for the involvement of schizophrenia risk genes in antipsychotic effects. *Molecular Psychiatry, 23*(3), 708–712. https://doi.org/10.1038/mp.2017.111
310. Readhead, B., Hartley, B. J., Eastwood, B. J., Collier, D. A., Evans, D., & Farias, R. (2018). Expression-based drug screening of neural progenitor cells from individuals with schizophrenia. *Nature Communications, 9*(1), 4412. https://doi.org/10.1038/s41467-018-06515-4
311. Xu, M., Lee, E. M., Wen, Z., Cheng, Y., Huang, W. K., Qian, X., et al. (2016). Identification of small-molecule inhibitors of Zika virus infection and induced neural cell death via a drug repurposing screen. *Nature Medicine, 22*(10), 1101–1107. https://doi.org/10.1038/nm.4184
312. Zhou, T., Tan, L., Cederquist, G. Y., Fan, Y., Hartley, B. J., Mukherjee, S., et al. (2017). High-content screening in hPSC-neural progenitors identifies drug candidates that inhibit Zika virus infection in fetal-like organoids and adult brain. *Cell Stem Cell, 21*(2), 274–283.e275. https://doi.org/10.1016/j.stem.2017.06.017
313. Watanabe, M., Buth, J. E., Vishlaghi, N., de la Torre-Ubieta, L., Taxidis, J., Khakh, B. S., et al. (2017). Self-organized cerebral organoids with human-specific features predict effective drugs to combat Zika virus infection. *Cell Reports, 21*(2), 517–532. https://doi.org/10.1016/j.celrep.2017.09.047

Modeling Inflammation on Neurodevelopmental Disorders Using Pluripotent Stem Cells

Beatriz C. Freitas, Patricia C. B. Beltrão-Braga, and Maria Carolina Marchetto

1 Introduction

Neurodevelopmental disorders (ND) impact the nervous system during its natural course of maturation, possibly lasting a lifetime. ND can have a wide range of consequences from intellectual disabilities (ID), speech, and communication like Autism Spectrum Disorders (ASD) [1]. Some specific syndromes displaying ND are related to specific genes (i.e., FXS, DS, RTT) while some are of polygenic or unknown etiologies (ex. idiopathic ASD). Additionally, literature shows that environmental factors can also lead to ND [2, 3]. Recent technological advances combining detailed clinical data with whole-genome sequencing have proposed better categorization of ASD by increasing diagnostic rates and syndrome classifications [4].

Impairment of cognitive function caused by ND is challenging to study in the neuroscience field due in part to the lack of good representation of human brain tissue samples during embryonic development and early postnatal years. In addition, existing animal models lack the complexity of structures and functions found in human brain and do not have the same genetic background [5]. Specifically regarding neuroinflammation, animal models are known to have different cytokine

B. C. Freitas
Laboratory of Disease Modeling, Department of Microbiology, Institute of Biomedical Sciences, University of São Paulo, São Paulo, SP, Brazil

P. C. B. Beltrão-Braga
Laboratory of Disease Modeling, Department of Microbiology, Institute of Biomedical Sciences, University of São Paulo, São Paulo, SP, Brazil

School of Arts, Sciences and Humanities, University of São Paulo, São Paulo, SP, Brazil

M. C. Marchetto (✉)
Department of Anthropology, University of California, San Diego, La Jolla, CA, USA
e-mail: marchetto@salk.edu; mcmarchetto@ucsd.edu

E. DiCicco-Bloom, J. H. Millonig (eds.), *Neurodevelopmental Disorders*, Advances in Neurobiology 25, https://doi.org/10.1007/978-3-030-45493-7_7

secretion profiles compared to humans [6, 7]. Cell reprograming technology is an attractive alternative to model human ND as it uses patients' own cells to generate immature neurons and glial cells allowing for the live study on human neurodevelopment in a dish. The use of human stem cell models has already brought significant progress to the field by comparing genetic models and patient-specific phenotypes [8]. For this review, we will focus on the development of stem cell research elucidating the role of inflammation in neurodevelopmental disorders (Fig. 1).

Animal models continuously provide crucial data on important neurodevelopmental hallmarks throughout years of research. They can be housed on a controlled environment in research facilities and have a shorter developmental period that facilitates study in a laboratory setting and can be genetically engineered for adding or removing genes of interest. Some of the major disadvantages of the animal model system are their differences in cortical structure and developmental timing compared to humans, differences in genome complexity, lack of reproducibility in findings across multiple murine models, and challenges with the interpretation of behavioral changes and its significance to human cognition. Animal models can also vary in complexity, and non-human primates have the potential to provide a more comprehensive approach for ND than rodent models [9]. Still, non-human primate research is limited by requirements of physical space and funding; and currently about 95% of animal model research is mainly performed using rodents.

Given the dynamic nature of NDs, longitudinal clinical data analysis and postmortem brain tissue collection could indicate critical periods associated with the onset of ND-associated deficits. Extensive longitudinal clinical data combined with postmortem samples used in one study concluded that the optimal stage of study autism is during prenatal life up to three years of life [10]. Human stem cells can offer data overlapping with the developmental timing and methods mentioned above, while taking into consideration patients' genetic background that contains

Fig. 1 Different models and methods applied to improve our understanding of neurodevelopmental disorders

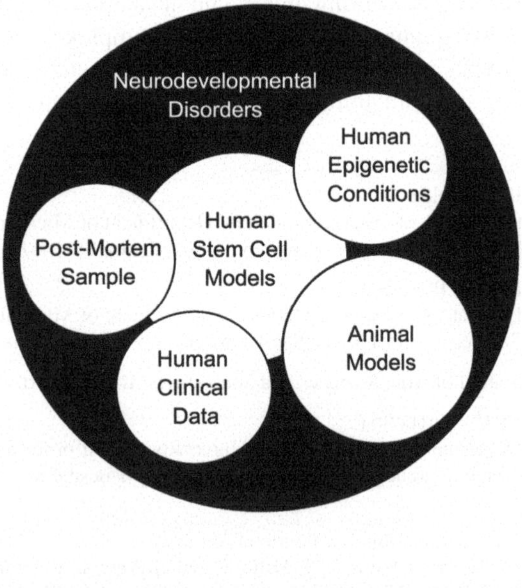

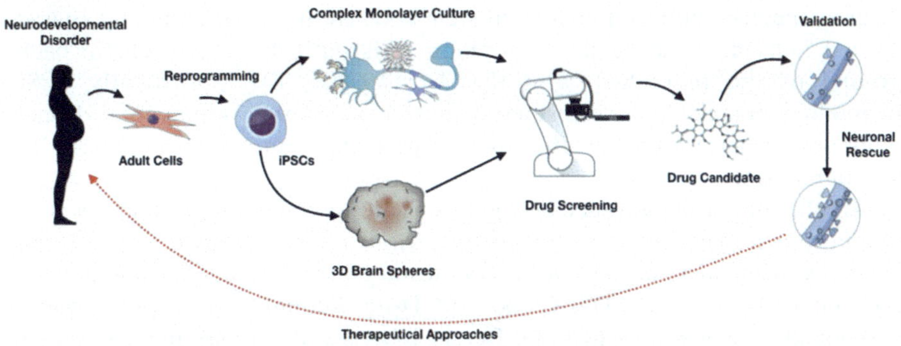

Fig. 2 Neurodevelopmental Disorders modeled by iPSCs. Cells are reprogrammed from adult somatic cells, differentiated into 2D complex cultures or 3D brain organoid models leading to drug screenings aiming to rescue cellular phenotypes turning into possible therapeutic options that incorporate information regarding cellular inflammatory response

gene variants that predispose individuals to autism [8]. Reprograming technology emerges as an essential method to perform validation experiments using data obtained from other model systems and with the potential to generate a relevant human model in a dish to study autism etiology and prognostics.

1.1 Stem Cells and Neurodevelopment Research

Stem cells are defined by the ability of self renewal and to develop into specialized cell types in the body. These cells can be found in adult individuals during their lifetime within specific tissues, such as the bone marrow, and most stem cells are already committed to a particular cell fate regarding their niche source (i.e., skin stem cells will give rise to skin cells). Embryonic stem cells (ES) are found during embryogenesis and are less biased towards a specific cell lineage and have the potential to differentiate into most cell types available. ES research brought several breakthroughs in science, such as the potential for effective macular regeneration treatment in clinical trials [11]. Some limitations of working with human ES are the restricted number of the lines that are currently eligible for use in NIH (National Institues of Health) funded research and the reduced genetic and ethnic background of the lines. Additionally, only a small fraction of the human ES lines available carry disease-specific mutations and even a smaller fraction carries mutations involved in neurodevelopmental conditions. Another caveat is that given the embryonic origin of the human ES lines, it is not possible to obtain any clinical information regarding the individual's disease onset or progression that could be relevant for the in vitro modeling and future therapies.

The advent of somatic reprogramming technology alleviated some of the constraints related to human ES cell research. It allowed for the generation of disease-specific induced pluripotent stem cells (iPSC) from clinically characterized individuals using donor somatic cells with the exact genetic background of the individual with the disease [12, 13]. Specifically, iPSC technology represented a substantial impact on neurodevelopmental disease modeling because it enables the study of patients' neuronal development live and in real time.

Another important advance brought by the reprogramming technology is the ability to test exposure to pathogens (ex. viral infection), environmental toxicants (ex. pesticides) or to perform drug testing on patient-derived and disease-relevant cells. Human FDA (Food and Drug Administration) pre-approved compounds can now be tested in different subtypes of neurons and toxicity can be also accessed in other tissues from the same individual, since iPSC can generate non-neuronal lines [14]. A clinical trial in a dish would cost a fraction of actual tests and could represent a complementary approach for traditional trials [15]. Current limitations for this approach are the need for improvements in the automation of stem cell culture and differentiation pipeline and for better high-throughput readouts of neuronal function. Incorporation of other brain niche cells including glia, endothelial cells, and immune system cells are in progress and will increase the complexity and improve the clinical relevance of the in vitro models [16].

Accurate in vitro maturation and maintenance of system complexity are important concerns when using stem cells for disease modeling of neurodevelopmental conditions. Regarding in vitro neurodevelopmental timing, studies on transcriptional profile of human postmortem brain during development indicated that human iPSCs-derived neurons reflect the developmental stage of fetal stem cells, therefore, mimicking in uterus development [17]. These results are supporting for the use of iPSC-derived neurons to study conditions that are known to develop during early brain development, such as ASD and ID.

Searching for a more complex system to study live human brain development researchers established a model, in which stem cells will form a tridimensional (3D) tissue that resembles early stages of the neural tube development (also called brain organoids). These 3D aggregates are differentiated for months in the presence of specific growth factors and can form different structures representative of the brain ventricles, neural progenitor layer, intermediate zone where neurons are migrating and cortical layers [18, 19]. One of the current challenges of brain organoid technology is the variability between methods available and their limited capabilities for high-throughput screenings (HTS) compared to the more traditional monolayer culture system [20–22]. Furthermore, cell viability and compound permeability can be variable depending on the composition and size of the brain organoid and more research needs to be done to standardize the culture methodologies and post-processing staining.

1.2 Inflammation Impact on Neurodevelopment

Immune system misbalances have been observed in ND. Particularly, immune system involvement has been associated with two major circumstances: (1) *Environmental*: maternal immune activation (MIA) during pregnancy or early childhood (ex. immune system activation after pathogen exposure) and; (2) *Genetic*: mutations in genes involved in immune system homeostasis [23]. In both cases, dysfunctional immune system during brain development was proposed to cause alterations on synapse pruning during a critical brain expansion period, affecting proper neuronal wiring.

Environmental Impact Inflammation has been associated with neurodevelopmental disorders in several animal models of Maternal Immune Activation (MIA) [24]. In these models, pregnant rodent females are exposed to immune system activation either by a viral or bacterial infection resulting in the developmental impairment of the progeny [25, 26]. MIA is one of the accepted methods for environmentally caused idiopathic autism, and cytokine release (ex. Interleukin-6, IL-6) has been shown to have a detrimental effect to the pups' developing brain [27, 28]. While cognitive dysfunctions such as sociability and communication—hallmarks of NDs—are challenging to model in rodents [29]; MIA models have reported a decrease in mice exploratory behavior, increased response to stressful situations, and altered processing of sensory information [30]. Maternal history of increased inflammation such as eczema, psoriasis, and asthma was associated with increased risk for both ASD and NDs [31]. Other risk factors apart from the immune deficiencies mentioned are gestational diabetes, obesity, and aging that were linked to an increased chance of a child to developing ASD [32, 33]. Other examples of early inflammatory challenges are Streptococcal infections, which can sometimes cause *PANDAS* (Pediatric Autoimmune Neuropsychiatric Disorders Associated with Streptococcal Infections). *PANDAS* are connected to obsessive-compulsive disorders (OCD) in children following infection [34, 35]. It is currently unknown if the patient's genetic background plays a role in determining the severity of the infection outcome.

Genetic Impact Several studies reported global expression differences in genes related to inflammatory response in ASD compared to controls [36–38]. Specifically, transcriptome analysis of postmortem brains revealed a significant association of genes related to glial cell activation and inflammation with ASD brains [39]. The proto-oncogene receptor, tyrosine kinase (*MET* gene) regulates cell immune functions among other roles and has been associated with ASD by different studies [40]. Single nucleotide polymorphisms (SNPs) were described at higher frequency in ASD and reduced expression of *MET* was shown in postmortem brains of ASD individuals [41]. SNPs affecting the complement component 4 (C4) gene function, located in the major histocompatibility complex (MHC) region, have also been associated with ASD [42, 43]. Additional components of the complement cascade (*C3, Masp1, Masp2, C3aR,* and *C5aR*) have been implicated in altered cortical development in ASD and play a role in microglial function and neuronal migration [44, 45].

1.3 Using Stem Cells to Study Neuroinflammation in ASD and ND

ASD is a ND with complex and polygenic etiologies, characterized by impairment in social interaction, communication with restricted and repetitive patterns of behavior [1]. ASD has an alarming incidence of 1 in every 59 individuals [46] and is a lifelong condition. Due to its complexity, the spectrum of severity can vary from individuals with mild to severe symptoms. This multifaceted variability represents a challenge to model idiopathic ASD, and usually, researchers rely on subgroups of ASD individuals with comparable clinical features (i.e., macrocephaly) [47, 48].

Anomalies in immune signaling pathways are reported for ASD [49]. Specifically, altered cytokine profiles and immunoglobulin G (IgG) levels were associated with intensified cellular immunity and general amplified neuroinflammation. Detection of misbalanced expression of inflammatory cytokines (IL-6, TNF-α, TGF-β, IL-17, and IL-2) was reported in ASD patients' blood [50]. Clinical data from ASD patients suggests increased neuroinflammation in patients occurs mainly during gestational periods [51, 52]. In addition, increased levels of the inflammatory cytokine IL-6 were detected in autistic brains and correlated to alterations on neural cell adhesion, migration, and synaptic formation [53, 54].

Stem cell modeling of ASD and genetic syndromes associated with ND phenotype are quite extensive: Idiopathic ASD [55, 56], Rett Syndrome [57], Timothy Syndrome [58, 59], Fragile X [60, 61], and Intellectual Disability [62]. Most models mentioned above explored the altered phenotypes of iPSC-generated neurons and not on cells involved in immune activation, such as astrocytes and microglia.

An exception is Rett Syndrome (RTT), a monogenetic progressive neurologic disorder that shares proposed mechanistic and core symptoms with ASD, with mutations in the X-linked gene *MeCP2* occurring in approximately 90% of patients [63–65]. Neurons were thought to be the most relevant cell type for RTT pathology due to the high expression levels of MeCP2 protein in these cells. Previous data from Mandel's group showed that in the MeCP2 null mouse model, restoration of MeCP2 in the mutant astrocytes exerted a non-cell-autonomous positive effect on mutant neurons in vivo [66], Based on these findings, Williams et al showed that wild type neurons co-cultured on human iPSC-derived mutant RTT astrocytes presented morphological and functional defects [67]. To date, the direct role of RTT-derived astrocytes on inflammation has not been extensively explored. Several studies demonstrated that the immune system is involved in RTT in early life. Microglia activation and/or proliferation and defective BDNF (brain-derived neurotrophic factor) signaling is described in RTT [68]. Maezawa et al. proposed that RTT resting microglia are sensitive to both immunological stimuli and neuronal/astrocytic signals causing neuroinflammation and, consequently, affecting brain development [69, 70].

The use of idiopathic patient-derived lines and iPSC-modeling brought significant progress to identifying pathways associated with specific disorders based on genomic data [71]. Reprogramming technology also increased our overall knowledge of the mechanism of action of the disease and improved therapeutic options [72]. However, only a few models mentioned focused on neuroinflammation related to NDs. Using idiopathic ASD lines, researchers identified increased secretion of IL-6 by astrocytes, and toxicity to neurons after co-culture [73]. Autoimmune disease like Aicardi–Goutières syndrome (AGS) is an neuroinflammatory disorder that affects newborn infants and usually results in severe mental and physical handicap, and can be caused by mutations on the *TREX1* gene. iPSC-derived astrocytes from AGS patients caused additional neurotoxicity through increased type I interferon secretion [74]. Childhood cerebral adrenoleukodystrophy (CCALD) is an immune disease associated with *ABCD1* gene function and can lead to dementia and progressive neurological deficits. iPSCs generated from patients with CCALD displayed significant gene expression differences related to neuroinflammation when compared to controls, and provided new leads for pathogenic mechanisms that can be explored in the future [75].

2 Effects of Zika Virus (ZIKV) Infection on Neurodevelopment

In 2015, Brazil went through an alarming increase of newborns with microcephaly caused by a generalized clinical condition identified as Congenital Zika Syndrome (CZS) [76–78]. The CZS involves developmental abnormalities in the eyes, craniofacial, musculoskeletal, pulmonary, and central nervous systems. Specifically in the central nervous system (CNS) the effects of Zika Virus (ZIKV) infection are devastating, due to the ZIKV neurotropism to neural cells [79].

The first postmortem brain tissues investigated revealed several brain malformations in addition to microcephaly, such as hydrocephaly, agyria, and abnormal histological findings like multifocal dystrophic calcifications, cortical displacement, and focal neuroinflammation [80]. Histological findings using biopsied human brains generally revealed the presence of gliosis, microcalcifications, and infiltrated macrophages and mononuclear cells, with glial cells diffusely positive for ZIKV reactivity by immunofluorescence [81]. Rodent brains infected with ZIKV showed the presence of inflammatory cells, specifically lymphocytes, polymorphonuclear cells (PMNs) near blood vessels along with neuronal malformation [82, 83]. Signs of cellular apoptosis and autophagy were also noticed in ZIKV-infected brain tissue, corroborating the idea of neurodevelopmental disruption [83, 84]. Although the analyses on human post mortem brain tissues revealed the presence of ZIKV and signs of neuronal disturbance and inflammation, critical evidence confirming that ZIKV was responsible for CZS came from in vitro modeling using iPSC and in vivo mouse models of ZIKV-crossing placental barrier [84–86].

In one study, ZIKV strains were isolated in Brazil from infected mothers. ZIKV infection in human iPSC-derived CNS cells (in both 2D and 3D models) revealed that neural progenitor cells were the main target for the virus and infected cells were dying [84]. Neural progenitor cell death was primarily triggered by caspase 3 activation leading to apoptosis and was further confirmed on animal models [84, 85, 87–91]. Moreover, ZIKV infection induced an immune response, mainly elicited by innate immunity against viruses by host type I interferon response (IFN-I), activated by Toll-like receptors 3 (TLR3), which impaired neurogenesis leading to cell death [87].

While infected brains provide the opportunity to follow focal immune cellular infiltration, in vitro 2D or 3D iPSC-derived CNS models and fetal brain cells offer the possibility to closely monitor cytokine production by different cell types. ZIKV infection of microglia isolated from human fetal brain induced high levels the immune mediators: interleukin 6 (IL-6), tumor necrosis factor alpha (TNF-a), interleukin 1b (IL-1b), and monocyte chemotactic protein 1 (MCP-1). These experiments reveled that microglia activation was also involved in neuroinflammation process during congenital ZIKV pathogenesis [92].

Another study infected 3D models of brain spheroids co-cultured with or without iPSC-derived microglia cells with Zika or Dengue virus. The spheroids-containing microglia produced a stronger neuroinflammatory response, increasing the expression of IL-6, IL-1β, tumor necrosis alpha (TNF-α), and chemokine (C-C motif) ligand 2 (CCL2), revealing that microglia is physiologically relevant to elicit CNS inflammation cascade and cell death [93]. Although the cytokine production intends to quell infection, its effects on CNS cells could induce detrimental effects to developing neurons in fetal brain. While still under debate, it has been suggested that ZIKV congenital infection could predispose to ASD, but to date there is no data proving the proposed link in the literature [94].

3 Final Remarks

While anomalies in immune signaling pathways have been reported for ASD and ND (ex. altered cytokine profiles in brain and blood), due to the complexity of these disorders, none of these alterations resulted in a biomarker panel for predicting ASD/ND prognosis. It is our expectation that patient-derived stem cell research will contribute to a better understanding of inflammatory component(s) of neurodevelopmental conditions, as it allows for the derivation of patients' immune cell types and the study of its mechanistic impact on neuronal function. Understanding the biological interactions between inflammagens and neurons in ASD or ND background will help developing hypothesis for better diagnostics tools and potential novel therapeutic interventions that will incorporate information regarding cellular inflammatory response.

References

1. American Psychiatric Publishing. (2013). *Diagnostic and statistical manual of mental disorders* (5th ed.). Washington, DC: American Psychiatric Publishing.
2. Ye, B. S., Leung, A. O. W., & Wong, M. H. (2017). The association of environmental toxicants and autism spectrum disorders in children. *Environmental Pollution, 227*, 234–242.

3. Vohr, B. R., et al. (2017). Neurodevelopment: The impact of nutrition and inflammation during preconception and pregnancy in low-resource settings. *Pediatrics, 139*(Suppl 1), S38–S49.
4. Yuen, R. K. C., Merico, D., Bookman, M., Howe, J. L., Thiruvahindrapuram, B., Patel, R. V., et al. (2017). Whole genome sequencing resource identifies 18 new candidate genes for autism spectrum disorder. *Nature Neuroscience, 20*(4), 602–611.
5. Zhao, X., & Bhattacharyya, A. (2018). Human models are needed for studying human neurodevelopmental disorders. *American Journal of Human Genetics, 103*(6), 829–857.
6. Mestas, J., & Hughes, C. C. (2004). Of mice and not men: Differences between mouse and human immunology. *Journal of Immunology, 172*(5), 2731–2738.
7. Tao, L., & Reese, T. A. (2017). Making mouse models that reflect human immune responses. *Trends in Immunology, 38*(3), 181–193.
8. Tamburini, C., & Li, M. (2017). Understanding neurodevelopmental disorders using human pluripotent stem cell-derived neurons. *Brain Pathology, 27*(4), 508–517.
9. Watson, K. K., & Platt, M. L. (2012). Of mice and monkeys: Using non-human primate models to bridge mouse- and human-based investigations of autism spectrum disorders. *Journal of Neurodevelopmental Disorders, 4*(1), 21.
10. Courchesne, E., Campbell, K., & Solso, S. (2011). Brain growth across the life span in autism: Age-specific changes in anatomical pathology. *Brain Research, 1380*, 138–145.
11. da Cruz, L., et al. (2018). Phase 1 clinical study of an embryonic stem cell-derived retinal pigment epithelium patch in age-related macular degeneration. *Nature Biotechnology, 36*(4), 328–337.
12. Takahashi, K., et al. (2007). Induction of pluripotent stem cells from adult human fibroblasts by defined factors. *Cell, 131*(5), 861–872.
13. Shi, Y., et al. (2017). Induced pluripotent stem cell technology: A decade of progress. *Nature Reviews. Drug Discovery, 16*(2), 115–130.
14. Deshmukh, R. S., Kovacs, K. A., & Dinnyes, A. (2012). Drug discovery models and toxicity testing using embryonic and induced pluripotent stem-cell-derived cardiac and neuronal cells. *Stem Cells International, 2012*, 379569.
15. Fermini, B., Coyne, S. T., & Coyne, K. P. (2018). Clinical trials in a dish: A perspective on the coming revolution in drug development. *SLAS Discovery, 23*(8), 765–776.
16. Gonzalez, D. M., Gregory, J., & Brennand, K. J. (2017). The importance of non-neuronal cell types in hiPSC-based disease modeling and drug screening. *Frontiers in Cell and Development Biology, 5*, 117.
17. Mertens, J., et al. (2016). Evaluating cell reprogramming, differentiation and conversion technologies in neuroscience. *Nature Reviews. Neuroscience, 17*(7), 424–437.
18. Lancaster, M. A., & Knoblich, J. A. (2014). Organogenesis in a dish: Modeling development and disease using organoid technologies. *Science, 345*(6194), 1247125.
19. Zhang, Z. N., et al. (2016). Layered hydrogels accelerate iPSC-derived neuronal maturation and reveal migration defects caused by MeCP2 dysfunction. *Proceedings of the National Academy of Sciences of the United States of America, 113*(12), 3185–3190.
20. Yoon, S. J., et al. (2019). Reliability of human cortical organoid generation. *Nature Methods, 16*(1), 75–78.
21. Lancaster, M. A., & Knoblich, J. A. (2014). Generation of cerebral organoids from human pluripotent stem cells. *Nature Protocols, 9*(10), 2329–2340.
22. Sherman, S. P., & Bang, A. G. (2018). High-throughput screen for compounds that modulate neurite growth of human induced pluripotent stem cell-derived neurons. *Disease Models & Mechanisms, 11*(2), 031906.
23. Estes, M. L., & McAllister, A. K. (2015). Immune mediators in the brain and peripheral tissues in autism spectrum disorder. *Nature Reviews. Neuroscience, 16*(8), 469–486.
24. Solek, C. M., et al. (2018). Maternal immune activation in neurodevelopmental disorders. *Developmental Dynamics, 247*(4), 588–619.
25. Hsiao, E. Y., et al. (2012). Modeling an autism risk factor in mice leads to permanent immune dysregulation. *Proceedings of the National Academy of Sciences of the United States of America, 109*(31), 12776–12781.

26. Weir, R. K., et al. (2015). Preliminary evidence of neuropathology in nonhuman primates prenatally exposed to maternal immune activation. *Brain, Behavior, and Immunity, 48*, 139–146.
27. Wu, W. L., et al. (2015). The interaction between maternal immune activation and alpha 7 nicotinic acetylcholine receptor in regulating behaviors in the offspring. *Brain, Behavior, and Immunity, 46*, 192–202.
28. Wu, W. L., et al. (2017). The placental interleukin-6 signaling controls fetal brain development and behavior. *Brain, Behavior, and Immunity, 62*, 11–23.
29. Chadman, K. K. (2017). Animal models for autism in 2017 and the consequential implications to drug discovery. *Expert Opinion on Drug Discovery, 12*(12), 1187–1194.
30. Patterson, P. H. (2002). Maternal infection: Window on neuroimmune interactions in fetal brain development and mental illness. *Current Opinion in Neurobiology, 12*(1), 115–118.
31. Croen, L. A., et al. (2018). Family history of immune conditions and autism spectrum and developmental disorders: Findings from the study to explore early development. *Autism Research, 12*(1), 123–135.
32. Keil, A., et al. (2010). Parental autoimmune diseases associated with autism spectrum disorders in offspring. *Epidemiology, 21*(6), 805–808.
33. Atladottir, H. O., et al. (2009). Association of family history of autoimmune diseases and autism spectrum disorders. *Pediatrics, 124*(2), 687–694.
34. Frick, L., & Pittenger, C. (2016). Microglial dysregulation in OCD, Tourette syndrome, and PANDAS. *Journal of Immunology Research, 2016*, 8606057.
35. Perez-Vigil, A., et al. (2016). The link between autoimmune diseases and obsessive-compulsive and tic disorders: A systematic review. *Neuroscience and Biobehavioral Reviews, 71*, 542–562.
36. Lanz, T. A., et al. (2013). Transcriptomic analysis of genetically defined autism candidate genes reveals common mechanisms of action. *Molecular Autism, 4*(1), 45.
37. Gupta, S., et al. (2014). Transcriptome analysis reveals dysregulation of innate immune response genes and neuronal activity-dependent genes in autism. *Nature Communications, 5*, 5748.
38. Ansel, A., et al. (2016). Variation in gene expression in autism spectrum disorders: An extensive review of transcriptomic studies. *Frontiers in Neuroscience, 10*, 601.
39. Voineagu, I., et al. (2011). Transcriptomic analysis of autistic brain reveals convergent molecular pathology. *Nature, 474*(7351), 380–384.
40. Campbell, D. B., et al. (2006). A genetic variant that disrupts MET transcription is associated with autism. *Proceedings of the National Academy of Sciences of the United States of America, 103*(45), 16834–16839.
41. Peng, Y., et al. (2013). MET receptor tyrosine kinase as an autism genetic risk factor. *International Review of Neurobiology, 113*, 135–165.
42. Warren, R. P., et al. (1994). Decreased plasma concentrations of the C4B complement protein in autism. *Archives of Pediatrics & Adolescent Medicine, 148*(2), 180–183.
43. Odell, D., et al. (2005). Confirmation of the association of the C4B null allele in autism. *Human Immunology, 66*(2), 140–145.
44. Gorelik, A., et al. (2017). Developmental activities of the complement pathway in migrating neurons. *Nature Communications, 8*, 15096.
45. Campbell, D. B., et al. (2007). Disruption of cerebral cortex MET signaling in autism spectrum disorder. *Annals of Neurology, 62*(3), 243–250.
46. Baio, J., et al. (2018). Prevalence of autism spectrum disorder among children aged 8 years - autism and developmental disabilities monitoring network, 11 sites, United States, 2014. *MMWR Surveillance Summaries, 67*(6), 1–23.
47. Ornoy, A., Weinstein-Fudim, L., & Ergaz, Z. (2016). Genetic syndromes, maternal diseases and antenatal factors associated with autism spectrum disorders (ASD). *Frontiers in Neuroscience, 10*, 316.
48. Rossignol, D. A., & Frye, R. E. (2012). A review of research trends in physiological abnormalities in autism spectrum disorders: Immune dysregulation, inflammation, oxidative stress, mitochondrial dysfunction and environmental toxicant exposures. *Molecular Psychiatry, 17*(4), 389–401.

49. Goines, P., & Van de Water, J. (2010). The immune system's role in the biology of autism. *Current Opinion in Neurology, 23*(2), 111–117.
50. Eftekharian, M. M., et al. (2018). Cytokine profile in autistic patients. *Cytokine, 108*, 120–126.
51. Pardo, C. A., Vargas, D. L., & Zimmerman, A. W. (2005). Immunity, neuroglia and neuroinflammation in autism. *International Review of Psychiatry, 17*(6), 485–495.
52. Carpita, B., Muti, D., & Dell'Osso, L. (2018). Oxidative stress, maternal diabetes, and autism spectrum disorders. *Oxidative Medicine and Cellular Longevity, 2018*, 3717215.
53. Li, X., et al. (2009). Elevated immune response in the brain of autistic patients. *Journal of Neuroimmunology, 207*(1-2), 111–116.
54. Wei, H., et al. (2011). IL-6 is increased in the cerebellum of autistic brain and alters neural cell adhesion, migration and synaptic formation. *Journal of Neuroinflammation, 8*, 52.
55. Liu, X., et al. (2017). Idiopathic autism: Cellular and molecular phenotypes in pluripotent stem cell-derived neurons. *Molecular Neurobiology, 54*(6), 4507–4523.
56. Marchetto, M. C., et al. (2017). Altered proliferation and networks in neural cells derived from idiopathic autistic individuals. *Molecular Psychiatry, 22*(6), 820–835.
57. Marchetto, M. C., et al. (2010). A model for neural development and treatment of Rett syndrome using human induced pluripotent stem cells. *Cell, 143*(4), 527–539.
58. Pasca, S. P., et al. (2011). Using iPSC-derived neurons to uncover cellular phenotypes associated with Timothy syndrome. *Nature Medicine, 17*(12), 1657–1662.
59. Yazawa, M., et al. (2011). Using induced pluripotent stem cells to investigate cardiac phenotypes in Timothy syndrome. *Nature, 471*(7337), 230–234.
60. Vershkov, D., et al. (2019). FMR1 reactivating treatments in fragile X iPSC-derived neural progenitors in vitro and in vivo. *Cell Reports, 26*(10), 2531–2539.
61. Abu Diab, M., & Eiges, R. (2019). The contribution of pluripotent stem cell (PSC)-based models to the study of fragile X syndrome (FXS). *Brain Sciences, 9*, 2.
62. Montani, C., et al. (2017). The X-linked intellectual disability protein IL1RAPL1 regulates dendrite complexity. *The Journal of Neuroscience, 37*(28), 6606–6627.
63. Acab, A., & Muotri, A. R. (2015). The use of induced pluripotent stem cell technology to advance autism research and treatment. *Neurotherapeutics, 12*(3), 534–545.
64. Amir, R. E., et al. (1999). Rett syndrome is caused by mutations in X-linked MECP2, encoding methyl-CpG-binding protein 2. *Nature Genetics, 23*(2), 185–188.
65. Liao, W., et al. (2012). MeCP2+/- mouse model of RTT reproduces auditory phenotypes associated with Rett syndrome and replicate select EEG endophenotypes of autism spectrum disorder. *Neurobiology of Disease, 46*(1), 88–92.
66. Lioy, D. T., et al. (2011). A role for glia in the progression of Rett's syndrome. *Nature, 475*(7357), 497–500.
67. Williams, E. C., et al. (2014). Mutant astrocytes differentiated from Rett syndrome patients-specific iPSCs have adverse effects on wild-type neurons. *Human Molecular Genetics, 23*(11), 2968–2980.
68. Theoharides, T. C., et al. (2015). Dysregulated brain immunity and neurotrophin signaling in Rett syndrome and autism spectrum disorders. *Journal of Neuroimmunology, 279*, 33–38.
69. Maezawa, I., et al. (2011). Does microglial dysfunction play a role in autism and Rett syndrome? *Neuron Glia Biology, 7*(1), 85–97.
70. Maezawa, I., & Jin, L. W. (2010). Rett syndrome microglia damage dendrites and synapses by the elevated release of glutamate. *The Journal of Neuroscience, 30*(15), 5346–5356.
71. Gilissen, C., et al. (2014). Genome sequencing identifies major causes of severe intellectual disability. *Nature, 511*(7509), 344–347.
72. Hettige, N. C., et al. (2018). Strategies to advance drug discovery in rare monogenic intellectual disability syndromes. *The International Journal of Neuropsychopharmacology, 21*(3), 201–206.
73. Russo, F. B., et al. (2018). Modeling the interplay between neurons and astrocytes in autism using human induced pluripotent stem cells. *Biological Psychiatry, 83*(7), 569–578.
74. Thomas, C. A., et al. (2017). Modeling of TREX1-dependent autoimmune disease using human stem cells highlights L1 accumulation as a source of neuroinflammation. *Cell Stem Cell, 21*(3), 319–331.

75. Wang, X. M., et al. (2012). The gene expression profiles of induced pluripotent stem cells from individuals with childhood cerebral adrenoleukodystrophy are consistent with proposed mechanisms of pathogenesis. *Stem Cell Research & Therapy, 3*(5), 39.
76. de Araujo, T. V. B., et al. (2016). Association between Zika virus infection and microcephaly in Brazil, January to May, 2016: preliminary report of a case-control study. *The Lancet Infectious Diseases, 16*(12), 1356–1363.
77. Calvet, G., et al. (2016). Detection and sequencing of Zika virus from amniotic fluid of fetuses with microcephaly in Brazil: a case study. *The Lancet Infectious Diseases, 16*(6), 653–660.
78. Rasmussen, S. A., et al. (2016). Zika virus and birth defects--reviewing the evidence for causality. *The New England Journal of Medicine, 374*(20), 1981–1987.
79. Alvarado, M. G., & Schwartz, D. A. (2017). Zika virus infection in pregnancy, microcephaly, and maternal and fetal health: What we think, what we know, and what we think we know. *Archives of Pathology & Laboratory Medicine, 141*(1), 26–32.
80. Mlakar, J., et al. (2016). Zika virus associated with microcephaly. *The New England Journal of Medicine, 374*(10), 951–958.
81. Schwartz, D. A. (2017). Autopsy and postmortem studies are concordant: Pathology of Zika virus infection is neurotropic in fetuses and infants with microcephaly following transplacental transmission. *Archives of Pathology & Laboratory Medicine, 141*(1), 68–72.
82. Bell, T. M., Field, E. J., & Narang, H. K. (1971). Zika virus infection of the central nervous system of mice. *Archiv für die Gesamte Virusforschung, 35*(2), 183–193.
83. Dowall, S. D., et al. (2016). A Susceptible mouse model for Zika virus infection. *PLoS Neglected Tropical Diseases, 10*(5), e0004658.
84. Cugola, F. R., et al. (2016). The Brazilian Zika virus strain causes birth defects in experimental models. *Nature, 534*(7606), 267–271.
85. Tang, H., et al. (2016). Zika virus infects human cortical neural progenitors and attenuates their growth. *Cell Stem Cell, 18*(5), 587–590.
86. Garcez, P. P., et al. (2016). Zika virus impairs growth in human neurospheres and brain organoids. *Science, 352*(6287), 816–818.
87. Dang, J., et al. (2016). Zika virus depletes neural progenitors in human cerebral organoids through activation of the innate immune receptor TLR3. *Cell Stem Cell, 19*(2), 258–265.
88. Onorati, M., et al. (2016). Zika virus disrupts phospho-TBK1 localization and mitosis in human neuroepithelial stem cells and radial glia. *Cell Reports, 16*(10), 2576–2592.
89. Shao, Q., et al. (2016). Zika virus infection disrupts neurovascular development and results in postnatal microcephaly with brain damage. *Development, 143*(22), 4127–4136.
90. Wu, K. Y., et al. (2016). Vertical transmission of Zika virus targeting the radial glial cells affects cortex development of offspring mice. *Cell Research, 26*(6), 645–654.
91. Yockey, L. J., et al. (2016). Vaginal exposure to zika virus during pregnancy leads to fetal brain infection. *Cell, 166*(5), 1247–1256.
92. Lum, F. M., et al. (2017). Zika virus infects human fetal brain microglia and induces inflammation. *Clinical Infectious Diseases, 64*(7), 914–920.
93. Abreu, C. M., et al. (2018). Microglia increase inflammatory responses in iPSC-derived human brainspheres. *Frontiers in Microbiology, 9*, 2766.
94. Vianna, P., Gomes, J. A., Boquett, J. A., Fraga, L. R., Schuch, J. B., Vianna, F. S. L., et al. (2018). Zika Virus as a Possible Risk Factor for Autism Spectrum Disorder: Neuroimmunological Aspects. *Neuroimmunomodulation, 25*, 320–327.

Astrocyte-Derived Exosomes in an iPSC Model of Bipolar Disorder

D. Attili, D. J. Schill, C. J. DeLong, K. C. Lim, G. Jiang, K. F. Campbell, K. Walker, A. Laszczyk, M. G. McInnis, and K. S. O'Shea

1 Introduction

Bipolar Disorder (BP) is a common psychiatric condition that affects 1–3% of the world's population [1]. BP patients typically experience vacillations in mood from severe episodes of mania to life-threatening depressions. The heritability of BP is estimated to be ~80% [2] and family studies support a strong genetic component. First degree relatives of BP probands have a sevenfold higher risk of BP [2], while twin studies have identified a concordance of 15% in dizygotic compared to a 70% or higher concordance in monozygotic twins. Genome-wide association studies (GWAS) have identified many loci, each with small effect size [3]; however, a recent combined sample of over 18,000 cases identified 19 genome-wide significant loci in or near genes in new pathways that may be involved in BP [4]. Considerable data suggest that BP is genetically and mechanistically heterogeneous, and it appears increasingly likely that the final common pathway(s) to disease can be reached by multiple routes.

D. Attili and D. J. Schill contributed equally to the work.

D. Attili · D. J. Schill · C. J. DeLong · K. C. Lim · G. Jiang · K. F. Campbell
K. Walker · A. Laszczyk
Department of Cell and Developmental Biology, The University of Michigan, Ann Arbor, MI, USA

M. G. McInnis
Department of Psychiatry, The University of Michigan, Ann Arbor, MI, USA

K. S. O'Shea (✉)
Department of Cell and Developmental Biology, The University of Michigan, Ann Arbor, MI, USA

Department of Psychiatry, The University of Michigan, Ann Arbor, MI, USA
e-mail: oshea@umich.edu

E. DiCicco-Bloom, J. H. Millonig (eds.), *Neurodevelopmental Disorders*, Advances in Neurobiology 25, https://doi.org/10.1007/978-3-030-45493-7_8

In addition to yet unknown signaling pathways, there is no consensus cell type, neurotransmitter or brain region consistently affected in bipolar disorder. One mechanism which may play a role in mood disorders is alterations in exosome function, which is involved in both normal cell–cell communication and in the transport of pathogenic proteins between cells in the nervous system. Some of these membrane-bound vesicles are generated in the endosome via the ESCRT pathway and are collectively known as extracellular vesicles or exosomes.

Exosomes are small 30–150 nm lipid-bound vesicles that are released from many cells of the body—particularly those of the CNS [5]. A major role for exosomes appears to be to collect and sequester or remove toxic or misfolded proteins from the cell. Some of these are proteins involved in neurodegenerative disease, including the prion protein in prion disease [6], phosphorylated tau and amyloid beta peptide in Alzheimer's disease [7], Tau and alpha-synuclein in Parkinson's disease [8–10], myelin proteins in multiple sclerosis [11], RNAs associated with proliferation, invasion and immune evasion in glioblastoma [12, 13], dysbindin in schizophrenia [14], SOD1 in ALS [15], and mutant HTT in Huntington's disease [16]. Both inhibition and stimulation of exosome release have been proposed as therapeutic approaches to these conditions [17]. The observation that current can stimulate vesicle release has increased interest in electroshock as a potential therapeutic approach to neurodegenerative diseases, while drug development has largely focused on clearing of exosomes from the cell, with the resulting concern of toxic build up in downstream tissues.

Exosomes also play critical roles in homeostasis, cell migration, cell to cell communication, and development by controlling the targeted release of signaling and patterning molecules between cells. Their molecular cargo can contain nucleic acids, lipids, and soluble proteins which may lack signal sequences. Exosomes can either be delivered to adjacent cells or transported to distal sites, suggesting that peripheral monitoring might be useful as a biomarker of disease [18, 19]. Alternatively, since they can cross the blood–brain barrier and are non-immunogenic, they might have a role as carriers of miRNAs, proteins, and therapeutic agents [20].

Recently, exosomal vesicles containing microRNAs were isolated from the cortex (BA9) of individuals with bipolar disorder (miR29, miR149) and schizophrenia (miR497) [21, 22], and from peripheral blood of bipolar disorder patients [23], suggesting that psychiatric disorders may be added to the list of conditions involving exosome transport.

In the current report, we have taken advantage of ability to reprogram somatic cells to pluripotency and then to neural subtypes to characterize exosomes derived from bipolar patient and control astrocytes.

2 Methods

Stem Cell Derivation Skin biopsies (3 mm) were taken from patients diagnosed with bipolar disorder and from control individuals with no psychiatric diagnosis (IRB HUM00043228). Briefly, patient fibroblasts were expanded, and reprogrammed to pluripotency by transfecting them with episomal expression vectors containing the reprogramming factors: SOX2, OCT4, KLF4, L-MYC, and LIN28 (Epi5, Thermo-Fisher). Clonal lines were selected, tested for karyotype, lineage differentiation capacity, for mycoplasma, and to insure that the episomal vectors were no longer present in the cells.

Astrocyte and Neuronal Differentiation Induced pluripotent stem cell (iPSC) lines were expanded to freezer stock, then differentiated into astrocytes following [24, 25]; (Fig. 1a). Briefly, iPSC were induced to form neuroepithelium by culture in DMEM/F12 containing 20% KOSR, 100 μm BME, 1× Glutamax, 2 μg/ml heparin, 10 μM SB431542 and 2.5 μM Dorsomorphin. After 5 days, medium was changed to neural induction medium (NIM): DMEM/F12 containing 1× N2, 2 μg/ml heparin, 1× Glutamax, SB431542 and Dorsomorphin, and cells grown for 2–3 days as embryoid bodies (EB). Next, they were plated on Matrigel for 7 days to form rosettes, which were then grown in suspension in NIM containing FGF2 and EGF (NMEF) for >70 days differentiation. Astrospheres were dissociated, then plated on Matrigel in NMEF. These proliferating, immature astrocytes can be expanded and then frozen for later analysis or matured using cytokines for an additional 1–3 weeks.

Forebrain neural progenitor cells (NPC) were differentiated from iPSC using dual SMAD inhibition (Dorsomorphin and SB431542; [26]). They were grown as EB in neural induction medium, NIM (1× N2, 1× Glutamax, 1× NEAA, pen/strep, and 20 ng/ml FGF2 in DMEM/F12) for 3 days. EBs were then plated on polyornithine/laminin to form rosettes, which were passaged after 6–8 days in NIM. They were manually picked, dissociated, and resuspended in NPC medium (1× B27 (without RA), 1× NEAA, 1× Glutamax, pen/strep, and FGF2 in Neurobasal medium) on Matrigel. After expansion and characterization, NPC were stored at −180 °C for future use. For neuronal differentiation, NPC were transferred to neuronal differentiation medium (NDM; BrainPhys medium with 1× SM1 supplement, 1× N2-A supplement, BDNF and GDNF (20 ng/ml)). To study exosome uptake by differentiating neurons, labeled exosomes were added to proliferating NPC or to NPC differentiated for 6 days in NDM.

Phenotype Analysis Expression of pluripotency factors and lineage markers was examined in iPSC and in astrocytes differentiated from them using q-PCR and immunocytochemistry. For immunocytochemistry, cells were fixed in 4% paraformaldehyde for 10 min, were washed and stored in PBS at 4 °C prior to exposure to primary antibodies (S100ß, GFAP, nestin, Aldh1l1, etc.) overnight at 4 °C, followed by secondary antibody conjugated to Cy3 or FITC. NPC were identified by expression of Pax6, Foxg1, Nestin, and Sox2, and neurons for NeuN and ßIII tubulin. Differentiation was assessed and cells photographed with a Leitz DMI8 inverted fluorescence microscope, or time-lapse video recorded in an Incucyte Zoom (Sartorius).

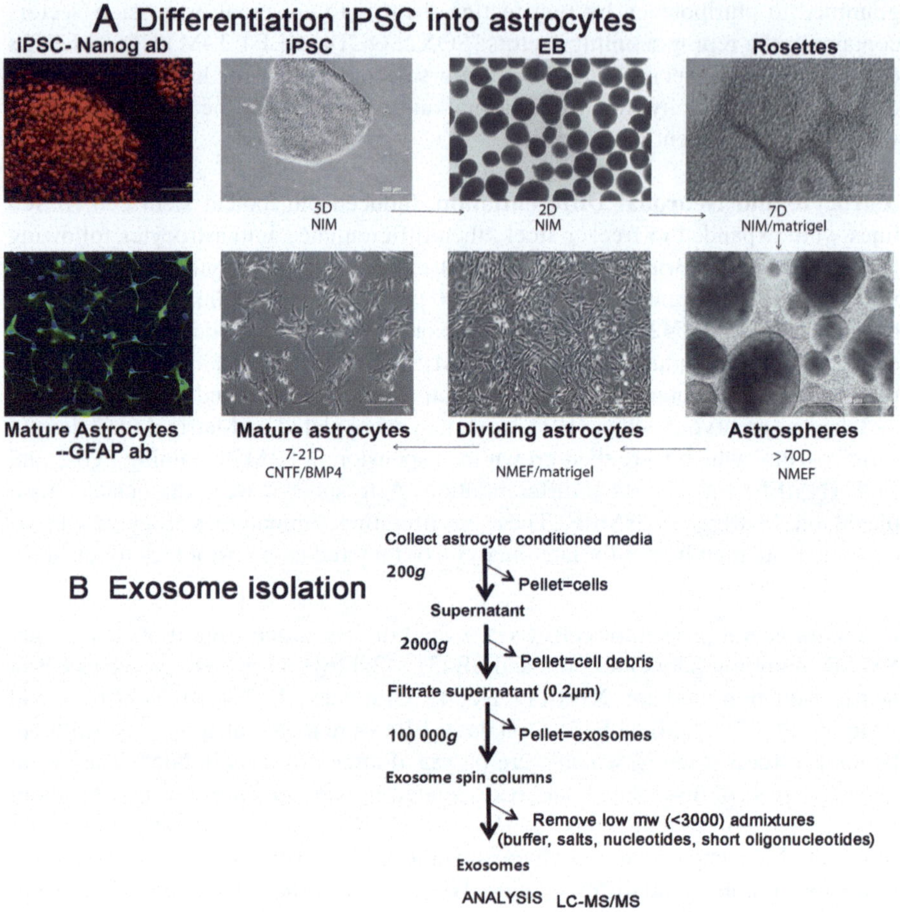

Fig. 1 (**a**) Derivation and characterization of bipolar disorder patient and control iPSC astrocytes. (**b**) Isolation of exosomes

For PCR analysis of astrocyte-restricted gene expression, RNAs were extracted, concentrations and quality measured, then PCR with gene specific primers (available on request) carried out.

Electron Microscopy

For transmission electron microscopy (TEM), astrocytes were fixed in 1% glutaraldehyde in phosphate buffer for 15 min at room temperature. They were rinsed in PBS, post-fixed in 1% osmium tetroxide, embedded in araldite and sectioned at 0.08 μm. They were viewed and photographed at 3000× in a JEOL JEM-1400+ TEM.

Exosome morphology was also examined in TEM following fixation 1% glutaraldehyde for 30 min at room temperature. A 20 μl sample was loaded onto formvarcarbon coated copper grids and allowed to stand for 15 min. Excess fluid was removed using filter paper and grids washed in distilled water (3×, 60 s). The sample was contrasted using a 50 μl drop of 4% uranyl acetate for 3 min. Samples were viewed and photographed at 15,000× in a JEOL JEM 400+ microscope operated at 80 kV.

RNAseq Analysis

RNA Extraction, RNAseq, and q-PCR Validation of mRNAs After 16 weeks of differentiation, total RNA was extracted from astrocytes using the RNAeasy kit mRNA Isolation Kit from Qiagen according to the manufacturer's instructions. RNA concentrations were calculated spectrophotometrically and samples were stored at −80 °C in 4 μg aliquots. RNA aliquots from three control and three BP-patient derived astrocyte samples were submitted to the Microarray Core Facility at the University of Michigan where RNAs were labeled and analyzed using the Illumina Hi-Seq 4000 platform. After hybridization, RMA was used to calculate expression values for each probeset. Expression of eight astrocyte-enriched mRNAs: ALDH1L1, AQP4, CD44, GLAST, GLUL, EAAT1, EAAT2, and S100ß was assessed in q-pcr to validate the RNAseq data.

Identification of Differentially Expressed Genes (DEG) Genes with a threshold fold change of 1.5 and a cut-off FDR of $\leq$0.05 were considered to be differentially expressed, and were selected for additional analysis using the Limma package of bioconductor implemented in the R statistical environment and using DAVID [27]. In addition to the analysis of DEG, the 200 most highly expressed genes in each group were selected, genes expressed by both groups identified and analyzed using hierarchical cluster analysis to identify pathways associated with these groups. Details of the RNAseq analysis are being prepared for publication elsewhere.

Exosome Isolation Exosomes were isolated from tissue culture supernatants from astrocytes grown to approximately 70% confluence by differential centrifugation (Fig. 1b). Supernatants were collected 48 h after medium change, spun at 200 Xg for 10 min, followed by 2000 Xg for 10 min, then passed through a 0.22 μm polyethersulfone (PES) filter to remove dead cells and cellular debris. The filtrate was then spun 100,000 Xg for 90 min. All centrifugation steps were carried out at 4 °C. After the final spin, the supernatant was removed and exosomes resuspended in ultrapure nuclease free deionized water (Thermo-Fisher), then passed through exosome spin columns (mw 3000, Invitrogen) to remove larger size vesicles, salts, and other impurities.

Nanoparticle Tracking Analysis (NTA) The size and concentration of isolated exosomes were examined using a NanoSight NS300 (Malvern Instruments) equipped with a 405 nm laser. A video of 5 × 60 s duration was taken at a frame rate of 30 frames/s, and particle movement analyzed using NTA software (NTA version 3.3. NanoSight, Ltd).

Western Blotting Exosomes (60–80 million per cell line) were lysed in RIPA buffer (Thermo-Fisher). Extracts were run on 4–12% polyacrylamide gels and proteins transferred to a nitrocellulose membrane. Blots were then incubated overnight with primary antibody (e.g., anti-CD9, 1:1000) at 4 °C, washed with TBST buffer, followed by HRP conjugated secondary antibody (1:500; ExoAB kit, System Biosciences) for 1 h at 25 °C. Detection was by enhanced SuperSignal West Pico PLUS Chemiluminescent Substrate (Thermo-Fisher).

Exosome Dye Labeling, Co-Culture, and Imaging Exosomes were labeled with the fluorescent membrane dye PKH26 (1:250, Millipore-Sigma) at a 1:1 (v/v) ratio to exosomes. After a 5 min incubation at room temperature, excess dye was removed by filtering the labeled exosomes through exosome spin columns (Invitrogen) following the manufacturer's instructions. Filtered, labeled exosomes (2–4 million) were added to cultures of differentiating neurons or neural progenitor cells for 10 min up to 6 days, cells fixed with 4% PFA, and imaged in a Leica DMi8 inverted fluorescence microscope or Incucyte Zoom live cell imaging culture microscope.

3 Results

iPSC Derivation, Characterization, Astrocyte Differentiation All iPSC lines were karyotypically normal, mycoplasma negative, negative for the episomal vectors, expressed pluripotency markers (NANOG, OCT4, SOX2, TRA160, SSEA4, and alkaline phosphatase) in PCR and using immunofluorescence localization of protein. SNP analysis identified no new CNVs from parental fibroblasts. Embryoid bodies (EB) differentiated in suspension culture expressed transcripts associated with all three germ lineages: mesoderm, endoderm, and ectoderm. Astrocytes derived from the iPSC lines were positive for astrocyte markers including: S100ß, CD44, GLUL, EAAT1, EAAT2, AQP4, and ALDH1l1. They responded to cytokine (BMP4 or CNTF) addition to the culture medium by spreading and increasing their expression of GFAP. They were able to take up glutamate and exhibited calcium currents. Finally, co-culture with iPSC-derived neurons or culture in astrocyte-conditioned medium supported neuronal survival and process outgrowth.

RNAseq Analysis 501 DEG were identified from a total of 22,761 measurable transcripts, comparing bipolar patient and control astrocyte mRNAs. Of the top 200 most highly expressed transcripts, 99 were expressed at high levels by both control and BP astrocytes, 99 were unique to controls, and 95 restricted to BP astrocytes. When functional annotation cluster analysis was carried out on these transcripts. Three significant (Benjamini corrected) clusters resulted (Table 1).

Of the top 200 transcripts, many were associated with gene ontology (GO) terms related to cellular component "Exosome" (Table 2). A number of these transcripts are involved in CNS development (ADCY1, ALK, CKB, EFNB1, GFRA1, GPM6A,

Table 1 Hierarchical cluster analysis of RNAs from BP and Control astrocytes identified three significant clusters (Benjamini correction)

Cluster	Term	*n*	Score	*p*
1	Extracellular exosome	47	9.88	5.0×10^{-13}
2	Extracellular space	30	7.14	1.8×10^{-9}
3	Proteinaceous ECM	9	1.95	2.6×10^{-3}

Table 2 Exosome associated transcripts

Controls	Both	Bipolar	Prev Assoc with BP
ADCY1	ADAMTS3	AKR1C1	ADCY1
ALK	CFH	ARHGDIB	APOD
APOD	CKB	GALNT5	CD74
CD74	CSRP1	GFRA1	CFH
CLEC3B	EDIL3	ITGB3	CTSH
CTSH	GALNT7	LOXL4	GNGG2
EFNB1	GAS6	MIR21	GPM6A
FREM2	GNG2	NDRG1	GPRC5B
F10	GPM6A	RAB27B	GPX3
GPX3	GPRC5B	SDC1	HLA-DPA1
HLA-DPA1	NUCB2	SLC26A4	HLA-DRA
HLA-DRA	OLFML3		HLA-DRB1
HLA-DRB1	PLCB1		PLCB1
HLA-DRB5	PRDX4		SERC5
NEBL	RND3		
	SERINC5		
	SLC20A2		
	SLIT2		
	SLK		
	SYTL4		
	UACA		
	WNT5B		

NDRG1, PLCB1, RND3, SDC1, SLIT2, UACA, WNT5B), diseases of the nervous system (APOD, CD74, CSRP1, FREM2, MIR21, SLC20A2), immune function (CTSH, HLA-DPA1, HLA-DRA, HLA-DRB1), signaling (GFRA1, GNG2, GPRC5B, MIR21, PLCB1), or with vesicle trafficking, membrane fusion or release. Remarkably, 15 have been previously associated with bipolar disorder (Table 2).

Exosome Characterization To begin to characterize the size and concentration of exosomes isolated from BP and C astrocyte culture medium, we utilized a NanoSight NS300 (Malvern Analytical) laser particle counter. Exosomes derived from BP or C astrocytes had similar mean particle sizes (124.6 $\pm$ 5.4 nm vs 125.2 $\pm$ 8.4 nm, $p \leq 0.43$; Fig. 2a) but were slightly more concentrated than controls (3.54×10^8 $\pm$

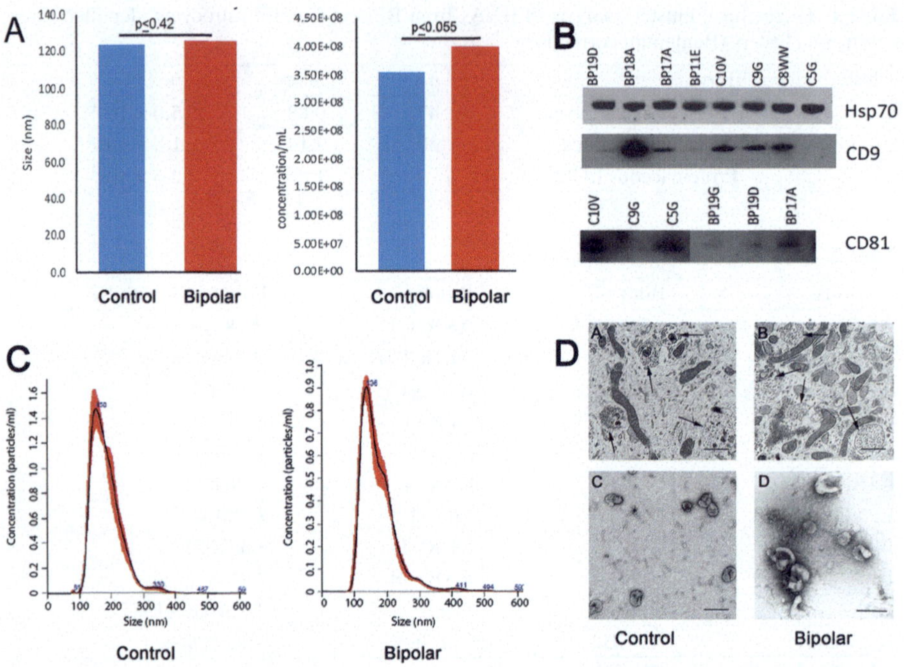

Fig. 2 Characterization of astrocyte-derived exosomes. (**a**) Mean particle size and concentration of exosomes derived from BP and C astrocytes. (**b**) Western blot of exosome markers CD9, CD81, and Hsp70 expression in exosomes from control (cell lines C10V, C9G, C8WW, C5G) and bipolar patient (lines BP19D/G, BP18A, BP17A, BP11E) astrocyte cell lines. (**c**) Representative NanoSight analysis of the size and concentration of astrocyte-derived exosomes from a control and a bipolar patient line. (**d**) TEM analysis of astrocyte cytoplasm illustrating the presence of many multivesicular bodies (**a**, **b**). (**c**, **d**) illustrates the typical cup shaped appearance of control and bipolar exosomes

0.15 vs $3.98 \times 10^8 \pm 0.64$, $p \leq 0.055$; C vs BP; Fig. 2a). To further confirm that the astrocyte-derived vesicles were in fact exosomes, we carried out western blot analysis of the exosome markers CD9, CD81, and Hsp70 [28]. Exosomes from several control and BP astrocyte lines show robust expression of CD9, CD81, and Hsp70 (Fig. 2b) and were also positive for other exosome proteins (not shown). These data suggest that BP and C astrocytes produce and release particles in the size range of exosomes (Fig. 2c, d), contain proteins typical of exosomes, and that there may be differences in cargo and concentration of exosomes from BP patients compared with controls.

TEM analysis indicated that isolated exosomes exhibited the typical cup shape (due to dehydration and loss of content during processing) and size range (Fig. 2d). In addition, ultrastructural analysis of the astrocyte cytoplasm identified a number of multivesicular bodies containing forming exosomes in both bipolar and control astrocytes (Fig. 2d).

Co-culture of Exosomes with Neural Progenitor Cells and Differentiating Neurons Exosome content can significantly impact the differentiation of recipient cells. It is currently unknown if astrocyte-derived exosomes from bipolar disorder patients differ in their cargo, or capacity to alter the development of recipient cells compared with those derived from undiagnosed individuals. To begin to investigate this question, we co-cultured labeled C and BP astrocyte-derived exosomes with control NPCs and performed live cell imaging using time lapse in the Incucyte Zoom (Fig. 3). Initially (time point 0 h), there were no labeled exosomes in the C NPCs (Fig. 3a, d). Imaging cells at 10 min intervals indicated that uptake of exosomes began by 10 min post-addition (data not shown), and after 4 h, C NPCs had taken up both C (Fig. 3b) as well as BP (Fig. 3e) exosomes. When we measured the area covered by red fluorescence (labeled exosomes) as an indication of uptake, after 4 h, similar areas (1789.1 mm^2 control and 1921.4 mm^2 BP) were covered by BP and control exosomes as indicated by the blue false coloring mask (Fig. 3c, f). This will be critical in comparing the uptake and localization of exosomes by recipient cells over time, as both the lipid bilayer and cargo are likely to differ between groups. These data demonstrate that control NPCs can take up exosomes derived from both BP and C astrocytes and this process can be visualized and quantified in real time.

To determine if uptake was as robust in neurons, we also differentiated control NPC for 6 days prior to adding exosomes (Fig. 4). After the addition of exosomes, neurons were grown for an additional 18 h in neuronal differentiation conditions, followed by fixation and imaging. After 6 days of differentiation, control neurons are beginning to extend neurites and were able to take up both C (Fig. 4a–c) and BP (Fig. 4d–f) exosomes. The majority of the exosomes were present in the cell body

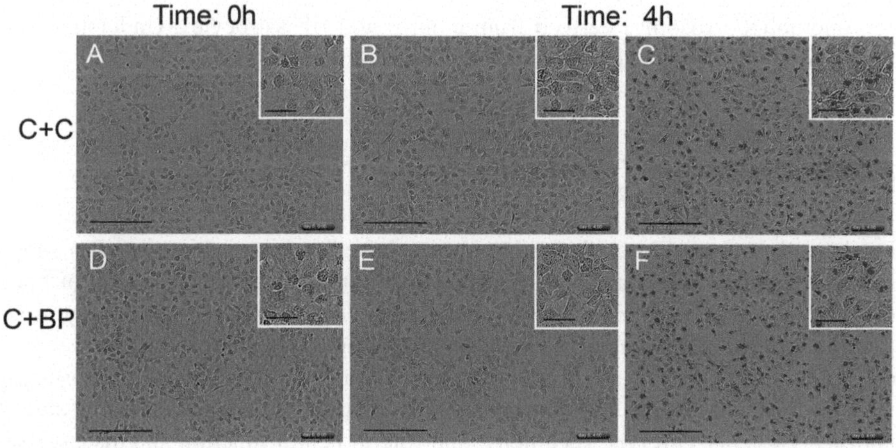

Fig. 3 Live cell imaging of control NPC cultured with BP or C astrocyte-derived exosomes (red) for 0 h (**a**, **d**) or 4 h (**b**, **e**). Blue masking at the 4-h time point (**c**, **f**) demonstrates the capability of the software to quantify exosome uptake over time. Scale bars = 200 μm, insets are 2×

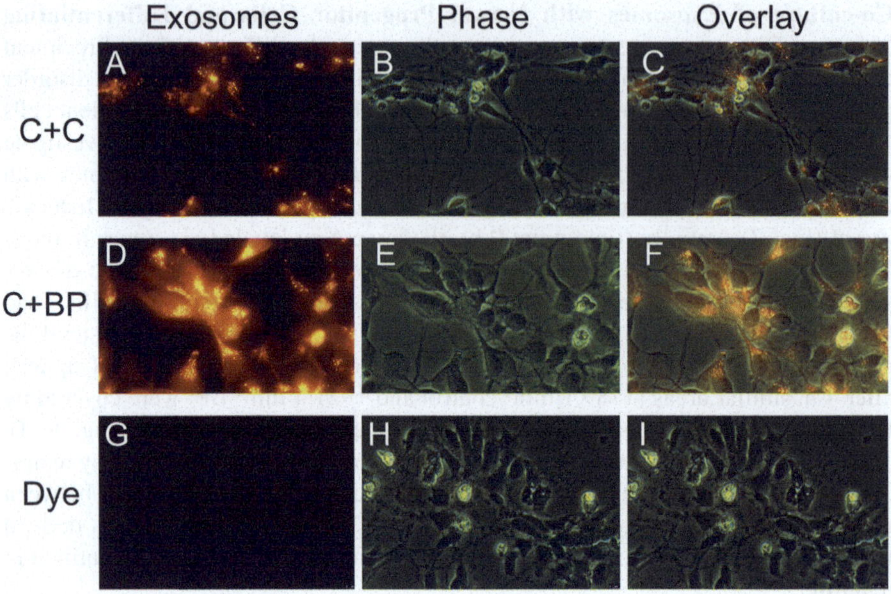

Fig. 4 Co-culture of BP and C exosomes with differentiating control forebrain glutamatergic NPC for 6 days illustrating the robust uptake of exosomes into the differentiating neurons. C+C = Control NPC plus Control exosomes (**a**–**c**); C+ BP = Control NPC plus BP exosomes (**d**–**f**), compared with dye only controls (**g**–**i**). Images were taken at 63×

but were also found in the cell processes (Fig. 4a–f), compared with dye only controls (Fig. 4g–i). These data suggest that both NPC and differentiating neurons can internalize exosomes derived from control and BP astrocytes. Understanding the sequelae of that uptake is in progress.

4 Discussion

Feeder cells—or media conditioned by them—have been widely employed to alternatively maintain pluripotency [29, 30] or to stimulate differentiation [31] of pluripotent stem cells. Astrocyte-conditioned medium (or co-culture) is routinely used to promote neuronal maturation for electrophysiological analysis [32], although the critical secreted factor(s) are not (yet) known. Neurons and glia produce exosomal vesicles [33] and release them into the CNS and into their tissue culture medium, but their contents, cell type specificity, and role in neural function and disease are just beginning to be elucidated.

Extracellular vesicles (EV) are critical mediators of cell–cell communication via their ability to target delivery of RNAs, proteins, and lipid into the aqueous microenvironment of the CNS, and then to cells. EV formation and release is regulated by the ESCRT machinery; exosomes are released when multivesicular bodies containing intraluminal vesicles (ILV) fuse with the plasma membrane releasing the ILV as exosomes into the extracellular space. Exosomes play an essential role in the transport of specialized cargos to target cells that then modify function in the acceptor cells [34]. Astrocyte-derived exosomes can support communication between glia and neurons in the absence of cell–cell contact or synaptic activity and have been suggested to mediate neuronal plasticity [35]. In addition to signaling, another function of exosomes in the CNS may be to relieve neurons of accumulated toxic content; pathology can result when there is an imbalance between the degradation and the sequestration and removal of insoluble protein aggregates [36], thus as exosome formation slows with age, AD symptoms increase [7]. Despite putative roles in disease, development, and tissue homeostasis, study of exosomes has been limited by the lack of unique markers, methods to isolate and study them, and their heterogeneity. Exosomes are typically defined by their size and endosomal origin, reflect the content of the cell of origin, but how exosomes interact with target cells, the mechanisms involved in cargo selection, secretion, and targeted uptake are largely unknown.

Most studies of CNS EV have been carried out on neurons or glia in culture, where exosomes are readily accessible. Neurons, astrocytes, microglia, and oligodendrocytes release exosomes [37, 38], which can modulate synaptic activity and alter neuronal morphology (review [39]), increase neurite outgrowth, and enhance regeneration [40], as well as trigger broad changes in the transcriptional profile of recipient neurons [41]. Exosome release also appears to play a critical role in many aspects of CNS development. As early as the fourth week of embryonic development, exosomes containing patterning molecules [42] are released from the primitive node to determine left-right asymmetry of the early post-implantation embryo. During the early development of the CNS, release of exosomes containing SHH ligands creates gradients that control the dorsal-ventral patterning of the nervous system [43], and maintain NPC proliferation [44]. Embryonic CSF has been shown to contain 1×10^{15} EV/ml [45], which may play an important role in controlling neurogenesis in neurogenic niches [46, 47]. Exosomes remain in the CSF of the adult brain, demonstrating the continued need for growth factor secretion and communication in the mature CNS. Finally, preferential localization of exosomes near dendrites of proliferative cortical pyramidal dendrites and Purkinje cells suggest an important role in development, plasticity, and synaptic function [44].

In the nervous system, exosome secretion can be triggered by stress [28], calcium influx, glutamatergic synaptic activity [48], and by KCl depolarization [37]. Astrocytes may control glutamate homeostasis in the nervous system by secreting and transporting the excitatory amino acid transporters EAAT1 and EAAT2 in exosomes [49]. It has been suggested that one mechanism underlying the success of electroconvulsive therapy in treatment-resistant CNS disease may be via its ability to release exosomes containing toxic proteins from neurons and glia—with the attendant concern they must then be cleared from the CNS, and the body.

Disease causing proteins have been detected in astrocyte-derived exosomes, which can fuse with neurons and cause neurotoxicity. Vesicle formation and sequestration of abnormal proteins have been considered both protective and causal in disease, but toxicity can result from proteasome dysfunction and failure to clear toxic proteins from neurons. Several lines of evidence suggest that there may be a role of altered exosome function in neuropsychiatric disorders, including BP. Bradshaw and Korth [50] hypothesize that there is overlap in the symptoms of neuropsychiatric disorders and those of early stages of neurodegenerative conditions, and that aberrant proteostasis may produce smaller or less toxic aggregates in brains of psychiatric patients. Differential expression of microRNAs in exosomes has been reported in the PFC of BP (miR29, miR149) compared with SZ (miR497) patients [21, 22]. Third, of the three miRNAs (miR21, miR222, miR612) identified in BP astrocytes in the current study, expression of miR21 was previously reported to be upregulated in astrocytes following CNS injury—both spinal cord injury [51], and following oxygen deprivation [52].

Finally, neurotoxic aggregates [53] of the protein products of two genes associated with schizophrenia and bipolar disorder, *DISC1* [54–56] and *DTNBP1* [57] have been shown to transfer to other cells [58] and have been identified in the cingulate cortex [59] of psychiatric patients. Dysbindin is a synaptic protein critical for vesicle tracking and exocytosis, while Disc1 is a multifunctional scaffolding protein which plays a role in many aspects of neural development via its ability to integrate signaling via multiple pathways, including cAMP and GSK3ß [60]. Interestingly, agents that increase cAMP activity/PKA signaling enhance proteasome activity [61, 62] and may attenuate proteasome dysfunction. One of these, rolipram increases cAMP activity by inhibiting PDE4B [63] and has been used clinically to treat schizophrenia and depression [64, 65]. Mutations in DISC1 that give rise to neuropsychiatric conditions are located in the binding sites for PDE4B thereby altering the binding of both PDE4B and GSK3ß to Disc1 [66], decreasing PDE4B activity, and increasing cAMP activity in specific cellular compartments [67]. SNPs in PDE4B have been associated with schizophrenia susceptibility [68, 69] in an isoform specific manner. In a mouse model, inhibiting both GSK3ß and cAMP pathways synergistically improved behavior [66], and may have implications for understanding the therapeutic capabilities of lithium [70]. Consistent with the ability of increased cAMP/PKA signaling to enhance proteasome pathway activity, treatment with rolipram promoted clearing of phosphorylated Tau and improved cognition [71] when delivered to transgenic animals with early but not late disease [72].

While it is unclear whether sequestration of abnormal proteins is neuroprotective or may become neurotoxic, exosomes may be useful in disease diagnosis once cell type-specific or activity-based "signature" cargos have been identified, as exosomes are present in CSF, serum, saliva, amniotic fluid, and urine. Another important use of exosomes may be in targeting treatments to specific cell types of the nervous system as they pass the blood–brain barrier and have surface antibodies that direct their fusion with desired target cells [73]. Ultimately, targeting the proteasomal pathway may represent a novel opportunity to understand and develop rational therapies for mood disorders [74].

Acknowledgements The work was supported by: The Heinz C. Prechter Fund, the Tam Fund, the Kelly Elizabeth Beld Memorial Fund, the University of Michigan Depression Center Eisenberg Translational Research Fund, and U19-MH106434. Assistance from members of the Joerg Lahan and the Carole Parent labs, Rebecca Tagett and the Bioinformatics group at the UM is gratefully acknowledged.

References

1. Goodwin, F. K., & Jamison, K. R. (2007). *Manic depressive illness: Bipolar disorders and recurrent depression*. Oxford: Oxford University Press.
2. Tsuang, M. T., & Faraone, S. V. (1990). *The genetics of mood disorders*. Baltimore: Johns Hopkins University Press.
3. Sklar, P., Ripke, S., Scott, L. J., Andreassen, O. A., Cichon, S., Craddock, N., et al. (2011). Large-scale genome-wide association analysis of bipolar disorder identifies a new susceptibility locus near ODZ4. *Nature Genetics, 43*, 977–983.
4. O'Dushlaine, C., Rossin, L., Lee, P. H., Duncan, L., Parikshak, N. N., Newhouse, S., et al. (2015). Psychiatric genome-wide association study analyses implicate neuronal, immune and histone pathways. *Nature Neuroscience, 18*, 199–209.
5. Lai, C. P., & Breakefield, X. O. (2012). Role of exosomes/microvesicles in the nervous system and use in emerging therapies. *Frontiers in Physiology, 3*, 228.
6. Vella, L. J., Sharples, R. A., Nisbet, R. M., Cappai, R., & Hill, A. F. (2008). The role of exosomes in the processing of proteins associated with neurodegenerative diseases. *European Biophysical Journal, 37*, 323–332.
7. Yuyama, K., Sun, H., Usuki, S., Sakai, S., Hanamatsu, H., Mioka, T., et al. (2015). A potential function for neuronal exosomes: Sequestering intracerebral amyloid-β peptide. *FEBS Letters, 589*, 84–88.
8. Giasson, B. I., Forman, M. S., Higuchi, M., Golbe, L. I., Graves, C. L., Kotzbauer, P. T., et al. (2003). Initiation and synergistic fibrillization of tau and alpha-synuclein. *Science, 300*, 636–640.
9. Shi, M., Kovac, A., Korff, A., Cook, T. J., Ginghina, C., Bullock, K. M., et al. (2016). CNS tau efflux via exosomes is likely increased in Parkinson's disease but not in Alzheimer's disease. *Alzheimers Dement, 12*, 1125–1131.
10. Shi, M., Liu, C., Cook, T. J., Bullock, K. M., Zhao, Y., Ginghina, C., et al. (2014). Plasma exosomal α-synuclein is likely CNS-derived and increased in Parkinson's disease. *Acta Neuropathologica, 128*, 639–650.
11. Galazka, G., Mycko, M. P., Selmaj, I., Raine, C. S., & Selmaj, K. W. (2018). Multiple sclerosis: Serum-derived exosomes express myelin proteins. *Multiple Sclerosis, 24*, 449–458.
12. Skog, J., Wurdinger, T., van Rijn, S., Meijer, D. H., Gainche, L., Sena-Esteves, M., et al. (2008). Glioblastoma microvesicles transport RNA and proteins that promote tumour growth and provide diagnostic biomarkers. *Nature Cell Biology, 10*, 1470–1476.
13. Tsilioni, I., Panagiotidou, S., & Theoharides, T. C. (2014). Exosomes in neurologic and psychiatric disorders. *Clinical Therapeutics, 36*, 882–888.
14. Straub, R. E., Jiang, Y., MacLean, C. J., Ma, Y., Webb, B. T., Myakishev, M. V., et al. (2002). Genetic variation in the 6p22.3 gene DTNBP1, the human ortholog of the mouse dysbindin gene, is associated with schizophrenia. *American Journal of Human Genetics, 71*, 337–348.
15. Boillee, S., Yamanaka, K., Lobsiger, C. S., Copeland, N. G., Jenkins, N. A., Kassiotis, G., et al. (2006). Onset and progression in inherited ALS determined by motor neurons and microglia. *Science, 312*, 1389–1392.

16. Gu, X., Li, C., Wei, W., Lo, V., Gong, S., Li, S. H., et al. (2005). Pathological cell-cell interactions elicited by a neuropathogenic form of mutant Huntingtin contribute to cortical pathogenesis in HD mice. *Neuron, 46*, 433–444.
17. Asai, H., Ikezu, S., Tsunoda, S., Medalla, M., Luebke, J., Haydar, T., et al. (2015). Depletion of microglia and inhibition of exosome synthesis halt tau propagation. *Nature Neuroscience, 18*, 1584–1593.
18. Narahari, A., Hussain, M., & Sreeram, V. (2017). MicroRNAs as biomarkers for psychiatric conditions: A review of current research. *Innovations in Clinical Neuroscience, 14*, 53–55.
19. Raghavan, V., Bhomia, M., Torres, I., Jain, S., & Wang, K. K. (2017). Hypothesis: Exosomal microRNAs as potential biomarkers for schizophrenia. *Medical Hypotheses, 103*, 221–225.
20. Lasser, C. (2015). Exosomes in diagnostic and therapeutic applications: biomarker, vaccine and RNA interference delivery vehicle. *Expert Opinion on Biological Therapy, 15*, 103–117.
21. Banigan, M. G., Kao, P. F., Kozubek, J. A., Winslow, A. R., Medina, J., Costa, J., et al. (2013). Differential expression of exosomal microRNAs in prefrontal cortices of schizophrenia and bipolar disorder patients. *PLoS One, 8*, e48614.
22. Choi, J. L., Kao, P. F., Itriago, E., Zhan, Y., Kozubek, J. A., Hoss, A. G., et al. (2017). miR-149 and miR-29c as candidates for bipolar disorder biomarkers. *American Journal of Medical Genetics Part B, 174*, 315–323.
23. Ceylan, D., Tufekci, K. U., Keskinoglu, P., Genc, S., & Ozerdem, A. (2020). Circulating exosomal microRNAs in bipolar disorder. *Journal of Affective Disorders, 262*, 99–107.
24. Krencik, R., & Zhang, S. C. (2011). Directed differentiation of functional astroglial subtypes from human pluripotent stem cells. *Nature Protocols, 6*, 1710–1717.
25. Santos, R., Vadodaria, K. C., Jaeger, B. N., Mei, A., Lefcochilos-Fogelquist, S., Mendes, A. P. D., et al. (2017). Differentiation of inflammation-responsive astrocytes from glial progenitors generated from human induced pluripotent stem cells. *Stem Cell Reports, 8*, 1757–1769.
26. Chambers, S. M., Fasano, C. A., Papapetrou, E. P., Tomishima, M., Sadelain, M., & Studer, L. (2009). Highly efficient neural conversion of human ES and iPS cells by dual inhibition of SMAD signaling. *Nature Biotechnology, 27*, 275–280.
27. da Huang, W., Sherman, B. T., Zheng, X., Yang, J., Imamichi, T., Stephens, R., Lempicki, R. A.. (2009). Extracting biological meaning from large gene lists with DAVID. *Current Protocols in Bioinformatics*. https://doi.org/10.1002/0471250953.bi1311s27
28. Hessvik, N. P., & Llorente, A. (2018). Current knowledge on exosome biogenesis and release. *Cellular and Molecular Life Sciences, 75*, 193–208.
29. Handyside, A. H., O'Neill, G. T., Jones, M., & Hooper, M. L. (1989). Use of BRL-conditioned medium in combination with feeder layers to isolate a diploid embryonal stem cell line. *Roux's Archives of Developmental Biology, 198*, 48–56.
30. Thomson, J. A., Itskovitz-Eldor, J., Shapiro, S. S., Waknitz, M. A., Swiergiel, J. J., Marshall, V. S., et al. (1998). Embryonic stem cell lines derived from human blastocysts. *Science, 282*, 1145–1147.
31. Kawasaki, H., Mizuseki, K., Nishikawa, S., Kaneko, S., Kuwana, Y., Nakanishi, S., et al. (2000). Induction of midbrain dopaminergic neurons from ES cells by stromal cell-derived inducing activity. *Neuron, 28*, 31–40.
32. Gallo, F., Morale, M. C., Avola, R., & Marchetti, B. (1995). Cross-talk between luteinizing hormone-releasing hormone (LHRH) neurons and astroglial cells: Developing glia release factors that accelerate neuronal differentiation and stimulate LHRH release from GT(1-1) neuronal cell line and LLHRH neurons induce astroglia proliferation. *Endocrine, 3*, 863–874.
33. Palay, S. L., & Palade, G. E. (1955). The fine structure of neurons. *Journal of Biophysical and Biochemical Cytology, 1*, 69–88.
34. Gauthier, S. A., Perez-Gonzalez, R., Sharma, A., Huang, F. K., Alldred, M. J., Pawlik, M., et al. (2017). Enhanced exosome secretion in Down syndrome brain – A protective mechanism to alleviate neuronal endosomal abnormalities. *Acta Neuropathologica Communications, 5*, 65.

35. Lafourcade, C., Ramirez, J. P., Luarte, A., Fernandez, A., & Wyneken, U. (2016). MiRNAs in astrocyte-derived exosomes as possible mediators of neuronal plasticity. *Journal of Experimental Neuroscience, 10*(S1), 1–9.
36. Takalo, M., Salminen, A., Soininen, H., Hiltunen, M., & Haapasalo, A. (2013). Protein aggregation and degradation mechanisms in neurodegenerative diseases. *American Journal of Neurodegenerative Disease, 2*, 1–14.
37. Faure, J., Lachenal, G., Court, M., Hirrlinger, J., Chatellard-Causse, B. B., Grange, J., et al. (2006). Exosomes are released by cultured cortical neurons. *Molecular and Cellular Neuroscience, 31*, 642–648.
38. Guescini, M., Genedani, S., Stocchi, V., & Agnati, L. F. (2010). Astrocytes and glioblastoma cells release exosomes carrying mtDNA. *Journal of Neural Transmission, 117*, 1–4.
39. Basso, M., & Bonetto, V. (2016). Extracellular vesicles and a novel form of communication in the brain. *Frontiers in Neuroscience, 10*, 127.
40. Lopez-Verrilli, M. A., Picou, F., & Court, F. A. (2013). Schwann cell-derived exosomes enhance axonal regeneration in the peripheral nervous system. *Glia, 61*, 1795–1806.
41. Rajendran, L., Bali, J., Barr, M. M., Court, F. A., Krämer-Albers, E. M., Picou, F., et al. (2014). Emerging roles of extracellular vesicles in the nervous system. *Journal of Neuroscience, 34*, 15482–15489.
42. Khalkhali-Ellis, Z., Galat, V., Galat, Y., Gilgur, A., Seftor, E. A., & Hendrix, M. J. C. (2016). Lefty glycoproteins in human embryonic stem cells: Extracellular delivery route and post-translational modification in differentiation. *Stem Cells and Development, 25*, 1681–1690.
43. Wessely, O., & De Robertis, E. M. (2002). Neural plate patterning by secreted signals. *Neuron, 33*, 489–491.
44. Coulter, M. E., Dorobantu, C. M., Lodewijk, G. A., Jacobs, F. M. J., Gaudin, R., & Walsh, C. A. (2018). The ESCRT-III protein CHMP1A mediates secretion of sonic hedgehog on a distinctive subtype of extracellular vesicles. *Cell Reports, 24*, 973–986.
45. Feliciano, D. M., Zhang, S., Nasrallah, C. M., Lisgo, S. N., & Bordey, A. (2014). Embryonic cerebrospinal fluid nanovesicles carry evolutionarily conserved molecules and promote neural stem cell amplification. *PLoS One, 9*, e88810.
46. Batiz, L. F., Castro, M. A., Burgos, P. V., Velasquez, Z. D., Munoz, R. I., Lafourcade, C. A., et al. (2016). Exosomes as novel regulators of adult neurogenic niches. *Frontiers in Cellular Neuroscience, 9*, 501.
47. Luarte, A., Cisternas, P., Caviedes, A., Batiz, L. F., Lafourcade, C., Wyneken, U., et al. (2017). Astrocytes at the hub of the stress response: Potential modulation of neurogenesis by miRNAs in astrocyte-derived exosomes. *Stem Cells International, 2017*, 1719050.
48. Lachenal, G., Pernet-Gallay, K., Chivet, M., Hemming, F. J., Belly, A., Bodon, G., et al. (2011). Release of exosomes from differentiated neurons and its regulation by synaptic glutamatergic activity. *Molecular and Cellular Neuroscience, 46*, 409–418.
49. Gosselin, R. D., Meylan, P., & Decosterd, I. (2013). Extracellular microvesicles from astrocytes contain functional glutamate transporters: Regulation by protein kinase C and cell activation. *Frontiers in Cellular Neuroscience, 7*, 251.
50. Bradshaw, N. J., & Korth, C. (2018). Protein misassembly and aggregation as potential convergence points for non-genetic causes of chronic mental illness. *Molecular Psychiatry, 24*(7), 936–951. https://doi.org/10.1038/s41280
51. Bhalala, O. G., Pan, L., Sahni, V., McGuire, T. L., Gruner, K., Tourtellotte, W. G., et al. (2012). microRNA-21 regulates astrocytic response following spinal cord injury. *Journal of Neuroscience, 32*, 17935–17947.
52. Ziu, M., Fletcher, L., Rana, S., Jimenez, D. F., & Digicaylioglu, M. (2011). Temporal differences in microRNA expression patterns in astrocytes and neurons after ischemic injury. *PLoS One, 6*, e14724.
53. Ottis, P., Bader, V., Trossbach, S. V., Kretzschmar, H., Michel, M., Leliveld, S. R., et al. (2011). Convergence of two independent mental disease genes on the protein level: Recruitment of

dysbindin to cell-invasive disrupted-in-schizophrenia 1 aggresomes. *Biological Psychiatry, 70*, 604–610.
54. Brandon, N. J., & Sawa, A. (2011). Linking neurodevelopmental and synaptic theories of mental illness through DISC1. *Nature Reviews Neuroscience, 12*, 707–722.
55. Craddock, N., O'Donovan, M. C., & Owen, M. J. (2006). Genes for schizophrenia and bipolar disorder? Implications for psychiatric nosology. *Schizophrenia Bulletin, 32*, 9–16.
56. Sawamura, N., Sawamura-Yamamoto, T., Ozeki, Y., Ross, C. A., & Sawa, A. (2005). A form of Disc 1 enriched in nucleus: Altered subcellular distribution in orbitofrontal cortex in psychosis and substance/alcohol abuse. *Proceedings of the National Academy of Sciences USA, 102*, 1187–1192.
57. Breen, G., Prata, D., Osborne, S., Munro, J., Sinclair, M., Li, T., et al. (2006). Association of the dysbindin gene with bipolar affective disorder. *American Journal of Psychiatry, 163*, 1636–1638.
58. Bader, V., Tomppo, L., Trossbach, S. V., Bradshaw, N. J., Prikulis, I., Leliveld, S. R., et al. (2012). Proteomic, genomic and translational approaches identify CRMP1 for a role in schizophrenia and its underlying traits. *Human Molecular Genetics, 21*, 4406–4418.
59. Leliveld, S. R., Bader, V., Hendriks, P., Prikulis, I., Sajnani, G., Requena, J. R., et al. (2008). Insolubility of disrupted-in-schizophrenia 1 disrupts oligomer-dependent interactions with nuclear distribution element 1 and is associated with sporadic mental disease. *Journal of Neuroscience, 28*, 3839–3845.
60. Lipina, T. V., Niwa, M., Jaaro-Peled, H., Fletcher, P. J., Seeman, P., Sawa, A., et al. (2010). Enhanced dopamine function in DISC1-L100P mutant mice: Implications for schizophrenia. *Genes, Brain, and Behavior, 9*, 777–789.
61. Lin, J. T., Chang, W. C., Chen, H. M., Lai, H. L., Chen, C. Y., Tao, M. H., et al. (2013). Regulation of feedback between protein kinase A and the proteasome system worsens Huntington's disease. *Molecular and Cellular Biology, 33*, 1073–1084.
62. Lokireddy, S., Kukushkin, N. V., & Goldberg, A. L. (2015). cAMP-induced phosphorylation of 26S proteasomes on Rpn6/PSMD11 enhances their activity and the degradation of misfolded proteins. *Proceedings of the National Academy of Sciences USA, 112*, E7176–E7185.
63. Houslay, M. D., Schafer, P., & Zhang, K. Y. (2005). Keynote review: Phosphodiesterase-4 as a therapeutic target. *Drug Discovery Today, 10*, 1503–1519.
64. Pietzcker, A., Muller-Oerlinghausen, B., & Kehr, W. (1979). Antipsychotic activity of rolipram in schizophrenic patients. A pilot study. *Naunyn-Schmiedeberg's Archives of Pharmacology, 308*(S1), R44.
65. Zeller, E., Stief, H. J., Pflug, B., & Sastre-y-Hernandez, M. (1984). Results of a phase II study of the antidepressant effect of rolipram. *Pharmacopsychiatry, 17*, 188–190.
66. Lipina, T. V., Wang, M., Liu, F., & Roder, J. C. (2012). Synergistic interactions between PDE4B and GSK-3: DISC1 mutant mice. *Neuropharmacology, 62*, 1252–1262.
67. Millar, J. K., Mackie, S., Clapcote, S. J., Murdoch, H., Pickard, B. S., Christie, S., et al. (2007). Disrupted in schizophrenia 1 and phosphodiesterase 4B: Towards an understanding of psychiatric illness. *Journal of Physiology, 584*(2), 401–405.
68. Feng, Y., Cheng, D., Zhang, C., Li, Y., Zhang, Z., Wang, J., et al. (2016). Association of PDE4B polymorphisms with susceptibility to schizophrenia: A meta-analysis of case-control studies. *PLoS One, 11*, e0147092.
69. Picard, B. S., Thomson, P. A., Christoforou, A., Evans, K. L., Morris, S. W., Porteous, D. J., et al. (2007). The PDE4B gene confers sex-specific protection against schizophrenia. *Psychiatric Genetics, 17*, 129–133.
70. Carlyle, B. C., Mackie, S., Christie, S., Millar, J. K., & Porteous, D. J. (2011). Co-ordinated action of DISC1, PDE4B and GSK3B in modulation of cAMP signaling. *Molecular Psychiatry, 16*, 693–694.
71. Myeku, N., Clelland, C. L., Emrani, S., Kukushkin, N. V., Yu, W. H., Goldberg, A. L., et al. (2015). Tau-driven 26S proteasome impairment and cognitive dysfunction can be prevented early in disease by activating cAMP-PKA signaling. *Nature Medicine, 22*, 46–53.

72. Zhu, C.-Y., Shen, Y., & Xu, Q. (2015). Propagation of dysbindin-1B aggregates: exosome-mediated transmission of neurotoxic deposits. *Neuroscience, 291*, 301–316.
73. Bryniarski, K., Ptak, W., Jayakumar, A., Pullmann, K., Caplan, M. J., Chairoungdua, A., et al. (2013). Antigen specific, antibody-coated exosome-like nanovesicles deliver suppressor T-cell microRNA-150 to effector T cells to inhibit contact sensitivity. *Journal of Allergy and Clinical Immunology, 132*, 170–181.
74. Kawikova, I., & Askenase, P. W. (2015). Diagnostic and therapeutic potentials of exosomes in CNS diseases. *Brain Research, 1617*, 63–71.

Using Two- and Three-Dimensional Human iPSC Culture Systems to Model Psychiatric Disorders

Kimberly M. Christian, Hongjun Song, and Guo-li Ming

1 Introduction

Psychiatric disorders are among the most challenging human diseases to understand at a mechanistic level due to the heterogeneity of symptoms within established diagnostic categories, the general absence of focal pathology, and the genetic complexity inherent in these mostly polygenic disorders. Each of these features presents unique challenges to disease modeling for biological discovery, drug development, or improved diagnostics. In addition, live human neural tissue has been largely inaccessible to experimentation, leaving gaps in our knowledge derived from animal models that cannot fully recapitulate the features of the disease, indirect measures of brain function in human patients, and from analyses of postmortem tissue that can be confounded by comorbid conditions and medication history.

Advances in cellular reprogramming of somatic cells have begun to bridge these gaps and hold the promise of being a novel translational tool to investigate the cellular and molecular mechanisms that contribute to pathology in disease-relevant cell

K. M. Christian (✉)
Department of Neuroscience, Perelman School of Medicine, University of Pennsylvania, Philadelphia, PA, USA

Mahoney Institute for Neurosciences, Perelman School of Medicine, University of Pennsylvania, Philadelphia, PA, USA
e-mail: kchristi@pennmedicine.upenn.edu

H. Song
Department of Neuroscience, Perelman School of Medicine, University of Pennsylvania, Philadelphia, PA, USA

Mahoney Institute for Neurosciences, Perelman School of Medicine, University of Pennsylvania, Philadelphia, PA, USA

Department of Developmental and Cell Biology, Perelman School of Medicine, University of Pennsylvania, Philadelphia, PA, USA

Institute for Regenerative Medicine, Perelman School of Medicine, University of Pennsylvania, Philadelphia, PA, USA

Institute for Epigenetics, Perelman School of Medicine, University of Pennsylvania, Philadelphia, PA, USA

E. DiCicco-Bloom, J. H. Millonig (eds.), *Neurodevelopmental Disorders*,
Advances in Neurobiology 25, https://doi.org/10.1007/978-3-030-45493-7_9

types. A landmark study in 2007 demonstrated that fibroblasts obtained from human adult donors could be reprogrammed into induced pluripotent stem cell (iPSC) lines [1], which can subsequently be differentiated into nearly any cell type in body. Human derived iPSCs (hiPSCs) are similar to human embryonic stem cells (hESCs) that derive from early blastocysts, both of which are capable of generating all the tissue and cell types in the body. Long considered the gold standard in stem cell-based research, there are several well-validated hESC lines that continue to be used in research but these are subject to more regulatory control. Although hiPSC technology has the potential to revolutionize our study of all human disease, there are distinct advantages for psychiatric and neurological disorders that include the ability to generate hiPSC lines that have the same genetic profile as the patient and the ability to generate a renewable source of specific populations of human neurons. In this chapter, we will discuss some of the key opportunities and challenges in hiPSC-based modeling of psychiatric disorders and highlight representative studies using either two- or three-dimensional (2D, 3D) cell cultures that illustrate the potential of each of these approaches to address some of the most critical, outstanding questions in the field. Although we are focusing on only a few studies to frame the discussion of the key issues, it is important to note that there has been a dramatic increase in hiPSC research in recent years and there are many valuable studies that we are unable to include in this chapter due to space limitations.

2 Genetics and Cohort Selection

Since the initial discovery showing the feasibility of generating hiPSCs, we have made great progress in optimizing reprogramming and differentiation protocols but both 2D and 3D iPSC-based studies are still labor-intensive and costly. For this reason, most studies often include a small number of subjects and therefore the choice of donors is critical. Typically, investigators have focused on clinically well-defined patients who are either idiopathic, for whom there is no known genetic risk factor, or those who harbor a causal or risk-associated genetic variant.

G.-l. Ming
Department of Neuroscience, Perelman School of Medicine, University of Pennsylvania, Philadelphia, PA, USA

Mahoney Institute for Neurosciences, Perelman School of Medicine, University of Pennsylvania, Philadelphia, PA, USA

Department of Developmental and Cell Biology, Perelman School of Medicine, University of Pennsylvania, Philadelphia, PA, USA

Institute for Regenerative Medicine, Perelman School of Medicine, University of Pennsylvania, Philadelphia, PA, USA

Department of Psychiatry, Perelman School of Medicine, University of Pennsylvania, Philadelphia, PA, USA

2.1 Monogenic Disorders

There are a few monogenic autism spectrum disorders (ASD) with known causal variants, which has been the focus of some of the first hiPSC-based studies of psychiatric disorders. For example, Rett syndrome, which is a result of mutations in the X-linked gene, *MeCP2*, is one such disorder and an early study revealed specific synaptic deficits in glutamatergic neurons derived from patient iPSCs, including reduced numbers of synapses, spines, and decreases in spontaneous synaptic currents [2]. This study was able to provide proof-of-concept support for the use of hiPSCs to model cellular pathology in 2D cultures using human neurons with disease-relevant mutations and to show that X-inactivation can occur upon neuronal differentiation following reprogramming. Furthermore, this study underscores the utility of hiPSC-based models to identify clear cellular phenotypes suitable for drug screening. The most prevalent monogenic cause of ASD is fragile X syndrome (FXS), which is the result of an expanded CGG trinucleotide repeat in the 5′ untranslated region of fragile X mental retardation gene 1 (*FMR1*). If the expansion is greater than 200 repeats, it leads to epigenetic silencing via CpG hypermethylation and a loss of the fragile X mental retardation protein (FMRP) expression. In one of the first studies to generate iPSC lines from FXS patients, instability of the expansion repeat length was observed after reprogramming, which correlated with methylation and FMRP expression levels [3]. Because the repeat expansion instability occurred even among hiPSC lines from the same subject, this study illustrates the importance of validating any expected genotype and monitoring any mosaicism in the gene of interest both in the fibroblast population before reprogramming and after reprogramming. As with the Rett syndrome study, any X-linked gene should also be monitored for X-inactivation, which can be altered during reprogramming but should be stable in clonal iPSC lines [4, 5]. Upon differentiation of the FXS hiPSC lines, morphological differences have been observed in FXS neurons that exhibit fewer neuronal processes and less complex branching, a phenotype consistent with the role of FMRP in mRNA transport from the nucleus to the dendrites and activity-dependent synaptic development [3, 6]. Additional studies have reported differentiation deficits, as well as impairments in synaptic function and presynaptic vesicle release [7–9]. A recent study used FXS hiPSC-derived neurons, as well as hESC-derived neurons in which the FMR1 gene had been knocked out, to investigate the role of FMRP in homeostatic plasticity mediated by retinoic acid signaling [10]. Importantly, the authors demonstrated that this form of plasticity is present in both hESC- or hiPSC-derived control human neurons, but compromised in FXS neurons. They did not, however, observe any of the overt phenotypes reported in other studies of deficits in differentiation and development and global synaptic dysfunction. Among the many methodological differences that could explain the discrepancy in results, the studies on homeostatic plasticity were predominantly performed in Ngn2-induced glutamatergic neurons that were co-cultured with rodent glial cells and/or rodent neurons. This approach suggests a more nuanced and cell-autonomous deficit in FXS neurons and perhaps that more severe phenotypes can be partly rescued by exposure to non-mutant glial or neural populations, even from other species.

2.2 Polygenic Disorders

In contrast to these monogenic disorders, most psychiatric disorders are polygenic. For disorders such as schizophrenia and bipolar disorder, there are hundreds of risk-associated genes that have been identified. In terms of the strength of association, genetic risk ranges from rare and highly penetrant variants to cumulative risk scores based on multiple common variants [11]. Structurally, risk variants can take many different forms, from discrete single nucleotide polymorphisms (SNPs) to large-scale copy number variations (CNVs) that include both duplications and deletions spanning from a few to hundreds of genes [12]. Adding further complexity, many of these risk variants are often associated with more than one disorder. For example, the 15q11.2–13.1 CNV has been associated with schizophrenia, ASD, and bipolar disorder [13]. Furthermore, some CNVs are associated with diametric risk for different psychiatric disorders based on gene dosage effects. For example, 22q11.2 microdeletions are associated with increased risk for schizophrenia, whereas duplications are associated with ASD [14], which is similar to the diametric risk associations identified for duplications and deletions in 15q11.2 [15].

Ultimately, one of the overarching goals of hiPSC-based research is to determine the degree of mechanistic convergence and divergence in the etiopathology within a single disorder and across clinically related disorders. This is particularly true for polygenic disorders in which patients will likely have distinct underlying genetic risk profiles. For schizophrenia, two of the earliest studies to perform comprehensive phenotyping of neurons differentiated from patient-specific hiPSC lines took different approaches to cohort selection and together were able to provide some information on mechanistic overlap. In one study, the investigators generated hiPSC lines from idiopathic schizophrenia patients [16]. Neurons generated from patient-specific hiPSC lines, as compared to lines generated from age- and sex-matched controls, exhibited several phenotypes including impaired migration, transcriptomic dysregulation, and synaptic deficits [16]. This study was among the first to show that consistent cellular phenotypes could be observed in neurons derived from a cohort of patients who did not share any known genetic or specific environmental risk factor.

Using a different strategy for cohort selection, another study focused on a rare mutation in *DISC1*, a gene that has been implicated in risk for multiple psychiatric disorders including schizophrenia and major depression [17, 18]. This mutation was identified in an American family, Pedigree H, in which some family members harbored a four base pair deletion in *DISC1*. hiPSC lines were generated from family members both with and without the mutation, which were analyzed together with another control line generated from an individual outside of the family to control for genetic background. Several robust cellular phenotypes were observed in forebrain glutamatergic neurons harboring the mutation including widespread transcriptional dysregulation and impaired synapse development and function [19]. Strikingly, several of the cellular phenotypes appeared to be similar to what was observed in the study of idiopathic SZ patients, suggesting a mechanistic convergence of the influence of different risk factors [20].

Importantly, by focusing on a discrete mutation in*DISC1* that is highly amenable to gene editing, investigators were able to generate pairs of isogenic cell lines by either correcting the mutation in patient lines or introducing the mutation in control lines. Investigation of these isogenic lines revealed that many of the cellular phenotypes could be rescued by repairing the mutation or induced by deleting the four base pairs in *DISC1* [19]. Although this study demonstrated the power of gene editing to reveal a causal relationship between a mutation and molecular/cellular phenotypes, it is important to note that this type of study cannot establish a causal relationship between a mutation and a specific disease, and that a direct mapping of cellular phenotype to clinical presentation of symptoms is unlikely. Even monogenic disorders are still considered heterogeneous at the level of gene expression and clinical symptomatology, illustrating the principle that even when a causal mutation and consistent cellular phenotypes have been identified, there can still be variability in how the disease manifests among individuals, likely involving other environmental, genetic, and epigenetic factors.

Ideally, to determine the causal role of any sequence variant in dysregulated cellular processes, it would be best to generate isogenic lines through targeted gene editing to either repair disease hiPSC lines or introduce the mutation in control lines and to show that the genotype is both required and necessary for the given phenotype. Focal- or mono-genetic variants with high penetrance are much easier to model with advanced gene editing strategies such as the recently developed CRISPR/CAS technologies. For disease-associated CNVs or high-risk polygenic scores affecting multiple genes, while it is relatively easy to generate hiPSC lines directly from patients that express the same genetic changes, it is nearly impossible to correct or introduce the variation in its entirety using current techniques, particularly for large CNVs that span hundreds of genes and diseases with polygenetic factors. On the other hand, manipulating a single gene within the CNV or a SNP is much more feasible, which may pinpoint the causal gene or genes for any observed cellular phenotype. There has been some success in identifying a single gene that may preferentially contribute to certain cellular phenotypes relevant to developmental disorders. For example, hiPSCs generated from the 15q11.2 microdeletion CNV carriers encompassing only four genes were differentiated toward a glutamatergic cortical lineage and deficits were observed in the earliest stages of development in the neural progenitors, which expressed aberrant adherens junctions and defects in apical polarity [21]. The authors identified*CYFIP1* as the gene within the 15q11.2 CNV that was responsible for the phenotype. This type of structural deficit in cortical progenitors could lead to dysregulated formation of the cerebral cortex and potentially contribute to the increased variability observed in some brain regions of schizophrenia patients [22] as well as localized irregularities in cortical structure observed in ASD patients [23].

2.3 *Pharmacotherapy Response*

In contrast to cohort selection based on genotype, an entirely different approach based on drug responsiveness was exemplified by a study in which the investigators generated iPSC lines from two groups of bipolar patients—those who were responsive to lithium treatment and those who were not [24]. Strikingly, they observed a hyperexcitability phenotype in neurons that could be selectively rescued with lithium, but only in the hiPSC lines derived from lithium-responsive patients. This finding is a remarkable illustration of how hiPSC models can capture some of the heterogeneity at the patient level in terms of treatment efficacy. A similar study used pharmacological responsiveness to selective serotonin reuptake inhibitors (SSRIs) to stratify patients with major depressive disorder, and found a serotonin-induced hyperactivity phenotype only in forebrain neurons derived from hiPSCs generated from patients who were resistant to SSRI treatment [25]. This "top-down" approach to patient stratification based on drug efficacy and treatment response has proven to be a fruitful strategy that has suggested the presence of basal differences in neuromodulatory pathways in a subset of patients, or selective rescue of phenotypes by a drug in cells derived from drug-responsive patients. Both of these open the door for hiPSC-based diagnostic screening and the potential to move toward a personalized medicine approach to better predict effective treatments for individuals.

In sum, these studies provide key examples of the importance of subject selection when designing an iPSC study of psychiatric disorders, regardless of whether 2D and/or 3D culture conditions will be used. For studies based on genetic variants, these should be verified before and after reprogramming and the possibility of isogenic controls should be considered when possible to allow for causal studies linking gene function to cellular phenotypes.

3 2D Models and Directed Differentiation

Following cohort selection, another important decision to be made when designing an hiPSC study is the cell type to be investigated. One of the key advantages of using 2D models is the ability to generate relatively homogenous populations comprised of a desired cell type (Fig. 1a). The earliest studies were not focused on generating highly specific cell types, but rather used less restrictive differentiation protocols to achieve populations enriched for a particular neurotransmitter subtype such as glutamatergic neurons. Typically, differentiation relies on introducing the same signaling molecules that are present during early development to pattern cells toward lineages of different germ layers and then specific tissues and regions [26]. Targeted differentiation protocols have dramatically improved since the introduction of hiPSC technology; however, achieving 100% purity for any specific cell type is still a challenge for the whole stem cell field. The success in generating a "pure" population of a particular cell type also depends on how the

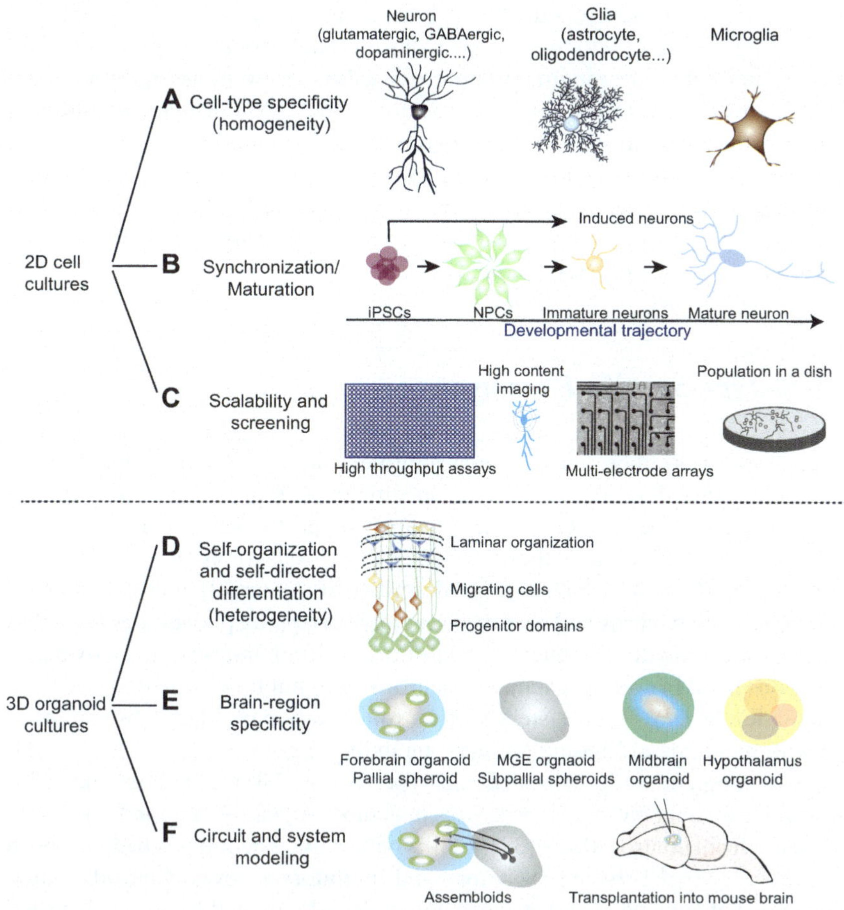

Fig. 1 2D and 3D cultures offer different advantages for modeling biological processes relevant to psychiatric disorders. 2D cell cultures allow for targeted differentiation of highly enriched populations of specific cell types (**a**) with efficient protocols and reproducible results; (**b**) cells can be timelocked during the differentiation process to perform assays in homogenous populations at specific developmental stages; (**c**) 2D cultures are amenable to scaling up to produce large quantities of cells and for high-throughput phenotypic assays; 3D organoid cultures allow for the self-directed fate specification of multiple cell types (**d**); (**e**) recapitulation of neural structures of specific brain regions; and (**f**) rudimentary modeling of neural circuits and axonal targeting

cell type is defined. For example, there are protocols that produce highly enriched populations of forebrain glutamatergic cortical neurons up to ~90% [19]. But this glutamatergic population is comprised of different subsets of cells expressing markers of individual cortical layers, which may have functionally different properties relevant to the disorder under investigation. Similarly, there are efficient protocols for generating GABAergic neurons, but protocols for generating a specific subtype of GABAergic cell, e.g. parvalbumin- or somatostatin-expressing neurons, are less well developed [27]. Advances in single cell biology are also beginning to reveal that there may be even more granularity in the potential classification of cells that are identified and cross-validated by transcriptomic, morphological, and/or electrophysiological analyses [28, 29]. Further complicating the picture is the fact that there seems to be a dynamic component to neurotransmitter expression and a switching phenomenon that can occur in the developing and adult brain [30]. Despite these caveats, monolayer cell cultures are the most efficient method to produce large quantities of relatively pure populations of target neural cell types for phenotypic analysis.

3.1 Cell-Type-Specific Phenotypes

Using a 2D approach, there have been several key findings in the field of hiPSC-based modeling revealing robust cell-type-specific phenotypes associated with particular disorders that have both provided data to support existing hypotheses and generated new mechanistic hypotheses based on unexpected biological insights. For schizophrenia, early studies such as those described above largely focused on excitatory glutamatergic neurons and observed several overlapping phenotypes related to synaptic plasticity and development. In addition, consistent transcriptional dysregulation was observed in cortical excitatory neurons from multiple patient hiPSC lines with a mutation in *DISC1*, consistent with its role as a hub of a large protein interactome with hundreds of binding partners, including transcription factors [19, 31]. Interestingly, a recent study identified cell-type-specific DISC1 interactomes with enrichment for distinct biological processes in neural progenitor cells and astrocytes differentiated from control iPSCs [32]. Cell-type-specific phenotypes had also been probed and compared between excitatory and inhibitory neurons from idiopathic schizophrenia patient iPSCs. For instance, the protocadherin pathway was identified as a locus of dysregulated gene expression only in GABAergic interneurons, but not in glutamatergic neurons [33]. Dysregulation of this pathway resulted in deficits in dendritic arborization and synapse development of interneurons, a result that was further validated with grafted human interneurons in the mouse cortex as well as in postmortem brain tissue [33].

Targeting brain regions and cell types implicated in cognitive impairments common to psychiatric disorders has led investigators to also focus on the hippocampus. The hippocampus is a critical site for many forms of learning and memory and is comprised of three canonical subregions, CA1, CA3, and dentate gyrus, which form a trisynaptic loop, as well as afferent connections from the entorhinal cortex and efferent projections to other cortical and subcortical regions. Subregion-specific differentiation protocols have been developed for both dentate gyrus [20] and CA3 neurons [34], allowing for the identification of cell-type-specific phenotypes, as well as co-culture systems that can begin to model one of the synaptic connections within the trisynaptic circuit. Selective deficits in spontaneous and evoked activity were observed in the idiopathic schizophrenia patient-derived CA3 cells [34], which also impacted synaptic activity in DG-CA3 co-cultures. In addition to finding a cell-type-specific deficit in this schizophrenia cohort, this study also demonstrates the possibility that some aspects of neural circuitry can be modeled in 2D cultures. Together, these studies reveal a context-dependent complexity of molecular and cellular phenotypes and we should be cautious when interpreting data from cell cultures with mixed neural cell populations.

3.2 Differentiation as a Phenotype

It is also important to recognize that aberrant differentiation itself may be a phenotype that could contribute to disease-relevant pathology. Brain imaging studies, for example, have reported a progressive reduction in both gray and white matter in schizophrenia patients [35, 36], suggesting that there may be increased neurodegeneration but also a deficit in the oligodendrocyte population contributing to the white matter volume. When comparing imaging results and hiPSC lines generated from six schizophrenia patients and age-matched healthy subjects, reduced white matter volume in vivo was correlated with less efficient differentiation of oligodendrocyte progenitors and overall fewer oligodendrocytes generated from the schizophrenia patient hiPSC lines [37]. This study demonstrates the value of targeted differentiation of a specific cell type to test a hypothesis based on patient observations. And this type of focused and hypothesis-driven investigation of a specific cell type is more tractable in homogeneous 2D cultures.

Cell-type-specific differentiation deficits have also been observed in hiPSC models for Timothy syndrome, a developmental disorder affecting heart, digit formation, and other physiological systems, as well as the nervous system, and is caused by mutations in *CACNA1c*. *CACNA1c* is a gene encoding the voltage-dependent cardiac L-type calcium channel and has been implicated as a risk factor for several psychiatric disorders, including schizophrenia and bipolar disorder. There is also a high co-morbidity of Timothy syndrome with ASD [38]. After differentiating Timothy syndrome hiPSCs toward a neural lineage, a differentiation phenotype was observed in the numbers of cortical neurons derived from the Timothy syndrome hiPSCs that expressed deep vs upper-layer cortical markers [39]. Specifically, there was an increase in cells expressing upper-layer markers, at the expense of cells expressing SATB2, a lower-layer marker that also identifies callosal projection neurons. This observation suggests that there may be a shift in the ratio of neurons projecting to subcortical vs callosal structures related to Timothy syndrome. Although this study was performed in 2D cell cultures, it may still provide clues as to potential structural changes in 3D brain architecture and indeed is consistent with several imaging studies of ASD patients showing reduced volume of corpus callosum but increased volume of frontal lobe cortical thickness [40].

3.3 *Optimization of Differentiation Protocols in 2D*

Efforts are ongoing to develop better differentiation protocols to generate highly enriched populations of neural cell types relevant to psychiatric and neurological disorders. Moving beyond the glutamatergic cortical neurons that were the focus of many of the early studies using patient hiPSCs, investigators are now targeting other neurotransmitter cell types and various brain regions that may contribute to specific symptomatology in patients. Targeted differentiation of dopaminergic neurons has been the focus of many groups due to the selective degeneration of this cell population in Parkinson's disease [41–43]. But it is well established that dopaminergic signaling is also altered in psychiatric disorders such as schizophrenia and addiction, and many antipsychotic drugs used in clinic target the dopamine system [44]. Thus, improved protocols to generate this type of neuron may shed light on the cellular phenotype and mechanisms underlying psychiatric disorders, as well as additional treatments. Likewise, there are focused attempts to optimize differentiation protocols to obtain relatively pure populations of GABA-expressing cortical neurons that arise from the medial ganglionic eminence in the ventral forebrain, a neural cell type affected in many disorders [45, 46]. In general, differentiation from hiPSC cultures follows the same relative timeline during human fetal brain development and GABAergic cells can take many weeks or months to differentiate in culture. Moreover, the efficiency is still relatively low for some of the latest appearing interneuron subtypes, such as parvalbumin-expressing neurons, a subtype that has been associated with excitatory–inhibitory balance and several psychiatric disorders including schizophrenia, bipolar disorder, and ASD [47, 48]. Other cell types that are currently being targeted for directed differentiation include hypothalamic neuropeptidergic neurons [49–51], cholinergic neurons [52], and various glial cell types [53]. With advanced differentiation protocols, hiPSC-based disease cellular models will provide a powerful system to characterize cell autonomous effects as well as cell-type-specific molecular and cellular changes that might contribute to disease pathology, to identify cell-type-specific biomarkers, and to specify the targets of drug action. Moreover, it is also possible to model simple neural networks when multiple neural cell types and glial cells are co-cultured.

4 Scalability, Screening, and Homogeneity

Another distinct advantage of iPSC-based 2D cell culture systems is the ability to scale up and generate large populations of neurons of a specific cellular age that can be grown in a higher-throughput format (Fig. 1c). This is advantageous for both diagnostic screening to identify phenotypes among several iPSC lines as well as for hypothesis-driven or unbiased screening of large compound libraries to rescue established phenotypes for drug discovery.

With respect to standardizing the culture conditions and cells, a major advantage of 2D cell cultures for large-scale screening is the fact that it is easier to synchronize cells during proliferation, differentiation, and postmitotic maturation. Antimitotic agents can be applied to block cell division and eliminate progenitor cells before reseeding to establish a postmitotic population for further differentiation. This ability to initiate a timestamped differentiation process and expose all cells in a monolayer to culture conditions equally allows for targeted investigations of specific stages of neuronal development at the population level (Fig. 1b). For example, one critical stage of neuronal development involves a change in the ratio of KCC2 (chloride exporter) to NKCC1 (chloride importer), and thus intracellular chloride levels, which mediates a polarity switch in GABAergic signaling from hyperpolarizing to depolarizing. In a Rett syndrome study, neurons differentiated from hiPSCs with an *MECP2* mutation exhibited a deficit in the expression of KCC2 [54]. As a result of the sustained low levels of KCC2 expression in these *MECP2* mutant neurons, GABA remained hyperpolarizing in these cells, reflecting aberrant neuronal maturation and a potential locus for widespread dysfunction caused by an excess of depolarizing GABA during an early critical period of development. Observation of this particular phenotype is greatly facilitated in 2D cultures in which the entire population of cells should be maturing at approximately the same time, making it easier to identify aberrations in developmentally and temporally regulated processes.

Once a clear and reproducible cellular phenotype has been discovered, unbiased screens to rescue the phenotype can be performed and effective hits could provide clues to the underlying biology and mechanisms. As a proof of principle for high-throughput drug screening on cellular phenotypes, a recent study using high-content imaging identified 108 compounds out of a library of over 4000 that modulate neurite growth [55]. Alternatively, screens can be performed to identify compounds that can modulate the expression of the target gene or a specific pathway linked to diseases. For example, among prominent risk genes associated with ASD, *SHANK3* has received attention because it encodes a scaffold protein that is highly expressed at the postsynaptic density and is also one of the genes affected by 22q13.3, or Phelan-McDermid deletion syndrome. One study employed a two-step approach to first identify compounds that increased SHANK3 expression in neurons differentiated from embryonic stem cells, which were then functionally validated in neurons derived from patient-specific hiPSCs with a SHANK3 haploinsufficiency [56]. This study highlights the potential utility of disease-relevant cellular models for developing specific disease-modifying treatments. A high-throughput screen has also been carried out to identify compounds that would increase FMRP expression in FXS patient cells, potentially by reversing the epigenetic silencing of the *FMR1* gene [57]. Using a fluorescence resonance energy transfer assay, several compounds, out of a library of approximately 5000, were identified that led to modest increases in FMR1 mRNA. While the increase was below the threshold of clinical relevance, the study demonstrated how unbiased screens may be useful to identify novel modulators of target pathways.

5 3D Models and Neural Architecture

Although the 2D cellular models described above have led to exciting breakthroughs and novel biological insight, there are clear limitations of this approach in modeling the cellular heterogeneity, neural architecture, and interconnected networks that are likely to be critical for understanding the mechanistic bases of psychiatric disorders. Several years after the initial study showing that adult somatic cells could be reprogrammed to pluripotency, laying the foundation for hiPSC-based disease modeling, a pioneering study published new methodology for generating complex 3D structures from self-organizing hiPSCs grown in a spinning bioreactor [58]. This initial report showed that this specialized culture condition led to the formation of 3D organoids that acquired structural properties reminiscent of different brain regions and related tissues. Using a limited differentiation protocol, these early organoids were allowed to follow intrinsic and stochastic patterning signals to generate disparate cell types and heterogeneous structures within a single organoid. Self-organization is a hallmark property of organoids and confirms the importance of cell–cell interactions in modulating the intrinsic capacities of cells to adopt particular fates and migration patterns (Fig. 1d). This exciting new technology development showed the promise of generating diverse cell types of neural tissue but it was difficult to predict which structural arrangements would emerge in each experiment. Whereas 2D cell culture approaches typically emphasize homogeneity allowing for targeted investigations of a single cell type, 3D organoid approaches promote cellular diversity and heterogeneity, as a function of the self-directed fate specification of progenitors and their progeny within a developing system. This reliance on self-organization introduces variability not only across hiPSC lines but also among organoids from the same line grown under the same conditions. Mechanistic studies, however, benefit from reproducibility and organoids present a unique challenge to ensure unbiased selection of sample regions for quantification and sufficient consistency to identify robust phenotypes and perform drug screens.

Since the initial reports showing highly heterogenous compound tissue organoids, many research groups have made concerted efforts to guide organoid differentiation toward specific brain regions of interest for use as models to investigate specific brain disorders [59, 60] (Fig. 1e). Among the first brain-region-specific organoids to be extensively characterized was a model of the developing neocortex [61]. At the time, it proved to be an optimal model to investigate microcephaly, which was the focus of global health crisis related to the reemergent Zika virus. Several other brain-region-specific models have since emerged including models of the cerebellum [62], hippocampus [63], midbrain [60, 64, 65], motor nerves with associated axon fascicles [66] and even assembloids, fusion of two brain-region-specific organoids [67, 68].

In addition to brain region specificity, organoids are also better suited to model some of the gross structural abnormalities associated with various psychiatric disorders, such as changes in brain volume observed in imaging studies of patients. For example, some of the reciprocal chromosomal rearrangements in CNVs in which both microdeletions and duplications have been associated with increased risk for psychiatric disorders have also been associated with reciprocal differences in brain volume. For CNVs such as 22q11.2, 15q11.2, and 16p11.2, an inverse correlation between gene dosage and brain volume in select brain regions has been observed [14, 69, 70]. Taking advantage of the organoid model, a recent study tested a specific hypothesis regarding brain volume abnormalities focused on the 16p13.11 microduplication CNV [71]. In a small group of patients with the 16p13.11 CNV, magnetic resonance imaging revealed a decrease in cortical volume. Strikingly, correlated to the imaging studies, smaller overall volumes were observed in cerebral organoids generated from those patients' hiPSCs, compared to those from unaffected family control hiPSCs. In addition, aberrant orientation of radial glial progenitors and deficits in neural progenitor cell proliferation were observed. After identifying dysregulation of the NFκB p65 pathway through transcriptomic analysis, pharmacological and genetic rescue targeting this pathway was able to ameliorate the proliferation phenotype. Therefore, well-designed experiments using the right approach to investigate specific phenotypes observed in patients, in this case using organoids to evaluate structural deficits, can lead to new biological insight and targets for drug discovery even from a small cohort of patients.

6 Spatiotemporal Influences on Epigenetics and Epitranscriptomics

In addition to the risk-associated genetic variants that may be associated with the etiopathology of various psychiatric disorders, we are just beginning to understand the degree to which gene expression can be dynamically modulated by environmental factors, epigenetic regulation, gene–gene interactions, and chromosomal architecture, as well as post-transcriptional [72] and post-translational modifications. All of these factors interact to further define the dynamic landscape of protein expression in a brain region and cell-type-specific manner and can influence susceptibility to psychiatric disorders. Brain-region-specific epigenetic profiles have been associated with psychiatric disorders and trait heritability [73], and cell-type-specific chromosomal conformations, for example, have been shown to be associated with schizophrenia risk loci in differentiating progenitor cells [74]. This complexity makes it extremely challenging to identify individual risk profiles and dissect the relative contributions of each component to the emergence of pathology and symptomatology. Brain organoids offer a more holistic approach to modeling the formation of neural systems and provide a platform to interrogate non-cell autonomous influences on the molecular dynamics of developmental processes. In comparison to fetal brain tissue at different gestational ages, brain-region-specific organoids can recapitulate transcriptional [61] and epigenetic signatures of fetal brain development [75], allowing for targeted investigations of early critical windows for long-lasting changes in gene expression.

7 Neural Systems as a Therapeutic Target

It has long been appreciated that the specific patterns of connectivity among neurons are fundamental to brain function. Indeed, several symptomatic domains are thought to involve anatomically distributed neural systems, such as the cortico-striatal-thalamic loops that are implicated in major depressive disorder, obsessive compulsive disorder, and substance abuse disorders [76]. Delineating relevant pathology in terms of neural circuits may be a better approach to understanding the cell and region-specific biological processes that play a causal role in psychiatric disorders, but these systems are more difficult to model in vivo due to potential species-specific differences and are also difficult to address therapeutically. Most pharmaceutical interventions will have off-target effects arising from the impact of drugs on functionally distinct systems that share a common molecular substrate. We need better strategies to target specific brain areas or to define a neural system in a way that distinguishes it from other circuitry. Although still in the early stages of protocol development, 3D cultures hold the promise of being able to dissect circuitry that may facilitate the development of rationally designed circuit-specific therapeutics.

However, we need to first be able to model components of neural circuitry, including accurate targeting of afferent and efferent projections of appropriate cell types. A major step in this direction are the recent studies showing that organoids, or neural spheroids, can be patterned to either a dorsal forebrain (pallium) or ventral forebrain (subpallium) identity and then assembled together to model the environment during fetal brain development when interneurons from the subpallium migrate and integrate into cortical circuits in the pallium [67, 68, 77]. These "assembloids" can approximate early migration events during fetal development and potentially reveal new phenotypes that would not be observable when analyzing region-specific organoids alone (Fig. 1f). Future studies may demonstrate the feasibility of fusing more mature brain regions together to model circuitry in the adult brain.

An alternative strategy to evaluate targeting and integration of hiPSC-derived neural cell types with appropriate presynaptic and postsynaptic partners is to transplant organoids to animal models to assess long-term survival, maturation, and integration. hiPSC-derived cerebral organoids transplanted to a cortical region in the mouse brain can survive for several months and show signs of successful integration into the local circuitry, including extensive axonal growth and evidence of presynaptic targeting of synapses on neurons in the host brain [78]. Importantly, these organoids became vascularized within the host brain, allowing circulating molecular signals to reach to the inner core of the organoid, which may promote long-term survival and maturation.

One challenge in the capacity of organoids to model relevant properties of neural systems is the inclusion of all relevant cell types. Following the same temporal sequence of cell type differentiation in human brain development, glial cells and inhibitory interneurons also emerge later in the developmental trajectory of hiPSC-derived brain organoids. To achieve a more representative distribution of cell types that exist in the mature brain, organoids must be cultured for longer periods of time, which is still a challenge. Efforts to encourage differentiation of astrocytes and oligodendrocytes in 3D cultures have been promising [79, 80]. As opposed to cell types that may arise from the same progenitor population and emerge at different timepoints, other critical populations may arise from a different embryonic origin and need to be co-cultured together with the organoids. Among these populations, microglia have garnered significant attention because these cells are the resident macrophages of the central nervous system. As resident immune cells, microglia are essential to model features of the neuroimmune axis that are not only important for infectious disease but are increasingly appreciated for their role in mediating long-term effects of stress and other factors that may contribute to psychiatric disorders [81]. Several groups have developed differentiation protocols for hiPSC-derived macrophages and microglia, which can be co-cultured in both 2D and 3D platforms [82] to better model the intact physiological environment.

8 Challenges and Future Directions

A theme that is emerging from the application of new technologies such as single cell biology and machine learning is that as we increase both the size of our datasets and the resolution of our analytical methods, some of the traditional categorical distinctions we have relied on for classification may be more fluid than previously thought. This can impact our understanding and investigation of psychiatric disorders at multiple levels. At the clinical level, there has been a growing movement to reconsider clinical diagnostic categories. In 2009, the NIMH proposed a new framework for the investigation of mental disorders based on Research Domain Criteria (RDoC). Currently, there are six recognized domains of human function that cover cognitive, social, and sensorimotor behaviors as well as affective and motivational states. The idea behind these domains is to provide another mode of interpretation for the constellation of symptoms exhibited by a patient, not to supplant the traditional diagnostic criteria. Although there are ongoing debates among clinicians about how and whether to implement this approach in the clinic [83], the RDoC framework can be informative in the design of iPSC studies during patient and cohort selection, or in trying to stratify and compare results across studies. At the cellular level, there is growing evidence that many of the bedrock criteria traditionally used to determine cell identity may be insufficient to distinguish among functionally discrete populations. Because organoids rely, at least in part, on self-organization to generate multiple constitutive cell types for a specific brain region, this approach can be less subject to bias about the importance of particular cell types. Organoids may be particularly useful for the identification and classification of human neural cell types based on single cell sequencing, morphology, and electrophysiological analyses in a more physiologically relevant context [84].

Defining, and achieving, cellular maturity is another challenge for the field. An important question is the extent to which organoids that recapitulate early brain developmental processes can also represent properties of a fully mature brain. For example, some organoid models have been shown to highly correlate with human fetal brain development at the structural, transcriptional, and epigenetic levels through the second trimester. These models may be able to capture dysregulation of early developmental processes that lead to the aberrant formation of neuronal networks, which could effectively form a neural substrate for increased susceptibility to psychiatric disorders. Indeed, there is an increasing appreciation for the idea that there is a developmental component for the majority of psychiatric disorders, which makes a developmental model highly relevant, even for later onset disorders. To identify effective therapies for adults, though, it is important to ask whether these approaches can also model features of the adult brain. Studies using treatment responsivity to stratify patient selection and identify cellular phenotypes suggest that 2D platforms may be an effective tool to generate populations of cells that can be used in diagnostic and therapeutic screening for adult patients. For organoids, although the global structure and transcriptional profiles mirror early brain development, this stage of development is also a period of heightened synaptogenesis and dendritic remodeling, which remain as core features of synaptic plasticity throughout life. As a model of disease-relevant biological processes, the 3D system can provide mechanistic insight into dysregulated plasticity that can affect neural function in the adult brain.

For both 2D and 3D hiPSC models of psychiatric disorders, it will become increasingly important to model the effects of environmental factors to approximate gene–environment interactions. Perturbagens may reveal risk gene-associated deficits in intracellular pathways in response to stress or inflexibility at the cellular or network level. Environmental factors may include exogenous toxins, drugs of abuse, or aberrant levels of endogenous hormones or neuromodulators. In addition to identifying biological processes that are affected at the cellular level by risk gene variants, there may be additional phenotypes that emerge in the presence of both a risk-associated mutation and exposure to an environmental stressor. And conversely, there may be some hiPSC lines that are not affected by the exogenous stressors, suggesting a resilience that could be investigated mechanistically. Finally, future studies should consider the possibility of sex differences when selecting a patient cohort and analyzing the data. Thus far, most studies have not reported any differences in cellular phenotypes in neural cells derived from male or female hiPSC lines, but sample sizes are typically small and some phenotypes may require a larger cohort size to detect. And given the literature on sexually dimorphic circuitry underlying different disorders and symptomatic domains, which can be further modulated by stress [85, 86], differences may be more likely to emerge at the structural level, which could be easier to detect in 3D cultures.

In sum, 2D and 3D cellular platforms from patient hiPSCs can provide complementary, and potentially confirmative, data to study psychiatric disorders. 2D cultures are advantageous when investigating a known specific cell type, scaling up for higher-throughput phenotypic or drug screens and for generating more mature neurons. 3D organoids, on the other hand, facilitate the investigation of specific brain regions over single cell types and can model features of neural structure, neuronal migration, and axonal targeting and interactions between different brain regions, to some extent. Both platforms are valuable and can be used at different stages of the investigation depending on the current state of knowledge about the biology of the disorder, the particular cohort being studied, and whether specific hypotheses are being tested or a more unbiased approach to discovery is warranted.

References

1. Takahashi, K., Tanabe, K., Ohnuki, M., Narita, M., Ichisaka, T., Tomoda, K., et al. (2007). Induction of pluripotent stem cells from adult human fibroblasts by defined factors. *Cell, 131*(5), 861–872.
2. Marchetto, M. C., Carromeu, C., Acab, A., Yu, D., Yeo, G. W., Mu, Y., et al. (2010). A model for neural development and treatment of Rett syndrome using human induced pluripotent stem cells. *Cell, 143*(4), 527–539.
3. Sheridan, S. D., Theriault, K. M., Reis, S. A., Zhou, F., Madison, J. M., Daheron, L., et al. (2011). Epigenetic characterization of the FMR1 gene and aberrant neurodevelopment in human induced pluripotent stem cell models of fragile X syndrome. *PLoS One, 6*(10), e26203.
4. Barakat, T. S., Ghazvini, M., de Hoon, B., Li, T., Eussen, B., Douben, H., et al. (2015). Stable X chromosome reactivation in female human induced pluripotent stem cells. *Stem Cell Reports, 4*(2), 199–208.
5. Tchieu, J., Kuoy, E., Chin, M. H., Trinh, H., Patterson, M., Sherman, S. P., et al. (2010). Female human iPSCs retain an inactive X chromosome. *Cell Stem Cell, 7*(3), 329–342.
6. Doers, M. E., Musser, M. T., Nichol, R., Berndt, E. R., Baker, M., Gomez, T. M., et al. (2014). iPSC-derived forebrain neurons from FXS individuals show defects in initial neurite outgrowth. *Stem Cells and Development, 23*(15), 1777–1787.
7. Telias, M., Kuznitsov-Yanovsky, L., Segal, M., & Ben-Yosef, D. (2015). Functional deficiencies in fragile X neurons derived from human embryonic stem cells. *The Journal of Neuroscience, 35*(46), 15295–15306.
8. Telias, M., Segal, M., & Ben-Yosef, D. (2013). Neural differentiation of Fragile X human embryonic stem cells reveals abnormal patterns of development despite successful neurogenesis. *Developmental Biology, 374*(1), 32–45.
9. Telias, M., Segal, M., & Ben-Yosef, D. (2016). Immature responses to GABA in fragile X neurons derived from human embryonic stem cells. *Frontiers in Cellular Neuroscience, 10*, 121.
10. Zhang, Z., Marro, S. G., Zhang, Y., Arendt, K. L., Patzke, C., Zhou, B., et al. (2018). The fragile X mutation impairs homeostatic plasticity in human neurons by blocking synaptic retinoic acid signaling. *Science Translational Medicine, 10*, 452.
11. Sullivan, P. F., Agrawal, A., Bulik, C. M., Andreassen, O. A., Borglum, A. D., Breen, G., et al. (2018). Psychiatric genomics: An update and an agenda. *The American Journal of Psychiatry, 175*(1), 15–27.
12. Bergen, S. E., Ploner, A., Howrigan, D., O'Donovan, M. C., Group CNVA, The Schizophrenia Working Group of the Psychiatric Genomics C, et al. (2019). Joint contributions of rare copy number variants and common SNPs to risk for schizophrenia. *The American Journal of Psychiatry, 176*(1), 29–35.
13. Malhotra, D., & Sebat, J. (2012). CNVs: Harbingers of a rare variant revolution in psychiatric genetics. *Cell, 148*(6), 1223–1241.

14. Lin, A., Ching, C. R. K., Vajdi, A., Sun, D., Jonas, R. K., Jalbrzikowski, M., et al. (2017). Mapping 22q11.2 gene dosage effects on brain morphometry. *The Journal of Neuroscience, 37*(26), 6183–6199.
15. Crespi, B. J., & Crofts, H. J. (2012). Association testing of copy number variants in schizophrenia and autism spectrum disorders. *Journal of Neurodevelopmental Disorders, 4*(1), 15.
16. Brennand, K. J., Simone, A., Jou, J., Gelboin-Burkhart, C., Tran, N., Sangar, S., et al. (2011). Modelling schizophrenia using human induced pluripotent stem cells. *Nature, 473*(7346), 221–225.
17. Chiang, C. H., Su, Y., Wen, Z., Yoritomo, N., Ross, C. A., Margolis, R. L., et al. (2011). Integration-free induced pluripotent stem cells derived from schizophrenia patients with a DISC1 mutation. *Molecular Psychiatry, 16*(4), 358–360.
18. Sachs, N. A., Sawa, A., Holmes, S. E., Ross, C. A., DeLisi, L. E., & Margolis, R. L. (2005). A frameshift mutation in disrupted in schizophrenia 1 in an American family with schizophrenia and schizoaffective disorder. *Molecular Psychiatry, 10*(8), 758–764.
19. Wen, Z., Nguyen, H. N., Guo, Z., Lalli, M. A., Wang, X., Su, Y., et al. (2014). Synaptic dysregulation in a human iPS cell model of mental disorders. *Nature, 515*(7527), 414–418.
20. Yu, D. X., Di Giorgio, F. P., Yao, J., Marchetto, M. C., Brennand, K., Wright, R., et al. (2014). Modeling hippocampal neurogenesis using human pluripotent stem cells. *Stem Cell Reports, 2*(3), 295–310.
21. Yoon, K. J., Nguyen, H. N., Ursini, G., Zhang, F., Kim, N. S., Wen, Z., et al. (2014). Modeling a genetic risk for schizophrenia in iPSCs and mice reveals neural stem cell deficits associated with adherens junctions and polarity. *Cell Stem Cell, 15*(1), 79–91.
22. Brugger, S. P., & Howes, O. D. (2017). Heterogeneity and homogeneity of regional brain structure in schizophrenia: A meta-analysis. *JAMA Psychiatry, 74*(11), 1104–1111.
23. Stoner, R., Chow, M. L., Boyle, M. P., Sunkin, S. M., Mouton, P. R., Roy, S., et al. (2014). Patches of disorganization in the neocortex of children with autism. *The New England Journal of Medicine, 370*(13), 1209–1219.
24. Mertens, J., Wang, Q. W., Kim, Y., Yu, D. X., Pham, S., Yang, B., et al. (2015). Differential responses to lithium in hyperexcitable neurons from patients with bipolar disorder. *Nature, 527*(7576), 95–99.
25. Vadodaria, K. C., Ji, Y., Skime, M., Paquola, A., Nelson, T., Hall-Flavin, D., et al. (2019). Serotonin-induced hyperactivity in SSRI-resistant major depressive disorder patient-derived neurons. *Molecular Psychiatry, 24*(6), 795–807.
26. Mertens, J., Marchetto, M. C., Bardy, C., & Gage, F. H. (2016). Evaluating cell reprogramming, differentiation and conversion technologies in neuroscience. *Nature Reviews. Neuroscience, 17*(7), 424–437.
27. Soliman, M. A., Aboharb, F., Zeltner, N., & Studer, L. (2017). Pluripotent stem cells in neuropsychiatric disorders. *Molecular Psychiatry, 22*(9), 1241–1249.
28. Tasic, B., Menon, V., Nguyen, T. N., Kim, T. K., Jarsky, T., Yao, Z., et al. (2016). Adult mouse cortical cell taxonomy revealed by single cell transcriptomics. *Nature Neuroscience, 19*(2), 335–346.
29. Holguera, I., & Desplan, C. (2018). Neuronal specification in space and time. *Science, 362*(6411), 176–180.
30. Spitzer, N. C. (2017). Neurotransmitter switching in the developing and adult brain. *Annual Review of Neuroscience, 40*, 1–19.
31. Porteous, D. J., Millar, J. K., Brandon, N. J., & Sawa, A. (2011). DISC1 at 10: Connecting psychiatric genetics and neuroscience. *Trends in Molecular Medicine, 17*(12), 699–706.
32. Wilkinson, B., Evgrafov, O. V., Zheng, D., Hartel, N., Knowles, J. A., Graham, N. A., et al. (2019). Endogenous cell type-specific disrupted in schizophrenia 1 interactomes reveal protein networks associated with neurodevelopmental disorders. *Biological Psychiatry, 85*(4), 305–316.

33. Shao, Z., Noh, H., Bin Kim, W., Ni, P., Nguyen, C., Cote, S. E., et al. (2019). Dysregulated protocadherin-pathway activity as an intrinsic defect in induced pluripotent stem cell-derived cortical interneurons from subjects with schizophrenia. *Nature Neuroscience, 22*(2), 229–242.
34. Sarkar, A., Mei, A., Paquola, A. C. M., Stern, S., Bardy, C., Klug, J. R., et al. (2018). Efficient generation of CA3 neurons from human pluripotent stem cells enables modeling of hippocampal connectivity in vitro. *Cell Stem Cell, 22*(5), 684–697.
35. Cannon, T. D., Chung, Y., He, G., Sun, D., Jacobson, A., van Erp, T. G., et al. (2015). Progressive reduction in cortical thickness as psychosis develops: A multisite longitudinal neuroimaging study of youth at elevated clinical risk. *Biological Psychiatry, 77*(2), 147–157.
36. Cropley, V. L., Klauser, P., Lenroot, R. K., Bruggemann, J., Sundram, S., Bousman, C., et al. (2017). Accelerated gray and white matter deterioration with age in schizophrenia. *The American Journal of Psychiatry, 174*(3), 286–295.
37. McPhie, D. L., Nehme, R., Ravichandran, C., Babb, S. M., Ghosh, S. D., Staskus, A., et al. (2018). Oligodendrocyte differentiation of induced pluripotent stem cells derived from subjects with schizophrenias implicate abnormalities in development. *Translational Psychiatry, 8*(1), 230.
38. Freitas, B. C., Trujillo, C. A., Carromeu, C., Yusupova, M., Herai, R. H., & Muotri, A. R. (2014). Stem cells and modeling of autism spectrum disorders. *Experimental Neurology, 260*, 33–43.
39. Pasca, S. P., Portmann, T., Voineagu, I., Yazawa, M., Shcheglovitov, A., Pasca, A. M., et al. (2011). Using iPSC-derived neurons to uncover cellular phenotypes associated with Timothy syndrome. *Nature Medicine, 17*(12), 1657–1662.
40. Pagnozzi, A. M., Conti, E., Calderoni, S., Fripp, J., & Rose, S. E. (2018). A systematic review of structural MRI biomarkers in autism spectrum disorder: A machine learning perspective. *International Journal of Developmental Neuroscience, 71*, 68–82.
41. Xue, Y., Zhan, X., Sun, S., Karuppagounder, S. S., Xia, S., Dawson, V. L., et al. (2019). Synthetic mRNAs drive highly efficient iPS cell differentiation to dopaminergic neurons. *Stem Cells Translational Medicine, 8*(2), 112–123.
42. Cobb, M. M., Ravisankar, A., Skibinski, G., & Finkbeiner, S. (2018). iPS cells in the study of PD molecular pathogenesis. *Cell and Tissue Research, 373*(1), 61–77.
43. Yan, Y., Yang, D., Zarnowska, E. D., Du, Z., Werbel, B., Valliere, C., et al. (2005). Directed differentiation of dopaminergic neuronal subtypes from human embryonic stem cells. *Stem Cells, 23*(6), 781–790.
44. McCutcheon, R. A., Abi-Dargham, A., & Howes, O. D. (2019). Schizophrenia, dopamine and the striatum: from biology to symptoms. *Trends in Neurosciences, 42*(3), 205–220.
45. Tao, Y., & Zhang, S. C. (2016). Neural subtype specification from human pluripotent stem cells. *Cell Stem Cell, 19*(5), 573–586.
46. Liu, Y., Liu, H., Sauvey, C., Yao, L., Zarnowska, E. D., & Zhang, S. C. (2013). Directed differentiation of forebrain GABA interneurons from human pluripotent stem cells. *Nature Protocols, 8*(9), 1670–1679.
47. Ferguson, B. R., & Gao, W. J. (2018). PV interneurons: Critical regulators of E/I balance for prefrontal cortex-dependent behavior and psychiatric disorders. *Frontiers in Neural Circuits, 12*, 37.
48. Volk, D. W., Sampson, A. R., Zhang, Y., Edelson, J. R., & Lewis, D. A. (2016). Cortical GABA markers identify a molecular subtype of psychotic and bipolar disorders. *Psychological Medicine, 46*(12), 2501–2512.
49. Wang, L., Meece, K., Williams, D. J., Lo, K. A., Zimmer, M., Heinrich, G., et al. (2015). Differentiation of hypothalamic-like neurons from human pluripotent stem cells. *The Journal of Clinical Investigation, 125*(2), 796–808.
50. Wang, L., Egli, D., & Leibel, R. L. (2016). Efficient generation of hypothalamic neurons from human pluripotent stem cells. *Current Protocols in Human Genetics, 90*, 21.

51. Merkle, F. T., Maroof, A., Wataya, T., Sasai, Y., Studer, L., Eggan, K., et al. (2015). Generation of neuropeptidergic hypothalamic neurons from human pluripotent stem cells. *Development, 142*(4), 633–643.
52. Hu, Y., Qu, Z. Y., Cao, S. Y., Li, Q., Ma, L., Krencik, R., et al. (2016). Directed differentiation of basal forebrain cholinergic neurons from human pluripotent stem cells. *Journal of Neuroscience Methods, 266*, 42–49.
53. Zheng, W., Li, Q., Zhao, C., Da, Y., Zhang, H. L., & Chen, Z. (2018). Differentiation of glial cells from hiPSCs: Potential applications in neurological diseases and cell replacement therapy. *Frontiers in Cellular Neuroscience, 12*, 239.
54. Tang, X., Kim, J., Zhou, L., Wengert, E., Zhang, L., Wu, Z., et al. (2016). KCC2 rescues functional deficits in human neurons derived from patients with Rett syndrome. *Proceedings of the National Academy of Sciences of the United States of America, 113*(3), 751–756.
55. Sherman, S. P., & Bang, A. G. (2018). High-throughput screen for compounds that modulate neurite growth of human induced pluripotent stem cell-derived neurons. *Disease Models & Mechanisms, 11*, 2.
56. Darville, H., Poulet, A., Rodet-Amsellem, F., Chatrousse, L., Pernelle, J., Boissart, C., et al. (2016). Human pluripotent stem cell-derived cortical neurons for high throughput medication screening in autism: A proof of concept study in SHANK3 haploinsufficiency syndrome. *eBioMedicine, 9*, 293–305.
57. Kumari, D., Swaroop, M., Southall, N., Huang, W., Zheng, W., & Usdin, K. (2015). High-throughput screening to identify compounds that increase fragile X mental retardation protein expression in neural stem cells differentiated from fragile X syndrome patient-derived induced pluripotent stem cells. *Stem Cells Translational Medicine, 4*(7), 800–808.
58. Lancaster, M. A., Renner, M., Martin, C. A., Wenzel, D., Bicknell, L. S., Hurles, M. E., et al. (2013). Cerebral organoids model human brain development and microcephaly. *Nature, 501*(7467), 373–379.
59. Amin, N. D., & Pasca, S. P. (2018). Building models of brain disorders with three-dimensional organoids. *Neuron, 100*(2), 389–405.
60. Qian, X., Jacob, F., Song, M. M., Nguyen, H. N., Song, H., & Ming, G. L. (2018). Generation of human brain region-specific organoids using a miniaturized spinning bioreactor. *Nature Protocols, 13*(3), 565–580.
61. Qian, X., Nguyen, H. N., Song, M. M., Hadiono, C., Ogden, S. C., Hammack, C., et al. (2016). Brain-region-specific organoids using mini-bioreactors for modeling ZIKV exposure. *Cell, 165*(5), 1238–1254.
62. Muguruma, K., Nishiyama, A., Kawakami, H., Hashimoto, K., & Sasai, Y. (2015). Self-organization of polarized cerebellar tissue in 3D culture of human pluripotent stem cells. *Cell Reports, 10*(4), 537–550.
63. Sakaguchi, H., Kadoshima, T., Soen, M., Narii, N., Ishida, Y., Ohgushi, M., et al. (2015). Generation of functional hippocampal neurons from self-organizing human embryonic stem cell-derived dorsomedial telencephalic tissue. *Nature Communications, 6*, 8896.
64. Jo, J., Xiao, Y., Sun, A. X., Cukuroglu, E., Tran, H. D., Goke, J., et al. (2016). Midbrain-like organoids from human pluripotent stem cells contain functional dopaminergic and neuromelanin-producing neurons. *Cell Stem Cell, 19*(2), 248–257.
65. Monzel, A. S., Smits, L. M., Hemmer, K., Hachi, S., Moreno, E. L., van Wuellen, T., et al. (2017). Derivation of human midbrain-specific organoids from neuroepithelial stem cells. *Stem Cell Reports, 8*(5), 1144–1154.
66. Kawada, J., Kaneda, S., Kirihara, T., Maroof, A., Levi, T., Eggan, K., et al. (2017). Generation of a motor nerve organoid with human stem cell-derived neurons. *Stem Cell Reports, 9*(5), 1441–1449.
67. Sloan, S. A., Andersen, J., Pasca, A. M., Birey, F., & Pasca, S. P. (2018). Generation and assembly of human brain region-specific three-dimensional cultures. *Nature Protocols, 13*(9), 2062–2085.

68. Bagley, J. A., Reumann, D., Bian, S., Levi-Strauss, J., & Knoblich, J. A. (2017). Fused cerebral organoids model interactions between brain regions. *Nature Methods, 14*(7), 743–751.
69. Niarchou, M., Chawner, S., Doherty, J. L., Maillard, A. M., Jacquemont, S., Chung, W. K., et al. (2019). Psychiatric disorders in children with 16p11.2 deletion and duplication. *Translational Psychiatry, 9*(1), 8.
70. Sonderby, I. E., Gustafsson, O., Doan, N. T., Hibar, D. P., Martin-Brevet, S., Abdellaoui, A., et al. (2018). Dose response of the 16p11.2 distal copy number variant on intracranial volume and basal ganglia. *Molecular Psychiatry, 25*(3), 584–602.
71. Johnstone, M., Vasistha, N. A., Barbu, M. C., Dando, O., Burr, K., Christopher, E., et al. (2019). Reversal of proliferation deficits caused by chromosome 16p13.11 microduplication through targeting NFkappaB signaling: an integrated study of patient-derived neuronal precursor cells, cerebral organoids and in vivo brain imaging. *Molecular Psychiatry, 24*(2), 294–311.
72. Kadumuri, R. V., & Janga, S. C. (2018). Epitranscriptomic code and its alterations in human disease. *Trends in Molecular Medicine, 24*(10), 886–903.
73. Rizzardi, L. F., Hickey, P. F., Rodriguez DiBlasi, V., Tryggvadottir, R., Callahan, C. M., Idrizi, A., et al. (2019). Neuronal brain-region-specific DNA methylation and chromatin accessibility are associated with neuropsychiatric trait heritability. *Nature Neuroscience, 22*(2), 307–316.
74. Rajarajan, P., Borrman, T., Liao, W., Schrode, N., Flaherty, E., Casino, C., et al. (2018). Neuron-specific signatures in the chromosomal connectome associated with schizophrenia risk. *Science, 362*, 6420.
75. Luo, C., Lancaster, M. A., Castanon, R., Nery, J. R., Knoblich, J. A., & Ecker, J. R. (2016). Cerebral organoids recapitulate epigenomic signatures of the human fetal brain. *Cell Reports, 17*(12), 3369–3384.
76. Fettes, P., Schulze, L., & Downar, J. (2017). Cortico-striatal-thalamic loop circuits of the orbitofrontal cortex: Promising therapeutic targets in psychiatric illness. *Frontiers in Systems Neuroscience, 11*, 25.
77. Birey, F., Andersen, J., Makinson, C. D., Islam, S., Wei, W., Huber, N., et al. (2017). Assembly of functionally integrated human forebrain spheroids. *Nature, 545*(7652), 54–59.
78. Mansour, A. A., Goncalves, J. T., Bloyd, C. W., Li, H., Fernandes, S., Quang, D., et al. (2018). An in vivo model of functional and vascularized human brain organoids. *Nature Biotechnology, 36*(5), 432–441.
79. Sloan, S. A., Darmanis, S., Huber, N., Khan, T. A., Birey, F., Caneda, C., et al. (2017). Human astrocyte maturation captured in 3D cerebral cortical spheroids derived from pluripotent stem cells. *Neuron, 95*(4), 779–790.
80. Marton, R. M., Miura, Y., Sloan, S. A., Li, Q., Revah, O., Levy, R. J., et al. (2019). Differentiation and maturation of oligodendrocytes in human three-dimensional neural cultures. *Nature Neuroscience, 22*(3), 484–491.
81. Salam, A. P., Borsini, A., & Zunszain, P. A. (2018). Trained innate immunity: A salient factor in the pathogenesis of neuroimmune psychiatric disorders. *Molecular Psychiatry, 23*(2), 170–176.
82. Muffat, J., Li, Y., Yuan, B., Mitalipova, M., Omer, A., Corcoran, S., et al. (2016). Efficient derivation of microglia-like cells from human pluripotent stem cells. *Nature Medicine, 22*(11), 1358–1367.
83. Ross, C. A., & Margolis, R. L. (2018). Research domain criteria: Cutting edge neuroscience or Galen's humors revisited? *Molecular Neuropsychiatry, 4*(3), 158–163.
84. Quadrato, G., Nguyen, T., Macosko, E. Z., Sherwood, J. L., Min Yang, S., Berger, D. R., et al. (2017). Cell diversity and network dynamics in photosensitive human brain organoids. *Nature, 545*(7652), 48–53.
85. Bangasser, D. A., & Valentino, R. J. (2014). Sex differences in stress-related psychiatric disorders: Neurobiological perspectives. *Frontiers in Neuroendocrinology, 35*(3), 303–319.
86. Tiwari, A., & Gonzalez, A. (2018). Biological alterations affecting risk of adult psychopathology following childhood trauma: A review of sex differences. *Clinical Psychology Review, 66*, 69–79.

Development of a 3-D Organoid System Using Human Induced Pluripotent Stem Cells to Model Idiopathic Autism

Jason W. Lunden, Madel Durens, Jonathan Nestor, Robert F. Niescier, Kevin Herold, Cheryl Brandenburg, Yu-Chih Lin, Gene J. Blatt, and Michael W. Nestor

1 Introduction

Autism spectrum condition (ASC) is a complex set of behavioral and neurological responses reflecting a likely interaction between autism susceptibility genes and the environment. Autism represents a spectrum in which heterogeneous genetic backgrounds are expressed with similar heterogeneity in the affected domains of communication, social interaction, and behavior. The impact of gene-environment interactions may also account for dsssifferences in underlying neurology and wide variation in observed behaviors. For these reasons, it has been difficult for geneticists and neuroscientists to build adequate systems to model the complex neurobiology causes of autism. In addition, the development of therapeutics for individuals with autism has been painstakingly slow, with most treatment options reduced to repurposed medications developed for other neurological diseases. Adequately developing therapeutics that are sensitive to the genetic and neurobiological diversity of individuals with autism necessitates personalized models of ASC that can capture some common pathways that reflect the neurophysiological and genetic backgrounds of varying individuals. Testing cohorts of individuals with and without autism for these potentially convergent pathways on a scalable platform for therapeutic development requires large numbers of samples from a diverse population. To date, human induced pluripotent stem cells (iPSCs) represent one of the best systems for conducting these types of assays in a clinically relevant and scalable way. The discovery of the four Yamanaka transcription factors (OCT3/4, SOX2,

J. W. Lunden · M. Durens · J. Nestor · R. F. Niescier · C. Brandenburg · Y.-C. Lin
G. J. Blatt · M. W. Nestor (✉)
Program in Neuroscience, Hussman Institute for Autism, Baltimore, MD, USA
e-mail: mnestor@hussmanautism.org

K. Herold
Program in Molecular Medicine, University of Maryland, School of Medicine, Baltimore, MD, USA

E. DiCicco-Bloom, J. H. Millonig (eds.), *Neurodevelopmental Disorders*, Advances in Neurobiology 25, https://doi.org/10.1007/978-3-030-45493-7_10

c-Myc, and KLF4) [1] allows for the induction of iPSCs from fibroblasts [2], peripheral blood mononuclear cells (PBMCs, i.e. lymphocytes and monocytes) [3, 4], or dental pulp cells [5] that retain the original genetics of the individual from which they were derived [6], making iPSCs a powerful tool to model neurophysiological conditions. iPSCs are a readily renewable cell type that can be developed on a small scale for boutique-style proof-of-principle phenotypic studies and scaled to an industrial level for drug screening and other high-content assays. This flexibility, along with the ability to represent the true genetic diversity of autism, underscores the importance of using iPSCs to model neurophysiological aspects of ASC.

iPSCs can be derived into most of the generalized categories of major cell types in the central nervous system. These include astrocytes [7–9], inhibitory neurons [10–12], and excitatory neurons [13–16]. Early neuronal work on iPSCs focused on 2-dimensional (2-D) differentiation into multiple neural cell types [17–19]. The addition of two separate antagonists to SMAD (a downstream effector to the TGFβ pathway) increased the yield of neural precursor cells (NPCs) from <10% to >80% [20]. This is commonly referred to as "dual-SMAD inhibition" and is used in the majority of cortical neuron derivation protocols. Additional protocols that have been developed allow differentiation towards an anterior or posterior fate, deep or superficial neuronal subtypes, dopaminergic, glutamatergic, pyramidal, gamma-aminobutyric acid (GABA)-ergic, parvalbumin (PV)-expressing, and somatostatin-expressing neurons [9, 11, 16, 21–27]. However, 2-D methods are unable to sufficiently mimic the complexity of 3-dimensional (3-D) neural tissues, particularly in terms of cytoarchitecture and a functional neural network that allows the study of electrophysiological properties [28, 29].

While monolayer cultures produce relatively uniform populations, multiple protocols have been developed to grow 3-D organoids that range in complexity from cortical rosettes to cerebral organoids possessing distinct forebrain, midbrain, and hindbrain characteristics (Fig. 1) [30]. Among the first 3-D neural cultures were

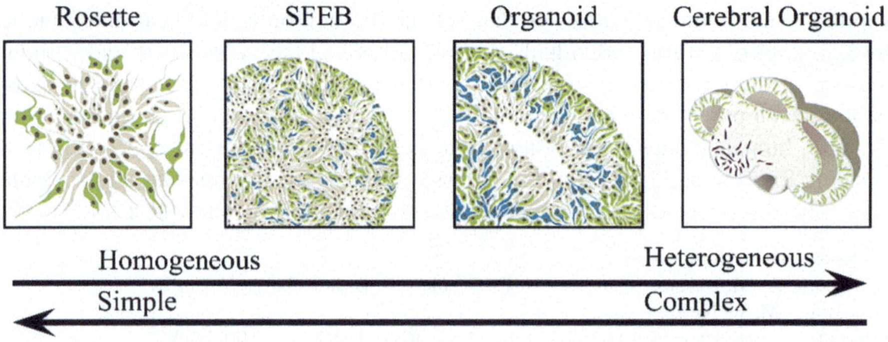

Fig. 1 iPSC culture systems range from homogeneous and simple to heterogeneous and complex. Neural rosettes typically consist of homogenous cell types in a 2-D culture that show uniform organization. Cerebral organoids consist of many cell types in a stochastically organized set of structures. SFEBs represent a good compromise between homogeneity and complexity, with relatively well controlled levels of uniformity (Adapted from [30])

serum free embryoid bodies (SFEBs) [31, 32]. When growing embryoid bodies (EBs) made from embryonic stem cells (ESCs) under serum free conditions (SF, i.e. without serum or growth factors) [33], the cells release autocrine signals that result in the generation of larger rosettes [32, 34] that retain a 3-D structure upon plating [30]. In early SFEB models, ESCs were differentiated into neural precursors by the addition of DKK1 and LEFTYA, inhibiting the WNT and Nodal pathways and driving self-renewal, ESC proliferation and differentiation [32]. During WNT/Nodal antagonization, neural precursors can also be exposed to WNT3a and SHH causing a direction towards a telencephalic fate (typically the pallial telencephalon) while being cultured in a V-bottom plate. The cellular aggregates of neural precursors in the V-bottom format stochastically self-organize into 3-D aggregations that can recapitulate structures in embryonic corticogenesis such as the rostral and caudal cortices, choroid plexus, olfactory bulb, and the retina [35, 36]. Indeed, cells in SFEBs were shown to express markers for the telencephalon, which grows into dorsal forebrain, or frontal cortex, but few or no markers for the mid and hindbrain [32].

SFEBs can be centrifuged in V-bottom plates to induce a quicker aggregation of neural precursors, iPSCs, or ESCs. By doing this while adding small molecules to the culture media, Mariani et al. [37] patterned "quick" SFEBs (SFEBq) into early dorsal telencephalon-like structures that express genes enriched in the subventricular zone and dorsal pallium. These structures are related to cortical neurogenesis, growth, and axon guidance [37]. This SFEBq system has been used to model individuals that have both autism and macrocephaly [38]. When SFEB-like systems are kept in culture for long periods of time (>90 days in vitro) they express markers for cortical lamination, deep layers, and a ventral zone including TBR1, BRN2, SATB2, and CTIP2 [29] as well as develop complex electrophysiological signatures [39]. Furthermore, SFEBs have functional neural networks that enable electrophysiological experiments to be performed [25, 35, 39, 40].

An SFEB transitions to a cerebral organoid upon acquiring discrete regional identities similar to a developing human brain, where interaction between these nascent regions is capable of influencing neural development [41, 42]. SFEBs can be induced to form neuroectoderm, followed by neuroepithelium, and subsequently embedded in Matrigel to form cerebral organoids [42]. During the first studies utilizing this method, the organoids grew beyond the limits allowed by the stationary diffusion of oxygen and nutrients, resulting in necrotic tissue in the center [41]. To compensate for this, bioreactors (small spinning containers filled with culture media) were used to enhance diffusion and ensure that the organoids remained in suspension. The gross morphology of these organoids suggests the presence of cerebral cortex, choroid plexus, retina, and meninges. Region-specific immunostaining found the presence of the markers PAX6, OTX1, and OTX2 indicating forebrain/midbrain developmental trajectories while adjacent cells expressed the hindbrain markers GBX2, KROX2,0 and PAX2, as well as others found in the hindbrain and retina [41]. This has been referred to as the "intrinsic protocol" because it relies on self-assembly without the addition of small molecules (growth factors, etc.) to artificially drive development towards a specific brain region or cell type [43]. One drawback of the intrinsic protocol and the culturing conditions for organoids is an increase in the sample-to-sample variability. Therefore a number of researchers have adopted the use of mini-bioreactors to decrease complications with quantitative analysis and increase reproducibility [43]. Taking embryoid bodies that are briefly treated with small molecules to initiate regional cell fate (i.e. ventral forebrain, hippocampus, hypothalamus, etc.) and placing these EBs in mini-bioreactors allowed in vitro development of all 6 cortical layers, and organoids expressing cells associated with hypothalamic development [43]. Furthermore, combining EBs with locked cell fate and mini-bioreactors reduced the heterogeneity that is seen when EBs develop under the intrinsic protocol in bioreactors [43]. Thus, cerebral organoids are an excellent model system to study human brain development.

2 iPSC Studies of ASC

So far, most ASC studies in human neurons have used 2-D cultures of iPSC-derived cortical neurons. Work on Rett Syndrome, a monogenetic form of ASC associated with a mutation of the MeCP2 gene, has yielded useful preclinical data. For these experiments, NPCs derived from iPSCs were grown in culture. Compared to controls, neurons derived from individuals with Rett Syndrome showed a reduced soma (cell body) size, spine density, and glutamatergic synapse number [44]. This was accompanied by a decrease in Ca^{2+} oscillation frequency, particularly in neurons expressing the synapsin protein, which implies a functional change in synaptic activity, and a decrease in the frequency and amplitude of spontaneous excitatory postsynaptic currents (EPSCs) [44]. In another monogenic iPSC line, TRPC6 mutations exhibited a 30% decrease in Ca^{2+} amplitude and a decrease in glutamate synapse number [5]. The overlap between these studies is not entirely surprising given that MeCP2 was found to act upstream of TRPC6 [5]. A recent study using monogenic iPSC lines examined an autism-related gene in schizophrenic individuals: CNTNAP2. In this study, a heterozygous proband, an unaffected heterozygous family member, and two controls were used [45]. Interestingly, there was a slight, but significant increase in CNTNAP2 mRNA in the CNTNAP2 deletion carriers as well as approximately a threefold increase in spike activity as measured by multielectrode arrays (MEA) [45]. In a study using a cohort of individuals with idiopathic autism, Derosa et al. [46] observed differential expression of genes in pathways related to GABA receptor signaling and interneuron (IN) differentiation. In addition, MEA and Ca^{2+} recordings from iPSC-derived cortical neurons revealed a decrease in spontaneous network-level spiking. Taken together these studies illustrate the utility in iPSC models for understanding the relationship between genetics and function in autism.

Some iPSC studies have utilized the neuropeptide insulin-like growth factor-1 (IGF-1), as evidence suggests that this growth factor may ameliorate some physiological deficits associated with ASC. This has led to a stage II clinical trial on IGF-1 as a treatment for ASC with results expected in 2020 [47]. Two independent MeCP2 models showed a reduction in glutamate synapses as measured by vesicular glutamate puncta. In addition, IGF-I administration restored vesicular glutamate puncta in both systems [5, 44], as well as restoring KCC2 expression in MeCP2 neurons, which resulted in a restoration of the Cl^- transmembrane gradient [48]. Thus, despite the limitations, there has been a sizable amount of work done with 2-D cultures. These phenotypes discovered using small cohorts can guide work transitioned to 3-D culture systems and scaled for drug discovery.

An example of how phenotypes in both 2-D and 3-D cultures of iPSC-derived neurons can be used to deepen our understanding of autism can be seen in a recent pair of studies. Idiopathic cohorts from ASC individuals with macrocephaly were used in either a 2-D culture or 3-D culture system [38, 49]. An analysis of cellular properties in a 2-D culture of cortical neurons found a decrease in synaptogenesis as measured by the protein synapsin 1, a decrease in vesicular glutamate transporter (VGLUT) positive puncta, decreases in GABA and GABAergic mRNA measured by an array of 84 GABA associated genes, but no change in the percentage of MAP2 positive neurons [49]. In contrast, Mariani et al., using 3-D organoids derived from four separate families measured many of the same properties and found opposite results. These include an increase in synaptogenesis as measured by the protein synapsin 1, more neurons as measured by MAP2, along with increases in GABA transcription factors, GABAergic enzymes, and vesicular GABA transporter (VGAT), but no change in VGLUT [38]. Beyond intrinsic differences in the genetics of the cohorts, the differences in these results highlight the need for more comparative studies within the same cohort between 2-D and 3-D iPSC culture systems (Table 1). Studies that employ an "apples to apples" 2-D and 3-D comparison will strengthen the argument put forth by us and others that 3-D iPSC-derived cortical neuronal cultures recapitulate more in vivo-like underlying biology with higher fidelity than 2-D models and as such are better for disease modeling and drug discovery.

Table 1 Major autism studies using either 2-D or 3-D iPSC models and the phenotypes uncovered in each case

2-D or 3-D	Phenotype	Results	IGF-I rescue	Source
2-D	CNTNAP2	↑ CNTNAP2 mRNA, ↑ neuronal spikes	Not tested	Flaherty et al. [45]
2-D	TRPC6/Rett	↓ Glu synapses, ↓ Ca^{2+} amplitude	Yes	Griesi-Oliveira et al. [5]
2-D	Rett	↓ Glu synapses, ↓ Ca^{2+} oscillation frequency	Yes	Marchetto et al. [44]
2-D	Macrocephaly	↓ synapsin 1, ↓ VGLUT, ↓ GABA, = MAP2	Yes	Marchetto et al. [49]
3-D	Macrocephaly	↑ synapsin 1, = VGLUT, ↑ GABA, ↑ MAP2	Not tested	Mariani et al. [38]
2-D	Rett	KCC2 loss prevents GABA switch from excitation to inhibition	Yes	Tang et al. [48]
2-D	Idiopathic ASC	↓ Network synaptic activity, ↓ Ca^{2+} oscillation frequency	Not tested	DeRosa et al. [46]

3 SFEBs Model Early Forebrain Characteristics

With the addition of appropriate media supplements while growing on a polytetrafluoroethylene (PFTE) cell culture insert, the protocol in this chapter allows SFEBs to acquire telencephalon/forebrain characteristics [39]. SFEBs created using these protocols express the neuronal markers NeuN, Tuj1, and MAP2, along with VGLUT, as early as days in vitro (DIV) 15 [40]. The expression of each of these markers increases between DIV 18 and DIV 30, with Tuj1 positive cells growing in an "inside out" ventricular zone motif, which resembles the development of the cortical subventricular zone [40]. Furthermore, network-wide Ca^{2+} oscillations, capable of being evoked by a tungsten microelectrode or by pharmacological intervention, and inhibited by tetrodotoxin (TTX), are observed in our SFEBs [40] (Table 2). This implies the presence of active glutamatergic synapses, as opposed to "silent" synapses potentially found in monolayer cultures that require pharmacological manipulation before activity can be measured [40, 50]. Finally whole cell patch clamp showed voltage-gated K^+ and Na^+ currents as well as action potentials demonstrating electrophysiological similarity to acute brain slices [40]. Additionally, SFEBs created using our protocol are approximately 100 μm in thickness, making them ideal for experiments that involve live-cell imaging or optogenetics.

Our SFEBs also express multiple markers associated with the caudal and medial ganglionic eminences (CGE and MGE, respectively), regions known to produce the cellular precursors of GABAergic INs. The transcription factors Dlx1/2, Dlx5, CoupTFI, and CoupTFII are known to enhance the migration of cortical INs derived from the CGE, while Nkx2.1 is a transcription factor that is known to induce GABAergic cell fate within the MGE [51, 52]. Neurons that migrate from the CGE are usually calretinin positive ($CalR^+$) [53], and $CalR^+$ neurons rarely originate in the MGE [54–56]. Nearly a third of neurons within the SFEBs created using our protocol are $CalR^+$, with 28% of $CalR^+$ neurons expressing the homeobox transcription factor Dlx5, and 37% of all cells expressing CoupTFII. Other cells expressed the IN precursor transcription factor Nkx2.1 [25]. Together these markers imply that within the GABA-positive population in SFEBs there is IN diversity reflecting the presence of cells that derive from the CGE and MGE. Furthermore, 61% of the total cells are VGLUT positive, and known to contain functional glutamatergic synapses [39]. Thus, SFEBs derived from this protocol contain many glutamatergic and GABAergic neurons. This creates an excellent environment to measure differences in the excitatory/inhibitory (E/I) balance, which has emerged as a potential hypothesis to understand some of the clinical phenotypes observed in ASC [36, 57–59].

Altered balance of excitatory and inhibitory inputs onto neurons (known as E/I imbalance) may explain some of the circuit-level changes observed in autistic brains [57, 60], and is associated with the functional impairment of INs [58, 61]. Impairment or loss of INs may contribute to disruption of coordinated activity of distant brain regions, language delays [62], and sensory processing problems frequently observed in ASC [63, 64]. Altered IN function is also linked to epileptic seizures [65], as INs are essential for preventing hyperexcitability in the brain [66], and epilepsy is com-

Table 2 Comparison of the differences between neuronal SFEB protocols (including SFEBq) with respect to major cell markers, percentages of neuronal/cell type techniques used to assay them, and major findings

Major findings	Cohorts	SFEBq	Cell types %	Cell markers	Assay techniques	Source
Spatial and temporal aspects of corticogenesis can be reproduced in vitro	Mice and human	Yes	82% Emx1, 83% Glutamate	Brn2, CamKIIa, Ctip2, Emx1, GluR1, Reelin, Tbr1, Tuj1, VGLUT	Immuno, e-phys, BrdU	Eiraku et al. [35]
Culturing with a sonic hedgehog inhibitor leads to intrinsic corticogenesis	Mice	No	10–75% Nestin, 35% VGLUT1, 35% VGAT	MAP2, Nestin, Nkx2.1, Otx1, Reelin, Tbr1, Tuj1, VGAT, VGLUT	Immuno, e-phys, RT-PCR, BrdU	Gaspard et al. [34]
Cerebral organoids can be grown via the intrinsic protocol by placing EBs in a bioreactor	Human iPSC	No	N/A	Ctip2, DCX, Foxg1, Fzd9, Nkx2.1, Pax6, Prox1, Reelin, Satb2, Sox2, Tbr1, Tbr2, Ttr, Tuj1	Immuno	Lancaster and Knoblich [41]
SFEBs can acquire telencephalon/ frontal cortex characteristics	Human iPSC	Yes	32% Tbr1, 42% Ctip2, 15% Brn2+, 10% Satb2	β3-tubulin, Ctip2, Dlx1/2, GAD-67, MAP2, N-cadherin, Nestin, PSD-95, Satb2, synapsin1, synaptophysin, Tbr1, Tbr2, Tle4	Immuno, microarray	(Mariani et al. [37]
SFEBs grown on a PFTE mesh maintain a 3-D structure while thinning, and have a neural network	Human iPSC	Yes	61% VGLUT, 47% β3-tubulin, 37% CoupTFII, 33% CalR, 27% GABA	DRAQ5, MAP2, NeuN, Nkx2.1, Pax6, Sox1, Topro-3, β3-tubulin, VGLUT, GABA	Immuno, e-phys	Nestor et al. [25, 40]
Human cortical spheroids can be generated from iPSCs	Human iPSC	No	78.8% β3-tubulin, 36.2% Neun, 7.6% GFAP	β3-tubulin, Brn2, Ctip2, EdU, GFAP, MAP2, Ncad, Neun, Pax6, Reelin, Satb2, Tbr1, Tbr2	Immuno, e-phys, Q-PCR	Pasca et al. [29]

(continued)

Table 2 (continued)

Major findings	Cohorts	SFEBq	Cell types %	Cell markers	Assay techniques	Source
EBs treated to induce cell fate can be grown in mini-bioreactors forming mini-organoids	Human iPSC	No	85% Sox2, 40% Pax6, 40% Otx2	β catenin, Ctip2, Fam107A, FoxA2, GABA, Hopx, Nestin, Nkx2.1, Nkx2.2, nNOS, NPY, OTP, Otx2, OXT, POMC, Ptprz1, PV, S100β, Sox2, SST, Tbr2, TH, Tuj1, VGLUT1	Immuno, e-phys, RNA-Seq	Qian et al. [43]
Mouse ESCs can form serum free, floating embryoid body aggregates	Mice	Yes	37% Bf1, 24% Pax6, 15% Nkx2.1	5-HT, Bf1, BrdU, E-Cad, Gsh2, Islet1, Nestin, Nkx2.1, Otx1, Pax2, Sox1, TH, Tuj1	Immuno, RT-PCR, Western Blot, FACS	Watanabe et al. [32]
Growing ES cells serum free can commit cells to a neuronal fate	Mice	No	N/A	β3-tubulin, CNPase, GABA, Nestin, Oct4, Sox1, Tau, TH, Tuj1	Immuno, RT-PCR	Ying et al. [33]

monly comorbid with ASC [65]. SFEBs created using our protocols represent a potentially useful platform for modeling E/I imbalance in a scalable system created from individuals with autism.

4 Drug Discovery: High-Throughput Assays in 3-D Culture

Multiplexing high-content assays together with 3-D culture has the potential to reduce development time and costs, and to decrease failure in drug discovery [67, 68]. Success rates for clinical trials have been declining for more than a decade, and around 90% of potential drugs that enter Phase I trials fail to reach the market [69]. Many of these studies start by screening compounds on in vitro primary cell systems that consist of isolated cell types. Human and animal cells grown in vivo have connections with multiple cell types, are surrounded by an extracellular matrix, and are continuously exposed to endocrine and neuronal signals that give cells within tissues different rates of proliferation, gene expression profiles, viability, apoptosis, and drug responses [28, 67]. 3-D tissues grown in vitro experience oxygen and nutrient gradients which are more likely to reflect in vivo physiology [67]. As a result, 2-D culture models have multiple differences in gene expression and cell behavior when compared to 3-D models, and these differences have led to clinical trials using an incorrect target, as the pathophysiological mechanism of those same cells grown in 3-D is entirely different [28, 67]. This has led many researchers to propose transitioning drug screening from 2-D to 3-D culture models, in part to decrease the number of false positive drug targets [18, 28, 36, 39, 67, 70].

High-content screening allows experiments to be done on multiple 3-D SFEBs, producing large amounts of useful data. The idea of extracting as much information as possible from each cell in an image, including temporal and spatial dynamics of cell constituents and processes led to the idea of high-content image screening [71]. High-throughput technology such as ThermoFisher's ArrayScan platform (ThermoFisher Scientific Waltham, MA) allows multiple 96-well plates, with each well containing an SFEB, to be scanned with a confocal microscope. Those images can be analyzed utilizing algorithms that search for neurite outgrowth, cell proliferation, apoptosis, and multiple other parameters. Fundamentally, the software can search for colocalization up to seven spectral channels, enabling the user to create additional programs to screen for cell dynamics that can be defined by localization of any combination of the available channels, such as heterogeneous combinations of transcription factors, receptor dimers or trimers.

One known limitation of this process is that a high cellular density prevents confocal imaging beyond 50 and 100 μm into spheroids and organoids, thus preventing the acquisition of cellular phenotypes deeper within the tissue [67, 70]. The scattering of light is caused by large differences in the refractive index in components within the cell membrane [72], which results in either incomplete data sets or false negative data points, limiting high-throughput microscopy [73]. To overcome this challenge, methods have been developed that homogenize tissue components to better match the refractive index, or by removing light scattering components of the tissue culture system [70, 74]. Despite this progress, direct application of these protocols to high-throughput methods has been limited [67, 70, 74]. We have developed a protocol to transfer SFEBs onto a Millipore 0.4 μm PFTE cell culture insert after 14 DIV, which thins SFEBs to ~100 μm by 30 DIV [40]. This is important because typical fluorescence from green fluorescent protein begins to scatter at approximately 75–100 μm in biological tissue due to Rayleigh scattering [75]. This scatter typically restricts imaging to the upper layers of organoids and makes advanced physiological techniques such as microphotolysis or optogenetics extremely difficult. In addition, we have adapted a protocol from Visikol Inc. to further improve light penetration for SFEB imaging.

Beyond immunostaining, our lab was the first to develop SFEB protocols for high-content screening by using MEA plates and Ca^{2+} imaging to assess real time reactions to pharmacological compounds. These approaches are particularly useful for dynamic processes. For instance, multiple biological events can be measured simultaneously, or the same event can be studied through multiple different assays. For example, an MEA plate can provide data on network burst activity, while Ca^{2+} imaging provides data on the activity of specific neurons within the plate [39, 46]. Subsequent immunohistochemistry or RNAseq can provide data on proteins or expression of the previously studied cells. The synergy between these methods increases the power and decreases the probability of false positive or negative results [68].

SFEBs represent a good balance between 2-D and 3-D. They are scalable and complex for high-throughput pharmacology. Although the iPSC field is in early development, the potential to model aspects of ASC using human neurons is promising [36]. The ability to develop human cortical tissue in vitro presents scientists with multiple new options to design experiments that integrate results from both rodent and clinical studies, which will result in greater clinical translatability.

5 Results

SFEBs derived from this protocol have telencephalon/forebrain characteristics [39] with active glutamatergic and GABAergic synapses. They express multiple neuronal markers including NeuN, Tuj1, and MAP2, with cells growing in an "inside out" ventricular zone motif, which parallels the cortical subventricular zone [40]. Furthermore, we have used these SFEBs for high-throughput experimental techniques including ArrayScan imaging, neuronal network activity, whole cell patch clamp, RNAseq, and Ca^{2+} imaging. By generating one or more 96-well plates per individual, these techniques have the statistical power to detect and quantify even small differences (Fig. 2).

6 Structure/Staining

As discussed earlier, SFEBs created using this protocol yield a robust population of GABAergic and glutamatergic neurons (~30% GABA; ~50% VGLUT, ~10 GFAP, ~10% neural progenitors/non-neuronal cells) [25, 39, 40]. SFEBs organize into an inner and outer zone that resembles a developing ventralized cortical plate (Fig. 3). MAP2 staining labels a large population of distinct neurons in all zones of the SFEB. In general, at DIV 60 these cells exhibit several primary branches from the cell body and limited secondary branch formation, giving them a qualitative appearance similar to that of DIV3-5 cultured mouse neurons. MAP2 staining is preserved in SFEBs cultured for 1 year on PFTE inserts but the morphological features of the outer zones are lost, which may be due to a technical limitation of the protocol. Robust MAP2 staining can be observed at the core of the SFEB, revealing the lack of a necrotic zone in this system. SFEBs created by this method have a generally uniform appearance and an outer edge where dendritic and axonal-like structures circle with embedded cell somas. SFEBs created using this protocol also robustly express the IN marker parvalbumin, which can be studied in detail by using tissue clearing techniques adapted from protocols developed by Visikol, Inc. This type of clearing combined with immunocytochemistry can reveal single axons and dendrites expressing parvalbumin in SFEBs (Fig. 3).

High-content approaches can be applied to SFEBs to study morphometric aspects of this system and to uncover potential phenotypes in SFEBs from individuals with

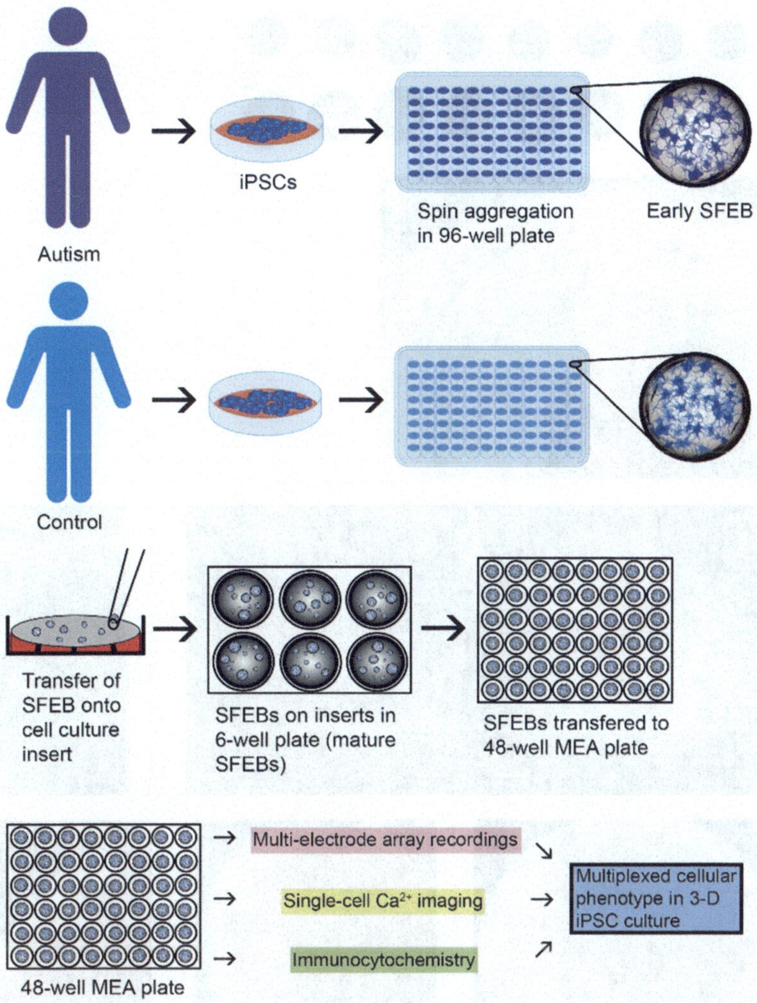

Fig. 2 Schematic illustrating the overall processing of iPSC into SFEBs in our system. iPSCs are created from individuals with and without autism and aggregated into 96-well, V-bottom plates. After aggregation and differentiation for 14 days, SFEBs are transferred onto Millipore PFTE cell culture inserts in 6-well plates and cultured for various days in vitro depending on the experimental question. After culturing, SFEBs are transferred to imaging plates or MEA plates and processed via various assays to build up a complex phenotype in 3-D culture

autism. An example of a high-content assay measuring VGLUT counts within SFEBs created from a control male individual without autism and a male with autism who shows low adaptive functioning, developmental delay and a normocephalic head circumference while harboring a mutation in RBFOX1, a candidate autism gene, can be seen in Fig. 4 [46]. There is a significant decrease in the number of VGLUT positive cells in the individual with autism, suggesting some potential

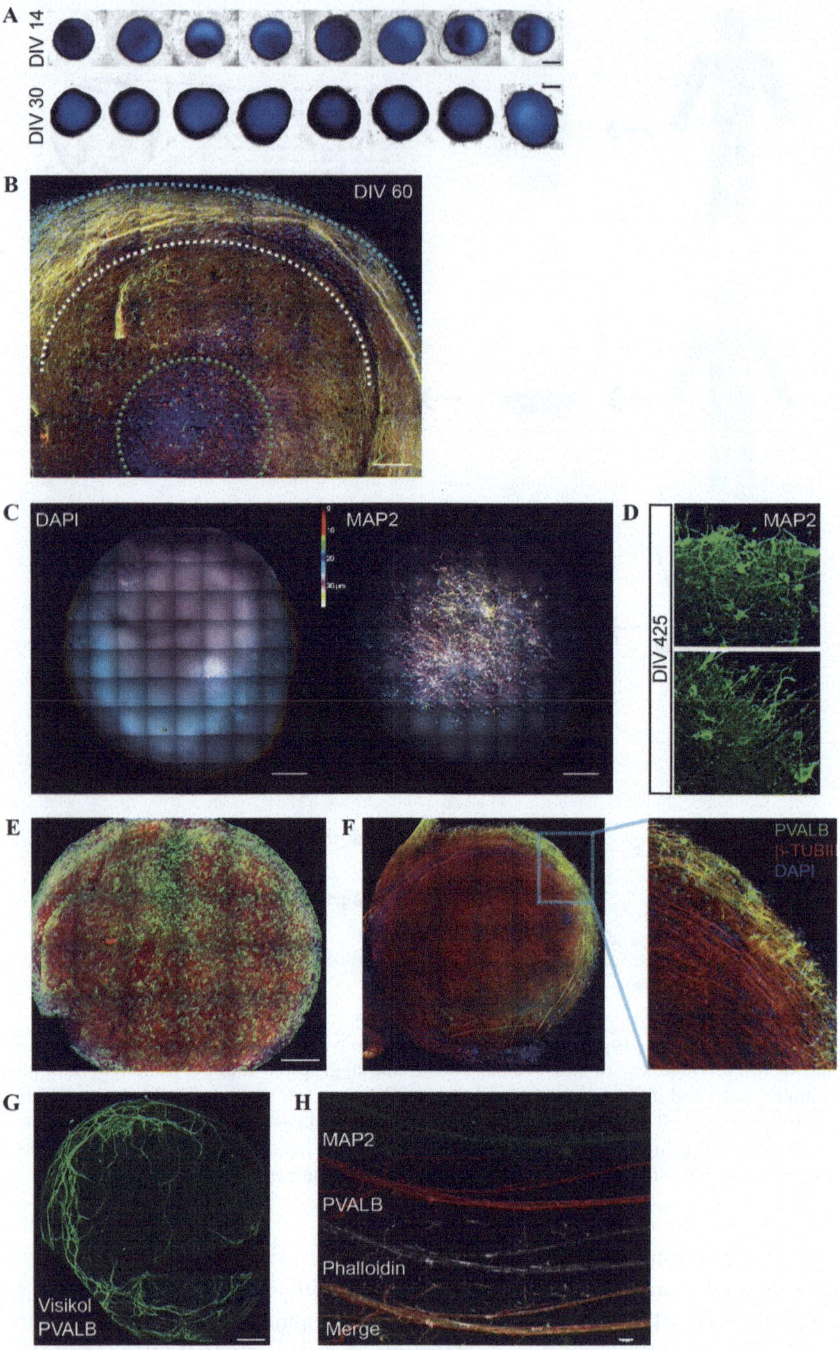

Fig. 3 (**a**) SFEBs stained for Hoechst 33342 and followed at DIV 14 (before transfer to cell culture insert), and 30. After transfer to insert, SFEBs increase in width from ~ 200 μm to ~800 μm (s.b. = 75 μm). (**b**) SFEBs matured to DIV 60 develop an inner core (green dashes) and an outer edge

early developmental shift in E/I balance. More cases and controls would need to be added to this assay before any conclusions about phenotypes can be asserted, but this example illustrates the usefulness of high-content morphometric assays in 3-D systems in developing cell-level and phenotypic-based hypotheses to test using autism disease models.

7 MEA

SFEBs display spontaneous local field potentials as assessed by MEA [39]. Inherent differences in spontaneous activity, bursting, and network activity between control and patient-derived SFEBs can also be examined. In addition, excitatory and inhibitory connections in SFEBs and their overall effect on activity can be tested by the addition of pharmacological inhibitors. We demonstrate that activity is decreased by the addition of (2R)-amino-5-phosphonovaleric acid (APV), an NMDA receptor antagonist [76], and cyanquixaline (CNQX), which is an AMPA receptor antagonist [77]. In contrast, inhibition can be examined by addition of picrotoxin (PTX) which blocks GABA receptors [78]. These initial studies indicate expression of these receptors and can be used as stepping stones to study synaptic plasticity and excitatory/inhibitory circuits in SFEBs.

Genetic and animal studies implicate synaptic dysfunction as a risk factor for many neurodevelopmental disorders including autism. Abnormalities in synaptic plasticity have been identified in animal models of Rett syndrome, Angelman syndrome, Tuberous sclerosis, and Shank3 disorders [79]. One form of synaptic plasticity is long-term potentiation (LTP), the prolonged strengthening of synaptic transmission in response to an increase in synaptic activity, which is considered to be one of the mechanisms underlying learning and memory [80]. Deficits in LTP can play a role in intellectual disability in neurodevelopmental disorders [79]. LTP produces increased signal transmission between two neurons that lasts for hours in vitro, and that can persist for periods of weeks or months in vivo (see [81] for a review). One form of LTP is dependent on the activation of NMDA receptors

Fig. 3 (continued) (white and blue dashes) that resembles a migratory zone approximating a developing cortical plate. Note that transfer to the culture inserts allows exchange of oxygen and media preventing the development of a necrotic core, common to other organoid models (red = MAP2, green = GABA, blue = DAPI; s.b.= 200 μm). (**c**) DAPI and MAP2 of SFEB on insert at DIV 60 staining show robust MAP2 expression throughout the core of the SFEB, indicating no necrotic core (s.b.= 500 μm). (**d**) SFEBs grown on PFTE culture inserts for 425 days show MAP2 staining at the outer edge but have lost the outer pattern observed at DIV 30–120 and take on a ragged edge morphology (s.b. = 25 μm). (**e**) SFEBs at DIV 60 show robust staining for both VGLUT (green) and GABA (red) as well as DAPI (blue) (s.b.= 200 μm). (**f**) Parvalbumin is expressed in the outer layers of SFEBs grown to DIV 60 and 120 on culture inserts (s.b. = 200 μm; inset s.b. = 50 μm). (**g**) Visikol protocol clears SFEBs to a greater extent than traditional fixation and staining methods. (**h**) Visikol staining can be used to isolate individual dendrites and axons within SFEBs (s.b. = 200 μm; inset s.b.= 10 μm)

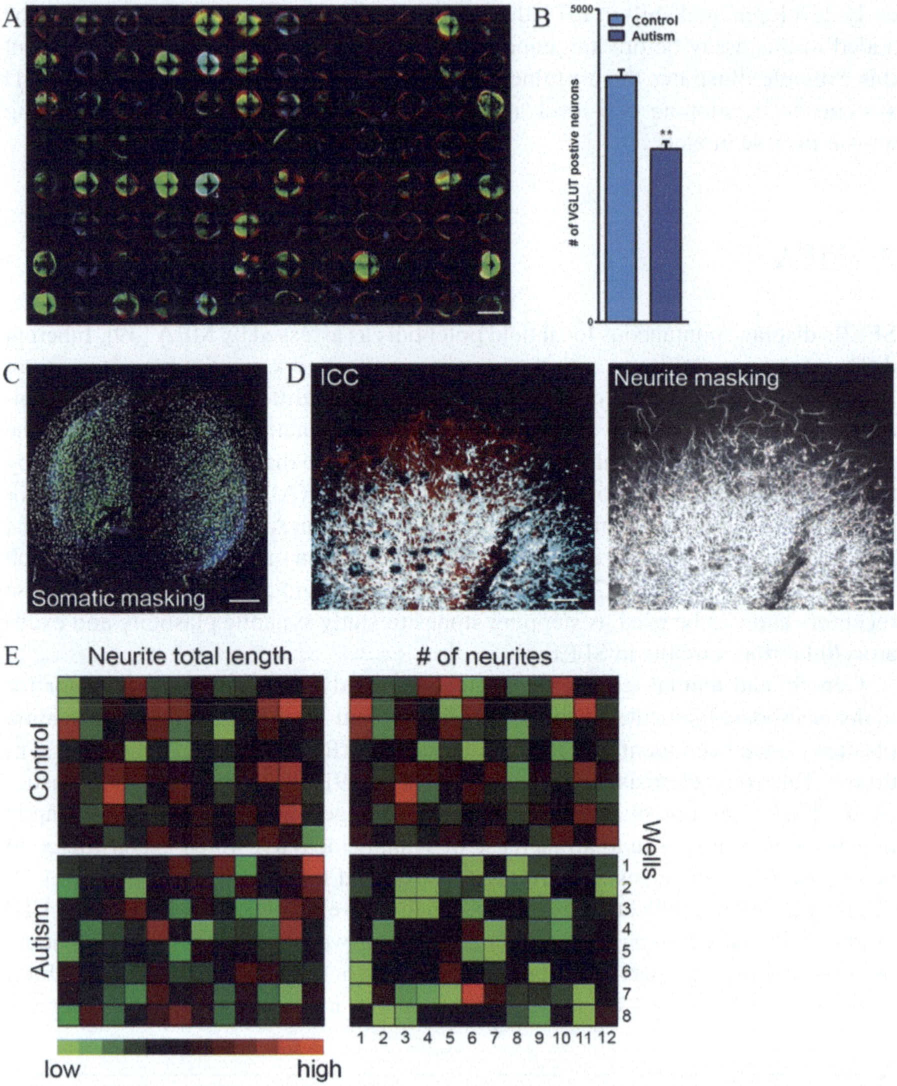

Fig. 4 (**a**) High-throughput montage of a single 96-well plate of SFEBs at DIV 120 (green = GABA; red = VGLUT; blue = hoechst 33,342; s.b.= 800 μm). High-content imaging can reveal well-to-well variability within and between individuals. (**b**) Example of how high-content imaging can reveal phenotypic differences. In this example of one individual with autism and one without, the number of VGLUT-expressing neurons within SFEBs is significantly decreased ($N = 96$ SFEBs/individual; $p < 0.001$; t-test). This difference may play a role in potentially driving differences in E/I balance-hypothesized to underlie some aspects of autism. (**c, d**) High-content imaging using systems such as the ArrayScan are used to mask both somatic and dendritic/axonal aspects of neurons within SFEBs in order to generate multiplexed data sets (s.b. = 200 μm). (**e**) High-throughput imaging provides large sample sizes, adding confidence to experimental effects. In this example, heat maps can be generated across wells and plates for morphometric measurements. The same individuals in panel (**b**) are examined for differences in total length of neurites per well and the number of neurites per neuron per well

(NMDAR), leading to Ca^{2+} influx and increased trafficking of AMPA receptors (AMPARs) to the postsynaptic membrane [80]. NMDAR-dependent LTP can be induced in neuronal cultures electrically via high-frequency stimulation or chemically (chemLTP) using a wide variety of compounds that increase NMDAR activation or levels of downstream effectors such as cAMP [82]. For example, the addition of glycine to cultured neurons has been shown to induce LTP [83]. Glycine acts on the NMDAR subunit GluN2A, producing AMPAR-mediated currents independently of NMDAR channel activation or glycine receptor activity [84]. Following attachment to a 48-well MEA plate, glycine induced nearly a twofold increase in SFEB firing rates ($p = 0.008$) (Fig. 5). Similar experiments can elucidate differences in LTP activation between ASC-derived cells and controls and can provide valuable insight into changes in synaptic plasticity in ASC. In addition, gene-editing techniques such as CRISPRi can be combined with the production of SFEBs to knockdown genes of interest and can be combined with electrophysiological assays.

8 Calcium Signaling

Obtaining MEA recordings is an immensely useful technique that provides a vast amount of information about the activity of neuronal populations with high temporal resolution. However, it cannot determine the activity, location, and cell type of individual neurons [85]. To supplement data obtained from MEA, calcium imaging can be used to visualize and quantify activity in individual cells. Neural activity is accompanied by rapid changes in the levels of intracellular Ca^{2+}. During action potential firing, Ca^{2+} enters the neuron and can be detected by calcium indicators. An example of this can be seen in the transfection of the genetically encoded GCamp6 protein [86] which can be transduced in SFEBs using AAV viruses. To assess firing in individual neurons within SFEBs, we infected the SFEBs with the calcium indicator pAAV9-CAG-GCamp6s-WPRE-SV40 (MOI = 500 K; Penn Vector Core; Addgene 100844-AAV9) 5–7 days after transferring to the MEA plate. SFEBs were incubated with the virus for 4 days, followed by a complete media change. The cells were imaged using the ThermoFisher ArrayScan (HCI) at 5× magnification. To facilitate analysis, cells were also treated with 100 ng/ml Hoechst 33342 to further distinguish cell somas and were imaged once per field. The sampling rate for each field was approximately 0.8 Hz. Analysis of Ca^{2+}-transients was performed with ArrayScan software, using the Hoechst images as a mask to assess spikes that are restricted to the cell body.

Using this technique, Ca^{2+} transients are visible in SFEBs 5–7 days after transduction. Calcium imaging can be used to further study local circuitry surrounding electrodes in an MEA plate and can complement extracellular recordings obtained using MEAs (Fig. 6).

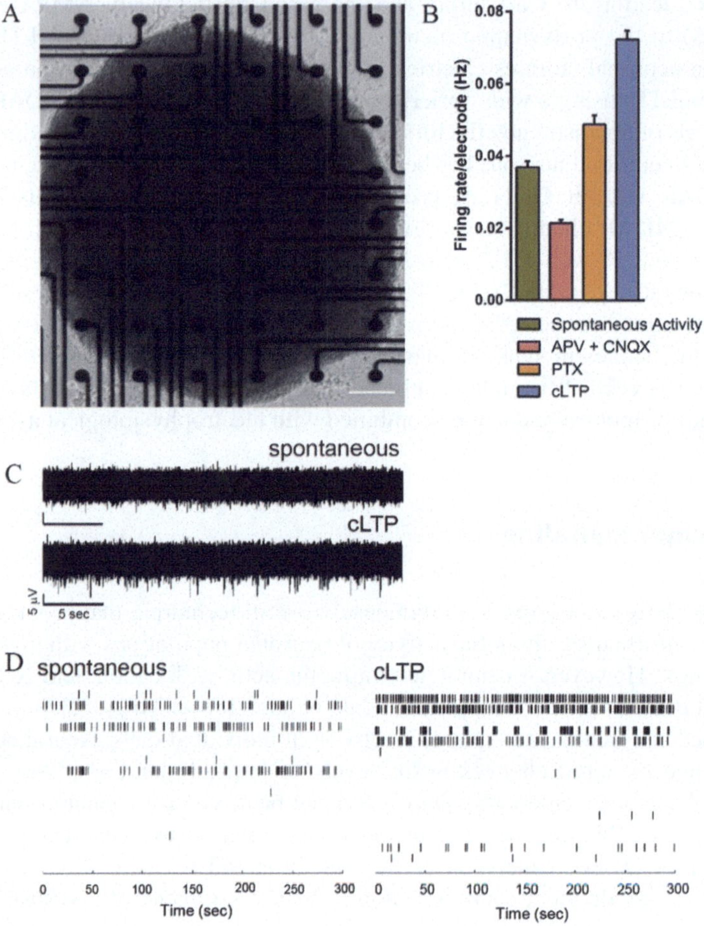

Fig. 5 (**a**) SFEB on MEA plate ready for recording. (**b**) Extracellular recordings obtained via MEA reveal changes in spike frequency in SFEBs in response to NMDAR/AMPAR blockade (50 μM APV + 25 μM CNQX) and GABA blockade (50 μM PTX). (**c**, **d**) Representative raw spikes and raster plots, respectively, demonstrate that SFEBs exhibit robust induction of long-term potentiation (LTP) with addition of 200 μm glycine ($N = 43$ SFEBs/three experiments)

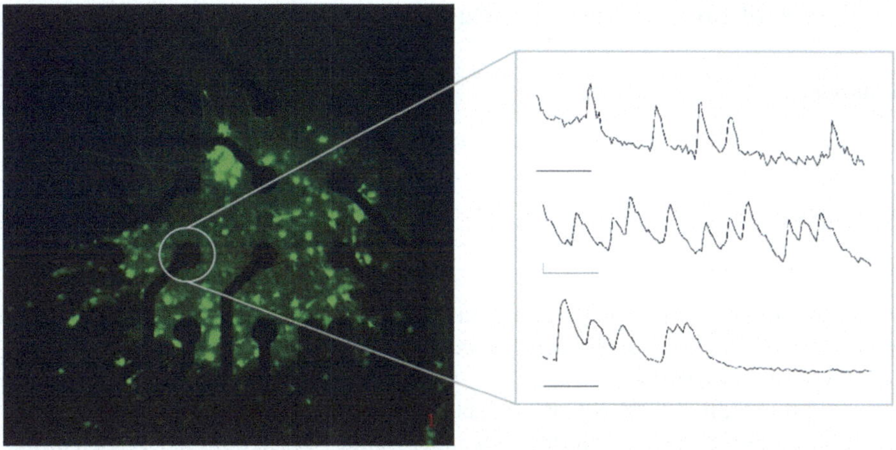

Fig. 6 Calcium imaging of SFEBs on MEAs using ThermoFisher ArrayScan (HCI) at 5× magnification. SFEBs were infected with AAV9.CAG.GCamp6s.WPRE.SV40 (MOI = 500 K; Penn Vector Core) and imaged after 7 days. Inset shows representative traces from individual neurons around MEA electrode (*N* = 32 SFEBs/two experiments)

Table 3 Drugs used in this study

	Manufacturer; catalog #	Final concentration (μM)	Stock concentration
APV	Tocris; Cat. No. 0106	50	100 m Min H_20
Picrotoxin	Tocris; Cat. No. 1128	50	100 m Min DMSO
CNQX	Tocris; Cat. No. 0190	25	100 m Min DMSO
Glycine	Sigma; PHR1111	200	500 mM in H_2O

Table 4 Antibodies used in this study

Name	Manufacturer	Host	Catalog number	Lot number	Concentration
Pan-albumin	Millipore	Mouse	MABI572	3041808	1:500
MAP2 (D5G1)	Cell Signaling Technology	Rabbit	8707S	1	1:200
β-Tubulin III	Sigma-Aldrich	Mouse	18578-100ul	054M4819V	1:200
PSD95	Synaptic Systems	Guinea pig	124014	N/A	1:500
Alexa fluor secondary antibodies	ThermoFisher	Donkey	N/A	N/A	1:200

9 Generation and Neural Induction of SFEBs from iPSCs

Equipment:

1. Countess™ II FL automated cell counter (ThermoFisher Scientific; CAT# AMQAF1000)

Materials:

1. SFEB generation
   a. EmbryoMax® 0.1% Gelatin Solution (Millipore Sigma; CAT# ES-006-B)
   b. StemPro™ Accutase™ cell dissociation reagent (ThermoFisher Scientific; CAT# A1110501)
   c. 35 mm culture dish (StemCell Technologies; CAT# 27100)
   d. Countess™ cell counting chamber slides (ThermoFisher Scientific; CAT#C10228)
   e. V-bottom 96 well plate (Evergreen Scientific; CAT# 222-8031-01V)
2. SFEB development
   a. Millipore 0.4 µm PFTE cell culture insert (Millipore Sigma; CAT# PICM0RG50)
   b. Corning™ Costar™ flat bottom cell culture plates (ThermoFisher Scientific; CAT#07-200-83)
   c. Wide-bore pipette tips, 200 µl (ThermoFisher Scientific; CAT# 14-222-730)
3. Medias (some medias adapted from [4]; for concentrations see Table 5)
   a. *mTeSR media*
      - mTeSR™1 (StemCell Technologies; CAT# 85850)
      - Human recombinant fibroblast growth factor-basic (FGF) (Stemgent/Reprocell; CAT# 03-0002)
      - CHIR99021 (Stemgent/Reprocell; CAT# 04-0004-10)
      - Thiazovivin (Stemgent/Reprocell; CAT# 04-0017)
      - PD0325901 (Stemgent/Reprocell; CAT# 04-0006)
      - Y27632 (ROCKi) (Stemgent/Reprocell; CAT# 04-0012-10)

   b. *Neural induction media (NIM)*
      - STEMdiff™ neural induction medium (StemCell Technologies; CAT# 05835)
      - FGF (Stemgent/Reprocell; CAT# 03-0002)
      - SB431542 (Stemgent/Reprocell;; CAT# 04-0010-10)
      - Thiazovivin (Stemgent/Reprocell; CAT# 04-0017)
      - Dorsomorphin (Stemgent/Reprocell; CAT# 04-0024)
      - Y27632 (ROCKi) (Stemgent/Reprocell; CAT# 04-0012-10)

   c. *Neural formation media (NF)*
      - Laminin (ThermoFisher Scientific; CAT# 23017015)
      - Forskolin (Sigma; CAT# F-6886)

- Recombinant brain derived neurotrophic factor (BDNF) (Peprotech; CAT# 450-02)
- Recombinant human neurotrophic factor-3 (NT3) (Peprotech; CAT# 450-03)
- Recombinant human beta-nerve growth factor (β-NGF) (Peprotech; CAT# 450-01)
- *N*-[*N*-(3,5-Difluorophenacetyl)-L-alanyl]-*S*-phenylglycine t-butyl ester (DAPT) (Stemgent/Reprocell; CAT# 04-0041)
- Heparin (Sigma; CAT# H3149)
- Crystalline bovine serum albumin (BSA) (Sigma; CAT# A-4161)
- Putrescine (Sigma; CAT# P7505)
- Progesterone (Sigma; CAT# P8783)
- Triiodothyronine (T3) (Sigma; CAT# T6397)
- Thyroxine (T4)(Sigma; CAT# T1775)
- Dulbecco's modified eagle medium/nutrient mixture F-12 plus L-glutamine (DMEM/F12) (ThermoFisher Scientific; CAT# 11320-033)
- Neurobasal media minus phenol red (ThermoFisher Scientific; CAT# 12348-017)
- 100× N2 supplement (ThermoFisher Scientific; CAT# 17502-048)
- B27 (50×) supplement minus vitamin-A (ThermoFisher Scientific; CAT# 17504-044)
- MEM non-essential amino acids (NEAA) (ThermoFisher Scientific; CAT# 11140-050)
- Insulin-transferrin-selenium-A (ThermoFisher Scientific; CAT# 51300-044)
- Penicillin-streptomycin (Pen/Strep) (ThermoFisher Scientific; CAT# 15140-122)
- N-acetyl cysteine (Sigma; CAT# A8199)

d. *Terminal differentiation media (TD)*

- Dulbecco's modified eagle medium/nutrient mixture F-12 plus L-glutamine (DMEM/F12) (ThermoFisher Scientific; CAT# 11320-033)
- Neurobasal media minus phenol red (ThermoFisher Scientific; CAT# 12348-017)
- 100× N2 supplement (ThermoFisher Scientific; CAT# 17502-048)
- B27 (50×) supplement minus vitamin-A (ThermoFisher Scientific; CAT# 17504-044)
- MEM non-essential amino acids (NEAA) (ThermoFisher Scientific; CAT# 11140-050)
- Insulin-transferrin-selenium-A (ThermoFisher Scientific; CAT# 51300-044)
- Penicillin-streptomycin (Pen/Strep) (ThermoFisher Scientific; CAT# 15140-122)
- N-acetyl cysteine (Sigma; CAT# A8199)
- Laminin (ThermoFisher Scientific; CAT# 23017015)
- Forskolin (Sigma; CAT# F-6886)

- Recombinant brain derived neurotrophic factor (BDNF) (Peprotech; CAT# 450-02)
- Recombinant human neurotrophic factor-3 (NT3) (Peprotech; CAT# 450-03)
- N-[N-(3,5-Difluorophenacetyl)-L-alanyl]-S-phenylglycine t-butyl ester (DAPT) (Stemgent/Reprocell; CAT# 04-0041)
- Heparin (Sigma; CAT# H3149)

Methods:

1. Generation of SFEBs from iPSCs

iPSCs are created on mouse embryonic fibroblasts (MEFs) (see: [40, 87]). In this protocol 96 well plates are filled with 20,000 cells per well, requiring a minimum of 1.92×10^6 healthy and undifferentiated iPSCs per cell line. Taking two confluent wells of an iPSC line, approximately 20–50 iPSC colonies (per well) must be cut into small chunks, and subsequently dissociated with Accutase. These iPSCs grow into SFEBs within 2 weeks of plating under this protocol.

a. Prepare a gelatinized 10 cm cell culture dish for MEF depletion. Add 7 ml of gelatin to the culture dish. This will be the MEF-depletion plate. Incubate for 1 h in 37 °C incubator under standard conditions.
b. Start with expanded clean non-differentiated colonies. Obtain an iPSC plate. On a 37 °C heat plate under a microscope, select well-formed undifferentiated colonies. To ensure a cell count of at least 1.92×10^6 pick from two wells of each line (recommended), or possibly one well of a very confluent iPSC layer.
c. Transfer the media with the selected colonies from the wells into one 15 ml conical tube per line, and spin at 800 rpm for 4 min. Immediately replace mTeSR+ ROCKi media in the selected well while cell picking from the same wells is repeated.

   Note: Optionally, conditioned mTeSR+ media can be prepared at this point. Use a filtered syringe to remove sloughed off cells, but retain beneficial growth factors that can be reused. A 50:50 ratio of conditioned to new mTeSR+ media will create a preferred environment for iPSCs.
d. Aspirate the used mTeSR+ media (or use it to generate conditioned mTeSR+) from 15 ml conical (Note: do not remove the cell pellet via suction).
e. Add 1 ml of Accutase (the accutase should be pre warmed to 37 °C). Carefully triturate the cells by pipetting up and down 3–5 times. Not triturating enough will prevent the iPSCs from breaking up, while triturating too hard may decrease cell survival. Transfer the 15 ml conical with iPSCs to the incubator for ~5 min. After the 5 min incubation add 2 ml of mTeSR+ ROCKi media to de-activate the Accutase and spin down at 800 rpm for 4 min.
f. Gently aspirate the accutase/mTeSR+ mix and keep the pellet.
g. Re-suspend the iPSC pellet in 1 ml mTeSR+ ROCKi media. Carefully triturate the cells by pipetting up and down 3–5 times. Add another 6 ml of mTeSR+ ROCKi media.

h. Take the MEF-depletion plate out of the incubator and aspirate the remaining gelatin fluid. Transfer the 7 ml of mTeSR+ ROCKi media with the iPSCs to the MEF-depletion plate and incubate at 37 °C for 45 min.
i. Following the incubation bring the MEF-depletion plate into a laminar flow hood. Carefully rinse the MEF-depletion plate with the plate media to capture all the individual iPSCs. The gelatin should retain large chucks of iPSCs and MEFs. Do not rinse too aggressively or MEFs and remaining iPSCs may contaminate the separated and selected iPSCs. These are now MEF-depleted iPSCs.
j. Transfer the MEF-depleted iPSCs to a 15 ml conical tube and spin down at 900 rpm for 3 min.
k. Remove the supernatant and re-suspend in 1 ml of NIM+ ROCKi to dissolve individual cells for counting.
l. Take a 10 μl aliquot of cell suspension for cell counting. Use the cell counting instrument (or hemocytometer) according to manufacturer's directions. Using an appropriate volume of NIM+ ROCKi media, dilute the remaining cells such that 100 μl will contain 20,000 cells.
m. Add 100 μl of the iPSC suspension to each well of a V-bottom 96-well plate. This will be easier with a multichannel pipette. These iPSC aggregates will form SFEBs.
n. After the appropriate number of wells are plated, spin the plate at 900 rpm for 3 min.
o. Next day feed all wells by removing 75% of the media and replacing it with 75 μl of NIM+ without ROCKi. Be careful not to pipette the SFEB. For the next 5 days perform a 50% feed (remove 50% of the media and add 50% NIM+). Each subsequent feed should dilute the concentration of ROCKi in the media. On the 6th day it is acceptable to perform 50% feeds every other day. By day 14, SFEBs should be in neural formation (NF) media on meshes.

Note: Start step 2 only after the MEF-depletion plate has been incubated for 45–50 min. The gelatin must be solidified before MEF-depletion steps in this protocol.

2. SFEB neural induction

Following 14 days of NIM treatment, the SFEBs can be transferred to mesh inserts. A Millipore 0.4 μm PFTE cell culture insert will allow the nutrients to diffuse from the bottom while the SFEB remains dry. The use of neural formation (NF) and terminal differentiation (TD) media will induce the SFEB to grow neural progenitors and neurons.

a. Place Millipore cell culture inserts into the desired wells of a 6-well plate and add 1.5 ml NF+ ROCKi media at least 1 h prior to the addition of SFEBs.
b. Get a 96-well plate with SFEBs grown in NIM media for 14 days. Using a 200 μl wide-bore pipette tip transfer the SFEB with 20 μl medium onto the insert.

Note: Millipore cell culture inserts have specialized structures that sit slightly off the bottom of the well and consist of a PFTE membrane attached to a plastic frame. This membrane is biocompatible and can efficiently sustain nutrient and oxygen transport to the SFEBs, which are placed on top. They are commonly used with organotypic hippocampal slice cultures created from mouse or rat.

c. A single insert can accommodate 8–12 SFEBs. Arrange the SFEBs carefully to ensure adequate space between them. Remove as much excess solution as possible from around each SFEB using a fine pipette tip (e.g. 200 µl tip).

   Note: It is acceptable to briefly disturb the SFEB as they will move on the cell culture insert as the solution is removed from around the SFEBs.

d. Maintain the cultures in 1 ml of NF+ media at 37 °C, 95% humidity, and 5% CO_2. Feeding will be performed every other day by removing 50% of the medium and replacing with three quarters fresh NF+ medium.

e. After 14 days of culture, switch to TD+ medium with a 50% medium change every other day. SFEBs can be maintained in this manner for more than 150 days; we have kept these cultures for up to 1 year. The SFEBs will grow to about 1000 µm in diameter and are typically 100–150 µm thick (Fig. 3).

10 SFEB Experiments

Depending on the desired outcome, the protocol will diverge at this point. If the goal is strictly to perform immunohistochemistry, this is an ideal point to fix the SFEBs with paraformaldehyde (PFA). Following fixation, SFEBs will be less likely to be damaged during removal from the insert. If the goal is to perform electrophysiology or Ca^{2+} imaging, the SFEBs must be carefully removed from the insert without PFA fixation. Following removal, the SFEBs must be attached to an MEA or comparable plate where they can recover for at least 7 days prior to subsequent experiments.

Different strategies must be used to dislodge the SFEBs from the insert depending on the length of time they have been growing. Before 30 DIV, it is possible to detach the SFEBs from the mesh by gently pressure ejecting media from a pipette. After a certain time, a cell scraper will be needed to minimize damage to the base of the SFEB where it attaches to the insert.

Materials:

1. All protocols

   a. 96-well U bottom glass plate (Thomas Scientific CAT# 6902-D09)
   b. Fisherbrand™ cell lifters (ThermoFisher Scientific CAT# 08-100-240)
   c. Wide-bore pipette tips, 200 µl (ThermoFisher Scientific; CAT# 14-222-730)
   d. Image-iT™ fixative solution (ThermoFisher Scientific CAT# FB002)
   e. Phosphate-buffered saline (PBS) (Sigma-Aldrich D8537-500ml)
   f. Triton X-100 (Sigma CAT# X100-100ML)
   g. Normal donkey serum (Jackson ImmunoResearch CAT# 017-000-121)
   h. PBS with 0.1% Triton X-100
   i. Blocking buffer: (PBS with 0.2% Triton X-100, 10% donkey serum)

Method:

1. SFEB Processing for MEA plates
   a. Detach the SFEBs from the insert by gently pressure ejecting the media using a 200 μl wide-bore pipette tip (Day 1–30) or use a cell lifter to detach the SFEB (after Day 30).
      Note: To ensure that the cell lifter does not damage neighboring SFEBs, the cell lifter must be carefully cut to the length of approximately the largest SFEB.
   b. When the SFEB is detached from the insert, use a wide-bore pipette tip to gently transfer the SFEB to a prepared MEA plate (see MEA protocol).
2. SFEB Immunohistochemistry
   a. Detach the SFEBs from the insert by gently pipetting the media using a 200 μl wide-bore pipette tip (Day 1–30) or use a cell lifter to detach the SFEB (after day 30).
      Note: To ensure that the cell lifter does not damage neighboring SFEBs, the cell lifter must be carefully cut to the length of approximately the largest SFEB.
   b. When the SFEB is detached from the insert, use a wide-bore pipette tip to gently transfer the SFEB to a 96-well U bottom glass plate and wash with 50 μl PBS per well.
      Note: SFEBs will be suspended in solution in the 96-well plates. This is important to allow full penetration of the fixative, the blocking solution, and antibodies.
   c. Fix SFEBs by adding 50 μl of 4% paraformaldehyde per well at room temperature (RT) for 30–45 min. Wash fixed SFEBs twice with 100 μl PBS, 3–5 min each wash.
      Caution: Wear appropriate personal protective equipment when handling PFA.
   d. Permeabilize and block with 0.1% Triton-X in PBS and 10% normal donkey serum (blocking solution; 50 μl/well) for 1 h at room temperature.
   e. Remove the blocking solution and treat SFEBs with 50 μl primary antibodies in blocking solution. Incubate at 4 °C overnight.
      Note: When storing the 96 well plate, wrap it in Parafilm to prevent the evaporation of solutions.
   f. Wash SFEBs three times (3–5 min/wash) in 100 μl PBS containing 0.1% Triton-X.
   g. Prepare secondary antibodies 1:1000 in blocking solution. Incubate SFEBs with 50 μl secondary antibodies for 1 h at room temperature. Wash SFEBs with 100 μl PBS, 3 times for 3–5 min each.
   h. Store 96-well plates with SFEBs at 4 °C in PBS until they are ready for imaging on the ArrayScan.

11 Visikol SFEB Processing

Note: The following protocol was adapted from the Visikol protocol for fixation and staining of NCI2170 Lung Cancer Organoids. Incubation times may be adjusted according to the size and shape of the organoid. This protocol can be used for deeper penetration of antibodies or preparing the SFEBs for immunohistochemical analysis. Antibodies used in this protocol can be seen in Table 4.

Materials:

1. All Protocols

   a. Visikol®HISTO-M™
   b. 4′,6-Diamidino-2-Phenylindole, Dihydrochloride (DAPI) (ThermoFisher Scientific CAT# D1306)
   c. Methanol (Sigma-Aldrich; CAT# 322415-100ML)
   d. Dimethyl sulfoxide (DMSO) (Sigma-Aldrich; CAT# 472301-100ML)
   e. PBS (Sigma-Aldrich D8537-500ml)
   f. PBST: (PBS with 0.1% Tween 20)
   g. PBS with 1% Triton X-100
   h. Blocking buffer: (PBS with 0.2% Triton X-100, 6% donkey serum, 10% DMSO and 0.05% sodium azide)
   i. Antibody buffer: (PBS with 0.2% Tween 20, 3% donkey serum, 5% DMSO, and 0.05% sodium azide)
   j. Penetration buffer: (PBS with 0.2% Triton X-100, 0.3 M glycine, and 20% DMSO)
   k. Fisherbrand™ cell lifters (ThermoFisher Scientific CAT# 08-100-240)
   l. ProLong™ glass antifade mountant (ThermoFisher Scientific; CAT# P36980)
   m. 18 X 18 square glass coverslips (Fisher Scientific; CAT#12-518-105H)

Method:

1. Fixation

   a. Fix SFEBs by aspirating media and adding approximately 3 ml 4% PFA on top of the cell culture insert in a 6 well dish, allowing the PFA to overflow into the bottom of the well. Press down on the plastic ends of the insert to ensure that the SFEBs are fully immersed, and incubate at RT for 45 min.
   b. Remove the PFA and wash the SFEBs 3 times with PBS for 5 min at RT on an orbital shaker. At this time, the SFEBs should detach from the insert. If not, the SFEB may be gently pried from the insert mesh using a cell scraper.
   c. Transfer each individual SFEB into a 96-well plate well using a wide-bore transfer pipette.

2. Immunostaining

   Note: for the following incubations, the SFEB should be fully immersed in the required reagent. Particular volumes are not important unless specified.

a. Wash SFEBs in 100% methanol for 2 min at RT.
b. Wash SFEBs in 20% DMSO/methanol at RT for 2 min.
c. Wash SFEBs in 100% methanol at RT for 2 min.
d. Wash SFEBs in PBS with 1% Triton-X100 at RT for 2 min.
e. Incubate SFEBs in penetration buffer for 60 min at RT with gentle shaking.
f. Block SFEBs in blocking buffer at 37 °C with gentle shaking for 4 h.
g. Incubate SFEBs in 100 μl of primary antibody in antibody buffer and incubate at 37 °C with gentle shaking overnight.

   Note: Use antibody dilutions according to the optimization of the target antibody, typically 1:100–1:1000.
h. Wash SFEBs in PBST 3 times, 5 min each time, at 37 °C with gentle shaking.
i. Incubate SFEBs in 100 μl of secondary antibody in antibody buffer and incubate at 37 °C with gentle shaking for 5 h.

   Note: Add 1.5 μm DAPI to the secondary antibody buffer for nuclear staining. Dilution of the secondary antibody to 1:200 is recommended.
j. Wash SFEBs in PBST 3 times, 5 min each time, at 37 °C with gentle shaking.

3. Tissue clearing and imaging

   a. If clearing is not desired:

      - Move the SFEB to a slide using a wide-bore transfer pipette.
      - Remove excess PBST with a Kimwipe.
      - Add 2 full drops of prolong glass mounting reagent directly on top of the SFEB.
      - Place an 18 × 18 mm glass #1.5 square coverslip on top of the SFEB, and allow to cure overnight. Note: This method will flatten the SFEB, allowing for imaging into the denser parts of the organoid.

   b. If clearing is desired:

      - Incubate SFEBs in 100% methanol for 2 min at RT with gentle shaking.
      - Remove methanol from the well and add 200 μl Visikol HISTO-M to the well. Clearing will now begin, with best results occurring after overnight incubation.

Note: For high-throughput imaging on the ArrayScan, SFEBs can be imaged directly in the 96-well plate with Visikol. For high resolution imaging, SFEBs can be mounted onto a slide by placing a #1.5 -18 × 18 mm square glass coverslip on top of a paper towel, and moving the Visikol-treated SFEB onto the coverslip using a wide-bore transfer pipette. Gently place a slide on top of the SFEB/Visikol, allowing excess Visikol to be absorbed by the paper towel. The coverslip with the SFEB will adhere to the slide. Seal the coverslip using nail polish, and proceed with imaging when cured.

12 Morphometric Analysis of SFEBs Using the ArrayScan XTI Neurite Detection Protocol

Image analysis was performed on ArrayScan XTI using the "neurite detection protocol." This protocol allows for quantification of neurite length, width, branch points, cross points, and associated puncta spots within networks of neurons (Fig. 4). Organoids and SFEB morphology can be interrogated at the well level, field level, and as single cells in a 96-well plate. Morphological analysis of neuronal processes within sub-groups of neurons can also be quantified. Acquisition times must be taken into account when performing high-throughput analysis on the ArrayScan XTI. Robust penetrating antibody stains should be used for high-throughput immunocytochemistry. Tissue clearing protocols should be used to help with antibody penetration, which in turn allows for increased quantification of SFEB morphology and decreased acquisition times. Exposure times for plates can be determined via the following equation: Exposure time = N channels × N of fields × exposure time per channel × Wells per plate.

Equipment

1. ArrayScan XTI High Content Imaging Platform (ThermoFisher Scientific)

Materials

1. Optically clear 96 well U-bottom plates (Corning; CAT # 4520)
2. Cellomics slide port (ThermoFisher Scientific; CAT # N0101111)

Method

1. SFEB image acquisition
   a. Channel 1 should harbor DAPI or Hoechst as a nuclear stain when using the "neurite detection protocol." This channel is used for autofocusing the wells.
   b. Channel 2 should be stained with a neuronal whole cell body and neurite marker (e.g. MAP2 or β-tubulin III).
   c. The neurite detection algorithm also has the ability to detect puncta located on cell bodies/soma and neuronal processes. Pre- and postsynaptic markers can be used in channels 3–5 for quantification of synaptic connections.
   d. Optically clear glass U-bottom 96-well plates are recommended when imaging SFEBs on the ArrayScan XTI. The concave in the U-bottom plate creates a restricted focal region for easier "form factor calibration" and decreased acquisition times.
   e. Optically clear plates also allow the ability to live-image SFEBs for analysis of cellular propagation/ viability, nuclear translocation of transcription factors, and cellular migration to name a few.
   f. A slide port is available for high quality images on a two-photon or spinning disk confocal, followed up by post hoc analysis on the ArrayScan XTI.

g. It is recommended to obtain Z-focus in multiple random fields throughout the 96-well SFEB plate. This will provide an overall whole plate average Z-plane of focus to overcome the heterogeneity in size of the SFEBs throughout the plate due to individual-to-individual variability between cell lines.
h. Once an overall Z-plane is established for the 96-well plate format, it should be entered into the form factor calibration module to hard set parameters into 96-well form factor for each iPSC line.
i. During "configure acquisition," autofocus should be set to "relaxed" and the "extended range of focus" box should be checked. Setting these parameters will widen the autofocus algorithm to obtain better Z-offset planes of focus, which is important when obtaining Z-projections within the SFEB.
j. Camera autofocus mode should be set to 4 × 4 binning during image acquisition, which will decrease scan times. Binning in all other channels can stay lower to preserve image quality.
k. In focus acquisition, set imaging mode Channel 1 to "wide field." All other consecutive channels should be set as confocal. DAPI or Hoechst channel should be set to represent the autofocus reference channel.
l. Random fields and wells should be imaged during "image acquisition" to determine if focusing parameters capture the Z-projection range of the SFEB.
m. During image acquisition, background fluorescence can be eliminated by decreasing "target percentile." Cell bodies with strong fluorescent signals should have a lower target percentile, whereas cells with weak signals should have a higher target percentile.

Note: This can be done by toggling over the background and then over the ROI to obtain pixel intensities located in the bottom left-hand corner of screen. If the background is not drastically different from fluorescent ROI, masking and demarcation will be difficult for thresholding algorithms, which potentially can cause a loss of data. Due to long acquisition times when simultaneously obtaining images and collecting data, it is recommended to use the "image acquisition" protocol to obtain server stored images before running a full plate through the data acquisition steps.

2. Image processing

   a. During image processing, background is removed via "background correction," and levels of brightness and contrast are adjusted.
   b. Using a "low-pass" filter during processing can better resolve cell bodies with weak fluorescent signals, or accentuate the fluorescence of cell bodies or neurites within dense networks.

3. Neurite detection assay ("nuclei and cell body object identification")

   a. Dynamic thresholding such as triangle or isodata methods should be used to demarcate cell bodies and neurites in the SFEB. Dynamic thresholding methods can better detect representative biology by accounting for antibody intensity variability (the difference in the pixel intensity profile of an image) throughout the thickness of the SFEB.

b. During nuclei and cell body validation, it is important to efficiently demarcate nuclei and cell bodies. Cell body masking will determine where neuronal projections are searched for and masked during neurite detection.
c. Some level of smoothing should be used in combination with the dynamic thresholding methods isodata and triangle. Pixel averaging performed by a smoothing algorithm allows for parallel neurites to be counted as separate entities.
d. For object segmentation the use of the "intensity method" is recommended to segment touching cell bodies on the bases of their fluorescent profile. A starting measurement for "intensity segmentation" can be found by using the measuring tool to obtain the radius of normal cell size and dividing by two, making this the segmentation value.

4. Validation of objects

   a. "Neurite identification modifier" can be adjusted to account for neurites with low fluorescent signals.
   b. "Direction length" should be set to account for more curved or straight neurons depending on morphology.
   c. "Point resolution" can be adjusted to account for cross points and branching throughout the neuronal processes in the SFEB.
   d. "Aggressive tracing" is used to account for neurons traveling in between and under other neurites; this is useful in thick woven neuronal networks as seen in SFEBs.
   e. It is also good practice to use "tracing without seeds." This will prevent neurons from only being traced when associated with a nuclear body, which is helpful when tracing organized networks of neurons within organoids and SFEBs.
   f. The "gap tolerance parameter" is used to set designated pixel distance between neuronal processes, determining whether a neuron can be traced as a continuous process.
   g. During object validation, thresholds for user-defined cellular criteria can be set to determine what is a "valid object" or a "rejected object." This can be done by clicking on object selection criteria.
   h. Object selection criteria allows for rejection of objects based on area, shape, and intensity.
   i. A scatter plot of all data points can be created by clicking on "data" box. This shows a representation of all valid objects in the sample.
   j. By clicking on a data point that is being accepted as a valid or rejected object, one can see where it falls among the distribution of all data points around the mean of the selection parameter.

This is a very elementary outline of starting points to use when creating high-content imaging assays on the ArrayScan to elucidate neuronal morphology in

SFEBs. With such methods, compound screens can be utilized to better fit pharmacological intervention to specific individual genotype and phenotype. High-throughput analysis of SFEBs yields large sample sizes across many individuals with cell counts per plates within the millions. Such a large sample size adds to the confidence of effect when screening SFEB morphology or pharmacological targets.

13 Recording Neuronal Activity in SFEBs Using Multielectrode Array and Calcium Imaging

Equipment:

1. Maestro MM MEA system (Axion Biosystems)
2. Axion Integrated Studio (AxIS) software (Axion Biosystems)
3. ArrayScan XTI High Content Imaging Platform (ThermoFisher Scientific)
4. HCS Studio Cell Analysis Software (ThermoFisher Scientific)

Medias:

Terminal Differentiation Media (TD)

- Dulbecco's Modified Eagle Medium/Nutrient Mixture F-12 plus L-Glutamine (DMEM/F12) (ThermoFisher Scientific; CAT# 11320-033)
- Neurobasal media minus phenol red (ThermoFisher Scientific; CAT# 12348-017)
- 100X N2 supplement (ThermoFisher Scientific; CAT# 17502-048)
- B27 (50X) supplement minus vitamin-A (ThermoFisher Scientific; CAT# 17504-044)
- MEM non-essential amino acids (NEAA) (ThermoFisher Scientific; CAT# 11140-050)
- Insulin-transferrin-selenium-A (ThermoFisher Scientific; CAT# 51300-044)
- Penicillin-streptomycin (Pen/Strep) (ThermoFisher Scientific; CAT# 15140-122)
- N-acetyl cysteine (Sigma; CAT# A8199)
- Laminin (ThermoFisher Scientific; CAT# 23017015)
- Forskolin (Sigma; CAT# F-6886)
- Recombinant brain derived neurotrophic factor (BDNF) (Peprotech; CAT# 450-02)
- Recombinant human neurotrophic factor-3 (NT3) (Peprotech; CAT# 450-03)
- N-[N-(3,5-Difluorophenacetyl)-L-alanyl]-S-phenylglycine t-butyl ester (DAPT) (Stemgent/Reprocell; CAT# 04-0041)
- Heparin (Sigma; CAT# H3149)

Materials:

1. MEA
   a. Cytoview MEA 48-well plates (Axion Biosystems; CAT# M768-tMEA-48B)
   b. 50% polyethylenimine solution (PEI) (Sigma-Aldrich; CAT# P3143)
   c. Boric acid (Sigma-Aldrich; CAT# B6768-500G)
   d. Sodium tetraborate (Sigma-Aldrich; CAT# 221732-100G)
   e. Mouse laminin (ThermoFisher Scientific; CAT# 23017-015)
   f. Leibovitz's L15 (ThermoFisher Scientific; CAT# 11415-064)
   g. UltraPure™ DNase/RNase-free distilled water (ThermoFisher Scientific; Cat# 10977015)
2. Calcium Imaging
   a. pAAV9-CAG-GCamp6s-WPRE-SV40 (Addgene; CAT#100844-AAV9)
   b. Hoechst 33342 (ThermoFisher Scientific; CAT# 66249)

Method:

1. MEA plate preparation (adapted from Axion Biosystems MEA plating protocols)
   a. Wash the MEA plates 3 times for 5 min with sterile H_2O under aseptic conditions to clean. Wash for 5 min with 75% ethanol and 2 min with 100% ethanol to sterilize the plate.
   b. Dry the plate inverted in an oven for 4–5 h at 50 °C to complete the sterilization process.
   c. Prepare boric acid solution by dissolving 3.10 g of boric acid and 4.75 g sodium tetraborate in 950 ml of dH_20. Adjust pH to 8.4 with 1M HCl. Bring volume up to 1 L.
   d. Prepare 0.2% PEI by dissolving 48 μl of 50% PEI in 12 ml of boric acid solution. Filter solution through 0.22 μm filter.
   e. Add 125 μl 0.2% PEI to each well and incubate 1 h at room temperature. Aspirate PEI and wash the wells 4 times with sterile distilled H2O. Air dry in the hood overnight.
   f. Add sterile dH_2O to area surrounding wells to avoid evaporation of laminin.
   g. Prepare 20 μg/ml of laminin in Leibovitz's L15 medium. Add 5–10 μl per well directly on top of electrode array. Do not cover grounding electrodes. Incubate for 2 h at 37 °C.
2. SFEB transfer to MEA plates
   a. Remove laminin from MEA plate. Do not let laminin dry on the MEA surface.
   b. Detach the SFEB from the cell culture insert using a cell lifter. Gently transfer one SFEB to each well of the MEA plate using a wide-bore pipette tip. Add 50 μl of TD with 10 μM ROCK inhibitor and incubate overnight at 37 °C.

   c. The following day replace media with 100 μl of TD. Allow the cells 5–7 days of recovery before measuring activity, changing 50% of the media every 2–3 days.

3. Recording neuronal activity in SFEBs Using MEA

   a. Setup Maestro MM and AxIS Software using the following parameters: temperature = 37 °C; High pass filter = 200 Hz; Low-pass filter = 2000 Hz; Adaptive spike detection threshold = 5.25× SD.
   b. Record activity for 20 min for each plate.
   c. Pharmacology may be performed by recording from SFEBs before and after drug administration, or by comparing drug administration between control and ASC lines. Drugs used in this study can be seen in Tables 3, 4, and 5. Recording was done using the same parameters as step 3.a. Wash out the drugs by doing two complete media changes with 100 μl DRM TD.
   d. Endpoint measures such as spike frequency, bursting, and synchrony can be obtained using the AxIS software. Raster plots and continuous traces can also be generated through the Neural Metric Tool and MATLAB.

4. Recording neuronal activity in SFEBs using calcium imaging

   a. Prepare a 48 well MEA plate with SFEBs as described above. Viral transduction is done 5–7 days after transferring SFEBs to the MEA plate.
   b. Prepare DRM TD media + pAAV9-CAG-GCamp6s-WPRE-SV40 at 500,000 gene copies/cell (assume 50,000 cells/SFEB). Make enough media for a volume of 40 μl/well.
   c. Remove all media from well. Replace with 40 μl of virus solution. Incubate overnight.
   d. The next day add DRM TD media at 100 μl/well.
   e. Do not complete media change for 96 h after addition of the virus. Imaging can be performed 5–7 days after addition of the virus.
   f. Add 100 ng/ml Hoechst 33342 to SFEBs prior to imaging.
   g. Image cells using ArrayScan XTI at 5× magnification. This magnification is sufficient to image the whole electrode array. To minimize phototoxicity, only image Hoechst 33342 once per field. For our experiment, exposure for GCamp6 was set to 400 ms. Sampling rate was approximately 0.8 Hz.
   h. Analyze changes in fluorescence intensity using the Hoechst channel to create regions of interest that are restricted to the cell bodies within the SFEB. The Target Activation module of HCS studio is recommended for this. Use a cell calcium analysis software package such as PeakCaller [88] to analyze the number of calcium spikes.

Table 5 Major cell culture medias, small molecules, and concentrations of stocks used in this study

mTeSR media	Manufacturer; catalog #	Final concentration	Stock concentration
mTeSR	Stem Cell Technologies; 85850	500 ml	500 ml
FGF	StemGent; 03-0002	40 ng/ml	10 μg/ml
CHIR	StemGent; 04-0004-10	10 μM	20 mM in DMSO
Thiazovivin	StemGent; 04-0017	1 μM	10 mM in DMSO
PD0325901	StemGent; 04-0006	1 μM	10 mM in DMSO
ROCKi (Y27632)	StemGent; 04-0012-10	10 μM	10 mM in DMSO
NIM media	*Manufacturer; catalog #*	*Final concentration*	*Stock concentration*
Neural induction media	Stem Cell Technologies; 05835	250 ml	250 ml
FGF	StemGent; 03-0002	40 ng/ml	10 μg/ml
SB431542	StemGent; 04-0010-10	10 μM	10 mM in DMSO
Thiazovivin	StemGent; 04-0017	1 μM	10 mM in DMSO
Dorsomorphin	StemGent; 04-0024	1 μM	10 mM in DMSO
ROCKi (Y27632)	StemGent; 04-0012-10	10 μM	10 mM in DMSO
ENB base media	*Manufacturer; catalog #*	*Final concentration*	*Stock concentration*
DMEM/F12 +L-glutamine	ThermoFisher; 11320-033	500 ml	500 ml
Neurobasal minus phenol red	ThermoFisher; 12348-017	500 ml	500 ml
100X N2 supplement	ThermoFisher; 17502-048	100%	100×
50X B27 supplement minus Vit A	ThermoFisher; 17504-044	100%	50×
NEAA	ThermoFisher; 11140-050	100%	100×
Insulin-transferrin-selenium-A	ThermoFisher; 51300-044	100%	100×
Pen/strep	ThermoFisher; 15140-122	100%	100×, 10,000 U/ml
N-acetyl cysteine	Sigma; A8199	5 μg/ml	5 mg/ml
SATO mix	*Manufacturer; catalog #*	*Final concentration*	*Stock concentration*
Crystalline BSA	Sigma; A-4161	10 mg/ml	
Putrescine	Sigma; P7505	1.6 mg/ml	
Progesterone	Sigma; P8783	19.87 μM	79.5 mM in 100% EtOH
T3 (Triiodothyronine)	Sigma; T6397	6 μM	12 mM in 0.1M NaOH
T4 (Thyroxine)	Sigma; T1775	5.2 μM	10.3 mM in 0.1M NaOH

(continued)

Table 5 (continued)

mTeSR media	Manufacturer; catalog #	Final concentration	Stock concentration
Neural formation media (NF)	*Manufacturer; catalog #*	*Final concentration*	*Stock concentration*
DRM media	See recipe		
SATO	See recipe		
Laminin	ThermoFisher; 23017015	1 μg/ml	1 mg
Forskolin	Sigma; F-6886	5 μM	2.1 mg/ml (5mM)
BDNF	Peprotech; 450-02	20 μg/ml	20 ng/ml
NT3	Peprotech; 450-03	50 μg/ml	20 ng/ml
β-NGF	Peprotech; 450-01	50 μg/ml	20 ng/ml
DAPT	StemGent; 04-0041	2 μM	10 mM in DMSO
ROCKi (Y27632)	StemGent; 04-0012-10	10 μM	10 mM in DMSO
Heparin	Sigma; H3149	2 μg/ml	2 mg/ml
Terminal differentiation media (TD)	*Manufacturer; catalog #*	*Final concentration*	*Stock concentration*
ENB base media	See recipe		
SATO	See recipe		
Laminin	ThermoFisher; 23017015	1 μg/ml	1 mg
Forskolin	Sigma; F-6886	5 μM	2.1 mg/ml (5mM)
BDNF	Peprotech; 450-02	20 μg/ml	20 μg/ml
NT3	Peprotech; 450-03	50 μg/ml	50 μg/ml
DAPT	StemGent; 04-0041	2 μM	10 mM in DMSO
ROCKi (Y27632)	StemGent; 04-0012-10	10 μM	10 mM in DMSO
Heparin	Sigma; H3149	2 μg/ml	2 mg/ml
DRM media	*Manufacturer; catalog #*	*Final concentration*	*Stock concentration*
DMEM/F 12 +L-Glutamine	ThermoFisher; 11320-033	500 ml	500 ml
Neurobasal minus phenol red	ThermoFisher; 12348-017	500 ml	500 ml
100× N2 supplement	ThermoFisher; 17502-048	100%	100×
50X B27 supplement minus Vit A	ThermoFisher; 17504-044	100%	50×
NEAA	ThermoFisher; 11140-050	100%	100×
Insulin-transferrin-selenium-A	ThermoFisher; 51300-044	100%	100×
Pen/strep	ThermoFisher; 15140-122	100%	100×, 10,000 U/ml
N-acetyl cysteine	Sigma; A8199	5 μg/ml	5 mg/ml

14 Conclusions

ASC reflects a likely interaction between complex heterogeneous genetic etiology and environmental factors leading to a spectrum of traits affecting social interaction, communication, and behavior. To date most therapeutic interventions for ASC are off label applications of medicines developed for depression, anxiety, ADHD, and other neuropsychiatric conditions. Given the diverse neurobiological pathways that can lead to ASC behaviors, it may be beneficial to develop treatments that are specific to ASC etiologies.

iPSC-derived SFEBs are a platform that can overcome potential phenotypic bottlenecks by generating many samples from a diverse population. SFEBs allow the scientist to multiplex complex physiological endpoints on a 3-D scalable platform to more accurately model the interaction between genetics and physiology. The ultimate result is a highly dynamic model that will lead to more accurate drug discovery for complex brain disorders. We have demonstrated here a system that makes some compromises between the complexity and heterogeneity of 3-D iPSC neural organoids and the simplicity and homogeneity of 2-D neural iPSC cultures. SFEBs represent an important biological middle ground between these two extremes. Furthermore, we have developed an experimental workflow that subjects SFEBs to complex and multiplexed physiological and cell-based assays in a high-throughput environment. The ultimate goal of the combination of this culture system with these assays is to more accurately model complex biology in order to approximate circuit-level changes that may be observed in relation to the genetic background of individuals with autism. In doing so, we have provided a platform for more accurate drug screening and discovery, and for a basic understanding of the relationship between genetics and cellular phenotypes in autism.

It is important to recognize that there is still a lot of work to be done in this fast-paced and burgeoning field within neuroscience and genetics. For example, there have been no studies to date that probe directly the translatability of human iPSC models to complex animal behavioral models of autism, or other neurodevelopmental conditions. This is a necessary step to validate drugs identified in screens using iPSC-based models and ask questions that utilize behavioral pharmacology. One major question for this field is what phase of human and animal development can be directly pegged to immature and mature iPSC-derived neurons and iPSC-derived organoids? We also recognize that there is a dearth of information on how to control for, or understand, well-to-well, plate-to-plate, experiment-to-experiment, and individual-to-individual variability commonly observed in iPSC-based disease models. This issue becomes acute when using 3-D iPSC-based models, which suffer from internal heterogeneity. Nonetheless, if proper controls are in place, and experiments using these 3-D models are carried out with a high level of rigor, focused on consistency and reproducibility, 3-D models and SFEBs in particular, represent a tremendous scientific step in our modeling and understanding of neurodevelopmental conditions such as autism. Here, we have provided a small snapshot of a phenotypic output from our SFEB system modeling autism. We hope that our discussion of SFEBs and our methods for generating them to model autism will spark researchers to continue to refine and develop more complex iPSC-based systems to model the neurophysiology underlying brain disorders.

Acknowledgements We would like to thank Dr. Rania Deranieh and Omotayo Oduwole for technical assistance. We also like to thank Dr. John Hussman and Elizabeth Benevides for comments and reviews of this manuscript.

References

1. Takahashi, K., & Yamanaka, S. (2006). Induction of pluripotent stem cells from mouse embryonic and adult fibroblast cultures by defined factors. *Cell, 126*, 663–676.
2. Takahashi, K., Tanabe, K., Ohnuki, M., Narita, M., Ichisaka, T., Tomoda, K., et al. (2007). Induction of pluripotent stem cells from adult human fibroblasts by defined factors. *Cell, 131*, 861–872.
3. DeRosa, B. A., Belle, K. C., Thomas, B. J., Cukier, H. N., Pericak-Vance, M. A., Vance, J. M., et al. (2015). HVGAT-mCherry: A novel molecular tool for analysis of GABAergic neurons derived from human pluripotent stem cells. *Molecular and Cellular Neurosciences, 68*, 244–257.
4. DeRosa, B. A., Van Baaren, J. M., Dubey, G. K., Lee, J. M., Cuccaro, M. L., Vance, J. M., et al. (2012). Derivation of autism spectrum disorder-specific induced pluripotent stem cells from peripheral blood mononuclear cells. *Neuroscience Letters, 516*, 9–14.
5. Griesi-Oliveira, K., et al. (2015). Modeling non-syndromic autism and the impact of TRPC6 disruption in human neurons. *Molecular Psychiatry, 20*, 1350–1365.
6. Abyzov, A., Mariani, J., Palejev, D., Zhang, Y., Haney, M. S., Tomasini, L., et al. (2012). Somatic copy number mosaicism in human skin revealed by induced pluripotent stem cells. *Nature, 492*, 438–442.
7. Caiazzo, M., Giannelli, S., Valente, P., Lignani, G., Carissimo, A., Sessa, A., et al. (2015). Direct conversion of fibroblasts into functional astrocytes by defined transcription factors. *Stem Cell Reports, 4*, 25–36.
8. Gupta, N., Henry, R. G., Strober, J., Kang, S.-M., Lim, D. A., Bucci, M., et al. (2012). Neural stem cell engraftment and myelination in the human brain. *Science Translational Medicine, 4*, 155ra137.
9. Krencik, R., Weick, J. P., Liu, Y., Zhang, Z.-J., & Zhang, S.-C. (2011). Specification of transplantable astroglial subtypes from human pluripotent stem cells. *Nature Biotechnology, 29*, 528–534.
10. Heinrich, C., Blum, R., Gascón, S., Masserdotti, G., Tripathi, P., Sánchez, R., et al. (2010). Directing astroglia from the cerebral cortex into subtype specific functional neurons. *PLoS Biology, 8*, e1000373.
11. Kim, D.-S., Ross, P. J., Zaslavsky, K., & Ellis, J. (2014). Optimizing neuronal differentiation from induced pluripotent stem cells to model ASD. *Frontiers in Cellular Neuroscience, 8*, 109.
12. Nicholas, C. R., Chen, J., Tang, Y., Southwell, D. G., Chalmers, N., Vogt, D., et al. (2013). Functional maturation of hPSC-derived forebrain interneurons requires an extended timeline and mimics human neural development. *Cell Stem Cell, 12*, 573–586.
13. Ambasudhan, R., Talantova, M., Coleman, R., Yuan, X., Zhu, S., Lipton, S. A., et al. (2011). Direct reprogramming of adult human fibroblasts to functional neurons under defined conditions. *Cell Stem Cell, 9*, 113–118.
14. Liu, M. L., Zang, T., Zou, Y., Chang, J. C., Gibson, J. R., Huber, K. M., et al. (2013). Small molecules enable neurogenin 2 to efficiently convert human fibroblasts into cholinergic neurons. *Nature Communications, 4*, 1–10.

15. Pang, Z. P., Yang, N., Vierbuchen, T., Ostermeier, A., Fuentes, D. R., Yang, T. Q., et al. (2011). Induction of human neuronal cells by defined transcription factors. *Nature, 476*, 220–223.
16. Shi, Y., Kirwan, P., & Livesey, F. J. (2012a). Directed differentiation of human pluripotent stem cells to cerebral cortex neurons and neural networks. *Nature Protocols, 7*, 1836–1846.
17. Karus, M., Blaess, S., & Brüstle, O. (2014). Self-organization of neural tissue architectures from pluripotent stem cells. *The Journal of Comparative Neurology, 522*, 2831–2844.
18. Pasca, S. P., Panagiotakos, G., & Dolmetsch, R. E. (2014). Generating human neurons in vitro and using them to understand neuropsychiatric disease. *Annual Review of Neuroscience, 37*, 479–501.
19. Zhang, S.-C., Wernig, M., Duncan, I. D., Brüstle, O., & Thomson, J. A. (2001). In vitro differentiation of transplantable neural precursors from human embryonic stem cells. *Nature Biotechnology, 19*, 1129–1133.
20. Chambers, S. M., Fasano, C. A., Papapetrou, E. P., Tomishima, M., Sadelain, M., & Studer, L. (2009). Highly efficient neural conversion of human ES and iPS cells by dual inhibition of SMAD signaling. *Nature Biotechnology, 27*, 275–280.
21. Espuny-Camacho, I., Michelsen, K. A., Gall, D., Linaro, D., Hasche, A., Bonnefont, J., et al. (2013). Pyramidal neurons derived from human pluripotent stem cells integrate efficiently into mouse brain circuits in vivo. *Neuron, 77*, 440–456.
22. Kadoshima, T., Sakaguchi, H., Nakano, T., Soen, M., Ando, S., Eiraku, M., et al. (2013). Self-organization of axial polarity, inside-out layer pattern, and species-specific progenitor dynamics in human ES cell-derived neocortex. *Proceedings of the National Academy of Sciences, 110*, 20284–20289.
23. Krencik, R., & Zhang, S.-C. (2011). Directed differentiation of functional astroglial subtypes from human pluripotent stem cells. *Nature Protocols, 6*, 1710–1717.
24. Maroof, A. M., Keros, S., Tyson, J. A., Ying, S.-W., Ganat, Y. M., Merkle, F. T., et al. (2013). Directed differentiation and functional maturation of cortical interneurons from human embryonic stem cells. *Cell Stem Cell, 12*, 559–572.
25. Nestor, M. W., Jacob, S., Sun, B., Prè, D., Sproul, A. A., Hong, S. I., et al. (2015). Characterization of a subpopulation of developing cortical interneurons from human iPSCs within serum-free embryoid bodies. *American Journal of Physiology. Cell Physiology, 308*, C209–C219.
26. Shi, Y., Kirwan, P., Smith, J., Robinson, H. P. C., & Livesey, F. J. (2012b). Human cerebral cortex development from pluripotent stem cells to functional excitatory synapses. *Nature Neuroscience, 15*, 477–486.
27. Woodard, C. M., et al. (2014). iPSC-derived dopamine neurons reveal differences between monozygotic twins discordant for Parkinson's disease. *Cell Reports, 9*, 1173–1182.
28. Abbott, A. (2003). Biology's new dimension. *Nature, 424*, 870–872.
29. Pasca, A. M., Sloan, S. A., Clarke, L. E., Tian, Y., Makinson, C. D., Huber, N., et al. (2015). Functional cortical neurons and astrocytes from human pluripotent stem cells in 3-D culture. *Nature Methods, 12*, 671–678.
30. Kelava, I., & Lancaster, M. A. (2016). Stem cell models of human brain development. *Cell Stem Cell, 18*, 736–748.
31. Sasai, Y. (2013). Next-generation regenerative medicine: Organogenesis from stem cells in 3-D culture. *Cell Stem Cell, 12*, 520–530.
32. Watanabe, K., Kamiya, D., Nishiyama, A., Katayama, T., Nozaki, S., Kawasaki, H., et al. (2005). Directed differentiation of telencephalic precursors from embryonic stem cells. *Nature Neuroscience, 8*, 288–296.
33. Ying, Q.-L., Stavridis, M., Griffiths, D., Li, M., & Smith, A. (2003). Conversion of embryonic stem cells into neuroectodermal precursors in adherent monoculture. *Nature Biotechnology, 21*(2), 183–186.
34. Gaspard, N., Bouschet, T., Hourez, R., Dimidschstein, J., Naeije, G., Van Den Ameele, J., et al. (2008). An intrinsic mechanism of corticogenesis from embryonic stem cells. *Nature, 455*, 351–357.

35. Eiraku, M., Watanabe, K., Matsuo-Takasaki, M., Kawada, M., Yonemura, S., Matsumura, M., et al. (2008). Self-organized formation of polarized cortical tissues from ESCS and its active manipulation by extrinsic signals. *Cell Stem Cell, 3*, 519–532.
36. Nestor, M. W., Phillips, A. W., Artimovich, E., Nestor, J. E., Hussman, J. P., & Blatt, G. J. (2016). Human inducible pluripotent stem cells and autism spectrum disorder: Emerging technologies. *Autism Research, 9*, 513–535.
37. Mariani, J., Vittoria, M., Palejev, D., Tomasini, L., Coppola, G., Szekely, A. M., et al. (2012). Modeling human cortical development in vitro using induced pluripotent stem cells. *Proceedings of the National Academy of Sciences, 109*, 12770–12775.
38. Mariani, J., Coppola, G., Zhang, P., Abyzov, A., Provini, L., Tomasini, L., et al. (2015). FOXG1-dependent dysregulation of GABA/glutamate neuron differentiation in autism spectrum disorders. *Cell, 162*, 375–390.
39. Phillips, A. W., Nestor, J. E., & Nestor, M. W. (2017). Developing HiPSC derived serum free embryoid bodies for the interrogation of 3-D stem cell cultures using physiologically relevant assays. *Journal of Visualized Experiments, 125*, 55799.
40. Nestor, M. W., Paull, D., Jacob, S., Sproul, A. A., Alsaffar, A., Campos, B. A., et al. (2013). Differentiation of serum-free embryoid bodies from human induced pluripotent stem cells into networks. *Stem Cell Research, 10*, 454–463.
41. Lancaster, M. A., & Knoblich, J. A. (2014). Generation of cerebral organoids from human pluripotent stem cells. *Nature Protocols, 9*, 2329–2340.
42. Lancaster, M. A., Renner, M., Martin, C.-A., Wenzel, D., Bicknell, L. S., Hurles, M. E., et al. (2013). Cerebral organoids model human brain development and microcephaly. *Nature, 501*, 373–379.
43. Qian, X., et al. (2016). Brain-region-specific organoids using mini-bioreactors for modeling ZIKV exposure. *Cell, 165*, 1238–1254.
44. Marchetto, M. C. N., Carromeu, C., Acab, A., Yu, D., Yeo, G. W., Mu, Y., et al. (2010). A model for neural development and treatment of Rett syndrome using human induced pluripotent stem cells. *Cell, 143*, 527–539.
45. Flaherty, E., Deranieh, R. M., Artimovich, E., Lee, I. S., Siegel, A. J., Levy, D. L., et al. (2017). Patient-derived hiPSC neurons with heterozygous CNTNAP2 deletions display altered neuronal gene expression and network activity. *NPJ Schizophrenia, 3*, 35.
46. Derosa, B. A., El Hokayem, J., Artimovich, E., Garcia-Serje, C., Phillips, A. W., Van Booven, D., et al. (2018). Convergent pathways in idiopathic autism revealed by time course transcriptomic analysis of patient-derived neurons. *Scientific Reports, 8*, 1–15.
47. Kolevzon, A. (2013). *A pilot treatment study of insulin-like growth factor-1 (IGF-1) in autism spectrum disorder*. Retrieved from https://clinicaltrials.gov/ct2/show/NCT01970345
48. Tang, X., Kim, J., Zhou, L., Wengert, E., Zhang, L., Wu, Z., et al. (2016). KCC2 rescues functional deficits in human neurons derived from patients with Rett syndrome. *Proceedings of the National Academy of Sciences, 113*, 751–756.
49. Marchetto, M. C., et al. (2017). Altered proliferation and networks in neural cells derived from idiopathic autistic individuals. *Molecular Psychiatry, 22*, 820–835.
50. Gomperts, S. N., Rao, A., Craig, A. M., Malenka, R. C., & Nicoll, R. A. (1998). Postsynaptically silent synapses in single neuron cultures. *Neuron, 21*, 1443–1451.
51. Flames, N., Pla, R., Gelman, D. M., Rubenstein, J. L. R., Puelles, L., & Marin, O. (2007). Delineation of multiple subpallial progenitor domains by the combinatorial expression of transcriptional codes. *The Journal of Neuroscience, 27*, 9682–9695.
52. Kanatani, S., Yozu, M., Tabata, H., & Nakajima, K. (2008). COUP-TFII is preferentially expressed in the caudal ganglionic eminence and is involved in the caudal migratory stream. *The Journal of Neuroscience, 28*, 13582–13591.
53. Gonchar, Y., & Burkhalter, A. (1999). Connectivity of GABAergic calretinin-immunoreactive neurons in rat primary visual cortex. *Cerebral Cortex, 9*, 683–696.

54. Anderson, S. A., Kaznowski, C. E., Horn, C., Rubenstein, J. L. R., & McConnell, S. K. (2002). Distinct origins of neocortical projection neurons and interneurons in vivo. *Cerebral Cortex, 12*, 702–709.
55. Valcanis, H., & Tan, S.-S. (2003). Layer specification of transplanted interneurons in developing mouse neocortex. *The Journal of Neuroscience, 23*, 5113–5122.
56. Wichterle, H., Garcia-Verdugo, J. M., Herrera, D. G., & Alvarez-Buylla, A. (1999). Young neurons from medial ganglionic eminence disperse in adult and embryonic brain. *Nature Neuroscience, 2*, 461–466.
57. Hussman, J. P. (2001). Suppressed GABAergic inhibition as a common factor in suspected etiologies of autism. *Journal of Autism and Developmental Disorders, 31*, 247–248.
58. Hussman, J. P. (2007). The gap between intention and action: Altered connectivity and GABA-mediated synchrony in autism. In E. B. Torres & C. Whyatt (Eds.), *Autism: The movement sensing perspective*. New York: CRC Press.
59. Selten, M., van Bokhoven, H., & Nadif Kasri, N. (2018). Inhibitory control of the excitatory/inhibitory balance in psychiatric disorders. *F1000Research, 7*, 23.
60. Lee, E., Lee, J., & Kim, E. (2017). Excitation/inhibition imbalance in animal models of autism spectrum disorders. *Biological Psychiatry, 81*, 838–847.
61. Cao, W., Lin, S., Qiang, X., Lan, D. Y., Yang, Q., Ying, Z. M., et al. (2018). Gamma oscillation dysfunction in mPFC leads to social deficits in neuroligin 3 R451C knockin mice. *Neuron, 97*, 1394.
62. Edgar, J. C., Khan, S. Y., Blaskey, L., Chow, V. Y., Rey, M., Gaetz, W., et al. (2015). Neuromagnetic oscillations predict evoked-response latency delays and core language deficits in autism spectrum disorders. *Journal of Autism and Developmental Disorders, 45*(2), 395–405. https://doi.org/10.1007/s10803-013-1904-x
63. Rojas, D. C., Maharajh, K., Teale, P., & Rogers, S. J. (2008). Reduced neural synchronization of gamma-band MEG oscillations in first-degree relatives of children with autism. *BMC Psychiatry, 8*, 1–9.
64. Wilson, T. W., Rojas, D. C., Reite, M. L., Teale, P. D., & Rogers, S. J. (2007). Children and adolescents with autism exhibit reduced MEG steady-state gamma responses. *Biological Psychiatry, 62*, 192–197.
65. Jacob, J. (2016). Cortical interneuron dysfunction in epilepsy associated with autism spectrum disorders. *Epilepsia, 57*, 182–193.
66. Zhu, Y. (2004). Chandelier cells control excessive cortical excitation: characteristics of Whisker-evoked synaptic responses of layer 2/3 nonpyramidal and pyramidal neurons. *The Journal of Neuroscience, 24*, 5101–5108.
67. Boutin, M. E., Voss, T. C., Titus, S. A., Cruz-Gutierrez, K., Michael, S., & Ferrer, M. (2018). A high-throughput imaging and nuclear segmentation analysis protocol for cleared 3-D culture models. *Scientific Reports, 8*, 1–14.
68. Esner, M., Meyenhofer, F., & Bickle, M. (2018). Live-cell high content screening in drug development. *Methods in Molecular Biology, 1683*, 149–164.
69. Smietana, K., Siatkowski, M., & Møller, M. (2016). Trends in clinical success rates. *Nature Reviews. Drug Discovery, 15*, 379–380.
70. Chen, Y. Y., Silva, P. N., Syed, A. M., Sindhwani, S., Rocheleau, J. V., & Chan, W. C. W. (2016). Clarifying intact 3-D tissues on a microfluidic chip for high-throughput structural analysis. *Proceedings of the National Academy of Sciences, 113*, 14915–14920.
71. Giuliano, K. A., DeBiasio, R. L., Dunlay, R. T., Gough, A., Volosky, J. M., Zock, J., et al. (1997). High-content screening: A new approach to easing key bottlenecks in the drug discovery process. *Journal of Biomolecular Screening, 2*, 249–259.
72. Egner, A., & Hell, S. (2006). Aberrations in confocal and multi-photon fluorescence microscopy induced by refractive index mismatch. In J. B. Pawley (Ed.), *Handbook of biological confocal microscopy* (pp. 404–413). New York: Springer.
73. Pampaloni, F., Ansari, N., & Stelzer, E. H. K. (2013). High-resolution deep imaging of live cellular spheroids with light-sheet-based fluorescence microscopy. *Cell and Tissue Research, 352*, 161–177.

74. Boutin, M. E., & Hoffman-Kim, D. (2015). Application and assessment of optical clearing methods for imaging of tissue-engineered neural stem cell spheres. *Tissue Engineering. Part C, Methods, 21*, 292–302.
75. Jacques, S. L. (2013). Optical properties of biological tissues: A review. *Physics in Medicine and Biology, 58*(11), R37–R61.
76. Morris, R. G. (1989). Synaptic plasticity and learning: selective impairment of learning rats and blockade of long-term potentiation in vivo by the N-methyl-D-aspartate receptor antagonist AP5. *The Journal of Neuroscience, 9*, 3040–3057.
77. Honoré, T., Davies, S. N., Drejer, J., Fletcher, E. J., Jacobsen, P., Lodge, D., et al. (1988). Quinoxalinediones: Potent competitive non-NMDA glutamate receptor antagonists. *Science, 241*, 701–703.
78. Crossman, A. R., Walker, R. J., & Woodruff, G. N. (1973). Picrotoxin antagonism of gamma-aminobutyric acid inhibitory responses and synaptic inhibition in the rat substantia nigra. *British Journal of Pharmacology, 49*, 696–698.
79. Zoghbi, H. Y., & Bear, M. F. (2012). Synaptic dysfunction in neurodevelopmental disorders associated with autism and intellectual disabilities. *Cold Spring Harbor Perspectives in Biology, 4*, a009886.
80. Malenka, R. C., & Bear, M. F. (2004). LTP and LTD: An embarrassment of riches. *Neuron, 44*, 5–21.
81. Cooke, S. F. (2006). Plasticity in the human central nervous system. *Brain, 129*, 1659–1673.
82. Molnar, E. (2011). Long-term potentiation in cultured hippocampal neurons. *Seminars in Cell & Developmental Biology, 22*, 506–513.
83. Musleh, W., Bi, X., Tocco, G., Yaghoubi, S., & Baudry, M. (1997). Glycine-induced long-term potentiation is associated with structural and functional modifications of alpha-amino-3-hydroxyl-5-methyl-4-isoxazolepropionic acid receptors. *Proceedings of the National Academy of Sciences of the United States of America, 94*, 9451–9456.
84. Li, L. J., Hu, R., Lujan, B., Chen, J., Zhang, J. J., Nakano, Y., et al. (2016). Glycine potentiates AMPA receptor function through metabotropic activation of GluN2A-containing NMDA receptors. *Frontiers in Molecular Neuroscience, 9*, 102. https://doi.org/10.3389/fnmol.2016.00102
85. Shew, W. L., Bellay, T., & Plenz, D. (2010). Simultaneous multi-electrode array recording and two-photon calcium imaging of neural activity. *Journal of Neuroscience Methods, 192*, 75–82.
86. Chen, T. W., Wardill, T. J., Sun, Y., Pulver, S. R., Renninger, S. L., Baohan, A., et al. (2013). Ultrasensitive fluorescent proteins for imaging neuronal activity. *Nature, 499*(7458), 295–300. https://doi.org/10.1038/nature12354
87. Sproul, A. A., Jacob, S., Pre, D., Kim, S. H., Nestor, M. W., Navarro-Sobrino, M., et al. (2014). Characterization and molecular profiling of PSEN1 familial Alzheimer's disease iPSC-derived neural progenitors. *PLoS One, 9*(1), e84547.
88. Artimovich, E., Jackson, R. K., Kilander, M. B. C., Lin, Y. C., & Nestor, M. W. (2017). PeakCaller: An automated graphical interface for the quantification of intracellular calcium obtained by high-content screening. *BMC Neuroscience, 18*(1), 72.

Index

E. DiCicco-Bloom, J. H. Millonig (eds.), *Neurodevelopmental Disorders*, Advances in Neurobiology 25, https://doi.org/10.1007/978-3-030-45493-7

V

W

Y

Z

The manufacturer's authorised representative in the EU is Springer Nature Customer Service Centre GmbH, Europaplatz 3, 69115 Heidelberg, Germany. If you have any concerns regarding our products, please contact ProductSafety@springernature.com

Printed and bound by CPI Group (UK) Ltd, Croydon, CR0 4YY
05/10/2026
02228126-0007